Flow Cytometry

New Developments

NATO ASI Series

Advanced Science Institutes Series

A series presenting the results of activities sponsored by the NATO Science Committee, which aims at the dissemination of advanced scientific and technological knowledge, with a view to strengthening links between scientific communities.

The Series is published by an international board of publishers in conjunction with the NATO Scientific Affairs Division

A Life Sciences B Physics	Plenum Publishing Corporation London and New York
C Mathematical and Physical Sciences D Behavioural and Social Sciences E Applied Sciences	Kluwer Academic Publishers Dordrecht, Boston and London
F Computer and Systems Sciences G Ecological Sciences H Cell Biology I Global Environmental Change	Springer-Verlag Berlin Heidelberg New York London Paris Tokyo Hong Kong Barcelona Budapest

NATO-PCO DATABASE

The electronic index to the NATO ASI Series provides full bibliographical references (with keywords and/or abstracts) to more than 30 000 contributions from international scientists published in all sections of the NATO ASI Series. Access to the NATO-PCO DATABASE compiled by the NATO Publication Coordination Office is possible in two ways:

- via online FILE 128 (NATO-PCO DATABASE) hosted by ESRIN, Via Galileo Galilei, I-00044 Frascati, Italy.
- via CD-ROM "NATO Science & Technology Disk" with user-friendly retrieval software in English, French and German (© WTV GmbH and DATAWARE Technologies Inc. 1992).

The CD-ROM can be ordered through any member of the Board of Publishers or through NATO-PCO, Overijse, Belgium.

Series H: Cell Biology, Vol. 67

Flow Cytometry

New Developments

Edited by

Alain Jacquemin-Sablon

Centre National de la Recherche Scientifique
Laboratoire de Cytométrie
B. P. 3
94801 Villejuif, France

Springer-Verlag
Berlin Heidelberg New York London Paris Tokyo
Hong Kong Barcelona Budapest
Published in cooperation with NATO Scientific Affairs Division

Proceedings of the NATO Advanced Study Institute "4th International Course on Flow Cytometry" held at Villejuif, France, from April 6–10, 1992

ISBN 3-540-54606-5 Springer-Verlag Berlin Heidelberg New York
ISBN 0-387-54606-5 Springer-Verlag New York Berlin Heidelberg

Library of Congress Cataloging-in-Publication Data
Flow cytometry : new developments / edited by Alain Jacquemin-Salon.
(NATO ASI series. Series H, Cell biology ; vol. 67)
"Proceedings of the NATO Advanced Study Institute "4th International Course on Flow Cytometry" held at Villejuif, France, from April 6-10, 1992 – T.p. verso. Published in cooperation with NATO Scientific Affairs Division." Includes bibliographical references and index.
ISBN 3-540-54606-5 (alk. paper)
ISBN 0-387-54606-5 (alk. paper)
1. Flow cytometry–Congresses. I. Jacquemin-Salon, Alain. II. North Atlantic Treaty Organization. Scientific Affairs Division. III. NATO Advanced Study Institute "4th International Course on Flow Cytometry" (1992 : Villejuif, France) IV: Series. [DNLM: 1. Flow Cytometry–congresses. QH 585.5.F56 F644 1992] QH585.5.F56F585 1993 574.87'028-dc20

Printed in Germany

Typesetting: Camera ready by authors
31/3145 - 5 4 3 2 1 0 – Printed on acid-free paper

FOREWORD

The International Flow Cytometry course, in the NATO Advanced Study Institute Series, was designed to address three major areas of interest in cell and molecular biology, (a) cell activation and biological response, (b) membrane-ligand interactions and cell identity and (c) nuclear components form and function. Data management, expert systems and cell sorting techniques that are subjects of concern to all aspects of flow cytometry, were also presented.

Cell activation is an early event in physiological processes that elicits a response of a cell to perform some specialized function. It is known that several cytolosic events are related to cell activation. The course focused on optical techniques that can be used in flow cytometry to study the cascade of processes that are involved. One example was demonstrated by the use of a peptide hormone for studying the activation of smooth muscle cells. The aim was to determine whether calcium and mitochondrial responses could be observed that are consistent wit the interaction between the signal an energy transduction pathways. Methods for detecting activation-induced changes in intracellular pH and membrane potential were addressed during the laboratory exercise. Although the process of cell activation is still not well understood, many of events that occur in tandem can be studied by flow cytometry and new fluorescent probes have recently been developed for that purpose.

The capability for following the migration of specific cell types through different parts of the body has recently become available by a process that involves labeling of cell membrane with fluorescent, non-toxic probes that are not easily degraded. If cells undergo division the fluorescent label is partitioned between the two daughter cells. In this way flow cytometric analyses can be used to "track" and follow the proliferative potential and isolation of specific subpopulations of cells. Attempts to identify the hemapoeitic stem cells in bone marrow by flow cytometry have been of particular interest. Some succes has been achieved and stem cells, capable of reviving a functional hemapoeitic system in lethally irradiated mice have been be obtained by cell sorting. Several new FCM techniques are being developed for obtaining viable, cell populations for functional assays.

The numerous functional activities that occur in the cell nucleus is response to intra and intercellular signals has invoked a large series of flow cytometric techniques. Changes in chromatin structure in response to cell cycle or physiological-related changes can be analyzed with metachromatic fluorochromes such as acridine orange following acid or heat denaturation of cells. Rates of DNA synthesis are also examined by analysis of BrdU-incorporation into DNA and subsequent flow cytometric analysis of cellular BrdU content using fluorescein-labeled antibodies to BrdU. Chromosome analysis and sorting of chromosomes from metaphase arrested cells are now possible. Techniques are also available to perform fluorescence *in-situ* hybridization (FISH) on sorted chromosomes. Applications of PCR and gene mapping on sort chromosomes are now possible. The advanced flow cytometry course provided the most up to data techniques available at present.

Co-Directors :

Dr. Alain Jacquemin-Sablon
CNRS, Laboratoire de Cytométrie, 94800 Villejuif, France

Dr. Harry A Crissman
Los Alamos National Laboratory, Los Alamos, NM 87545, USA

Acknowledgements

Generous support form the following Institutions and Companies is gratefully acknowledged :

- the *Centre National de la Recherche Scientifique (CNRS)*, the *Association pour la Recherche sur le Cancer (ARC, Président : Jacques Crozemarie)*

- Becton-Dickinson
- Bio-Rad
- Boehringer Mannheim France
- Chemunex (Partec)
- Cohérent
- Coultronics
- Cytolab
- Leica
- Ortho Diagnostic Systems
- Sanofi Recherche
- Spectra-Physics
- Zeiss

We also wish to thank :

- Martine Bardissa
- Claude Boucheix
- Sophie Lafosse
- Dominique Laird
- Linda Pritchard
- Arlette Vervisch

for their expert assistance during the whole Course.

Table of Contents

PART I : Cell Activation and Biological Response Modifiers

PART II : Membrane - Ligand Interactions and Cell Identity

PART I

Cell Activation and Biological Response Modifiers

CYTOSOLIC EVENTS RELATED TO CELL ACTIVATION:

Potential Interaction between Signal and Energy Transduction Pathways in Smooth Muscle Cells

Bruce D. Jensen
Zynaxis Cell Science, Inc.
371 Phoenixville Pike
Malvern, PA 19355-9603

Introduction

A number of peptide hormones have been shown to induce metabolic and respiratory stimulation in hepatocytes and other mammalian cell types (Balaban and Blum, 1982), demonstrating that there is necessary intercommunication between the signal and energy transduction pathways within the cell. What has not been clear is whether the communication mechanism between these two important pathways is direct or indirect. The answer to this question could have significant impact on our understanding of the regulatory mechanism(s) for cellular energy production. The regulation of cellular energy production has been a major focus of research for many years, with primary candidates for the mechanism(s) of control including the relative levels of various respiratory and phosphorylation substrates (Balaban and Blum, 1982; Rutter and Denton, 1988). Other candidates which have recently gained importance are rate limiting calcium sensitive mitochondrial dehydrogenases (M^c^Cormack and Denton, 1990). It has also been suggested that calcium dependent changes of mitochondrial volume may play a role in metabolic regulation of hepatocytes (Halestrap, 1989). These latter categories of potential control mechanisms are particularly interesting because many hormones are known to induce pronounced rapid intracellular calcium transients of magnitudes consistent with stimulation of mitochondrial dehydrogenases (Tsien et al., 1985; Berk et al., 1987). Further, this type of control mechanism could potentially provide a direct avenue for communication between the signal and energy transduction pathways.

NATO ASI Series, Vol. H 67
Flow Cytometry
Edited by A. Jacquemin-Sablon

The goal of this investigation was to characterize the activation of smooth muscle cells by the peptide hormone 8-arginine-vasopressin (AVP) and to determine whether mitochondrial responses could be observed which are consistent with an interaction between the signal and energy transduction pathways. A number of endogenous and exogenously added probes were used to resolve the sequence of optically detectable events which take place during the cellular activation process. The flow cytometer was employed in this investigation because it possesses an unique combination of characteristics which make it possible to simultaneously time resolve multiple parameters in a cell population.

A rapid biphasic calcium transient was found to be the first optically detectable cellular activation event. This response has been well characterized in the literature (Monck et al., 1988) and is part of the V1 receptor mediated signal transduction pathway. The calcium transient was followed closely by a biphasic right angle light scatter response. Careful characterization of this response demonstrated that it may be consistent with a receptor dependent mitochondrial volume change which could play a role in cellular energy regulation.

Materials and Methods

Material - all cell culture materials were obtained from Gibco Laboratories (Grand Island, NY). Valinomycin, nigericin and antimycin A were obtained from Sigma Chemicals (St. Louis, MO). 8-Arginine-Vasopressin and vasopressin antagonist (SKF 100273) were synthesized at SmithKline and French Laboratories (King of Prussia, PA). Indo-1/AM, BCECF/AM and ionomycin were obtained from Molecular Probes, Inc. (Eugene, OR). Nylon mesh for filtering cell suspensions was obtained from Small Parts, Inc. (Miami, FL). All other chemicals were obtained from various sources and were always reagent grade or better.

Preparation of A10 Cells - Vascular smooth muscle cells, derived from rat thoracic aorta (Kimes and Brandt, 1976), were obtained from the American Type Culture Collection (A10 - ATCC CRL 1476). The cells were subcloned by laser isolation of a single cell using a Meridian ACAS 470 interactive laser cytometer, followed by propagation into a pure culture. Cells were routinely grown in a humidified air/CO_2 incubator (7.5% CO_2) in Dulbecco's Modified Eagle Medium supplemented with 20% fetal bovine serum, and were passaged weekly. Log phase cells were used in all experiments. Cells were harvested by trypsin/EGTA treatment and washed twice in Experimental Buffer Solution (EBS) containing 130mM NaCl, 4mM KCl, 12.5mM Na-MOPS, 12.5mM H-MOPS, 1mM $CaCl_2$, 0.5mM $MgCl_2$, 5mM Glucose and 0.1% bovine serum albumin at pH 7.4. All experimental measurements were performed in this medium. Cells were routinely filtered through 40 μm nylon mesh to eliminate clumps immediately before each experimental measurement was begun.

Loading A10 Cells with Indo-1 - Cells were suspended at a concentration of 10^7/ml in EBS medium and Indo-1/AM was added from a 1 mM stock solution in DMSO to a final concentration of 5μM. The cell suspension was incubated at 37°C in a temperature regulated water bath for 30 minutes. The cells were washed twice in EBS medium and were then resuspended at $5x10^5$/ml. The cell suspension was stored at 4°C until needed. A slow rate of dye leakage was observed which did not interfere with flow cytometric measurements. However, before bulk fluorometric measurements were begun, each specimen was spun down in a centrifuge and resuspended in fresh medium to eliminate interfering external dye signals. Each experimental sample was brought to and incubated at 37°C for 10 minutes before measurements were begun.

Measurement of Cytosolic Free Ca^{2+} Concentration and 90° Angle Light Scatter in A10 Cells - Calcium and light scatter responses were monitored simultaneously using a SLM Aminco SPF-500C spectrofluorometer controlled remotely from an IBM XT computer. The excitation monochrometer was set at 355nm and the emission monochrometer was rapidly moved under computer control from 355nm (light scatter) to 400nm (calcium bound Indo-1 signal). The calcium measurement technique was calibrated according to Grynkiewicz et al. (1985) for single wavelength detection. Briefly, at the end of each calcium experiment the cells were lysed with 0.05% triton-X 100. The 1 mM extracellular calcium concentration saturated the Indo-1, leading to a maximum fluorescence intensity (Fmax). 5 mM EGTA was then added and the pH was adjusted with NaOH until it exceeded 8 units. This resulted in a minimum fluorescence intensity for the probe (Fmin). The calcium concentration was then calculated using these data as described in the quoted reference.

Loading A10 Cells with BCECF for pH Measurement - Cells were suspended at a concentration of 10^7/ml in EBS medium and BCECF/AM was added from a 1 mM stock solution in DMSO to a final concentration of 1μM. The cell suspension was then treated as described above for Indo-1 loading.

Fluorometric Measurement of Cytosolic pH in A10 Cells - Cytosolic pH was monitored ratiometrically by rapidly scanning the excitation monochrometer between 430nm and 500nm and ratioing the emitted intensities at 530nm. The pH response of the BCECF fluorescence intensity was calibrated by clamping the intracellular pH at the extracellular value using the K^+/H^+ exchanging ionophore, nigericin, in a high potassium medium (Thomas et al., 1979; Musgrove et al., 1986). Briefly, multiple aliquots of BCECF loaded cell were suspended in a medium adjusted to various pH values which contained 130mM KCl, 4mM NaCl, 25mM MOPS/NaOH, 1mM $CaCl_2$, and 0.5mM $MgCl_2$. 10μM nigericin was added to each sample and they were allowed to incubate at room temperature for 10 minutes prior to fluorometric analysis. The relationship between the fluorescence intensity and the cytosolic pH was found to be linear from pH 6.7 to 8.4.

Flow Cytometer Configuration and Calcium Measurements - All measurements were performed on a Coulter EPICS 753 triple laser flow cytometer (Coulter Electronics, Hialea, FL) equipped with a high sensitivity flow tip. The front laser was adjusted for an output power of 80 milliwatts at ultraviolet wavelengths (351 and 363 nm). The rear laser was routed around the dye box and adjusted for an output power of 100 milliwatts at 488 nm. The front and rear laser beams were separated in space by 70 μm at the sample stream, allowing the ultraviolet and visible signals to be separated in time. Measurements were triggered by light scatter in the ultraviolet by placing a 365 nm bandpass filter in front of the forward light scatter photodetector. A 440 nm longpass dichroic filter was used to split the emitted signals. The off axis photomultiplier tube was filtered with a 405 nm bandpass filter used to collect the fluorescence signal from calcium bound Indo-1. The on axis photomultiplier tube was filtered with a 485 nm bandpass filter and was used to collect both 488 nm right angle scattered light (RALS) and the calcium free (485 nm) unbound Indo-1 fluorescence signal. A gated amplifier was employed so that the time delayed ultraviolet excited fluorescence signals could be matched up with the undelayed visible RALS signal. Real time analog ratioing was performed on the fluorescence signals from the bound and free Indo-1, followed by digitization and storage in a correlated fashion. All calcium measurements using flow cytometry were monitored as the ratio of the calcium bound and free Indo-1 signals. No attempt was made to calibrate these responses, since their purpose was only to confirm the activation event.

Preparation and Mitochondrial Measurements - Mitochondria were prepared according to previously described methods (Jensen et al., 1986). The intramitochondrial volume and rates of respiration and phosphorylation were also measured as previously described.

Results

Optically Detectable Cellular Activation Events Induced by Vasopressin - Typical calcium and pH responses induced by 100nm AVP are illustrated in figure 1A. The calcium transient is characterized by a very rapid spike response into the micromolar concentration range followed by a return to a stable level approximately 50nm above the equilibrium level. The pH response develops much more slowly than the calcium response, peaking at -0.025 pH units below the equilibrium level. The peak response occurs approximately 1 minute after the peak calcium response. In this cell line, under the present experimental conditions, the pH does not recover to or above the equilibrium level within 5 minutes. Figure 1B shows the right angle light scatter response from a bulk suspension of A10 cells exposed to 100nm AVP. The light scatter response lags behind the calcium response following approximately the same time course as the pH response. However, the light scatter response recovers much like the calcium response, only over a longer time course, consistent with the time lag in onset.

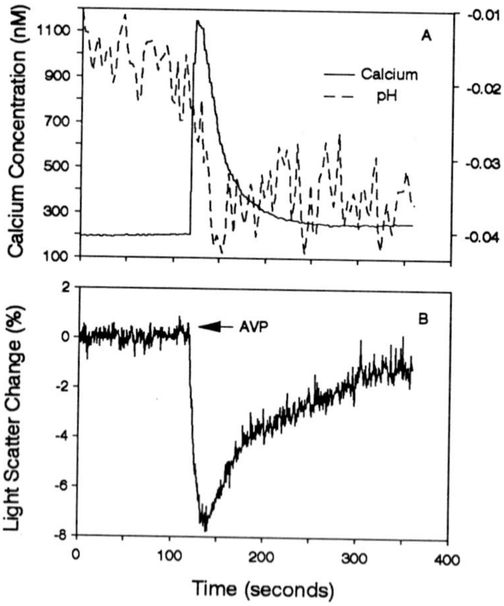

Figure 1 - AVP (100nM) induced responses in A10 smooth muscle cells.

(A) Calcium and pH
(B) Right angle light scatter

The Effect of Respiratory Inhibition on the Light Scatter Response - In an attempt to determine whether mitochondria play a role in the AVP induced light scatter response measurements were performed in the presence and absence of a respiratory inhibitor. Figure 2A shows the control response characterized by rapid onset, peaking within 1 minute followed by partial recovery within 5 minutes. In figure 2B the cells were incubated with 10μM antimycin A for 2 minutes before addition of AVP. The peak response was similar to control, but the recovery phase was completely obliterated. This demonstrates that mitochondria play a role in the scatter response, but whether the role is direct or indirect is not clear.

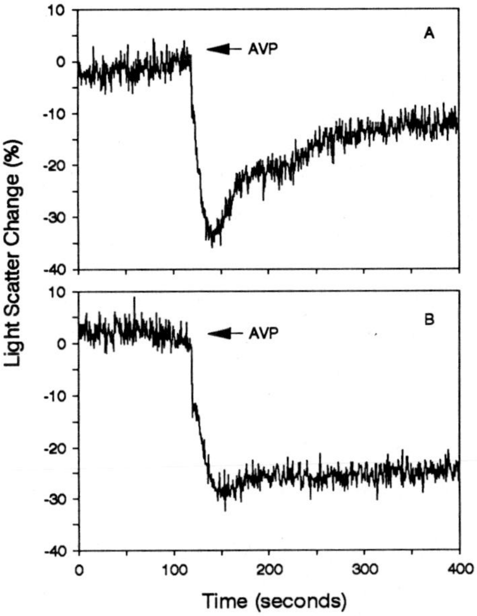

Figure 2 - Effect of respiratory inhibition on AVP induced light scatter response.

(A) Control
(B) 10μM antimycin A

Structural and Metabolic Effects of Ionophores on Isolated Mitochondria - Since the light scatter response could be due to mitochondrial volume changes induced by calcium (Halestrap, 1989), an independent means of affecting mitochondrial volume was sought to see if similar light scatter changes could be induced by known mitochondrial volume changes. Table 1 shows the effects of the ionophores nigericin (K^+/H^+ exchanger) and valinomycin (K^+ uniporter), at the indicated levels, on the volume of isolated rat liver mitochondria. The ionic selectivity of each ionophore cause the volume to increase in the case of valinomycin and decrease with nigericin.

TABLE 1

Effects of Valinomycin and Nigericin on the Volume of Rat Liver Mitochondria

	Control	Valinomycin (2ng/mg)	Nigericin (0.4ng/mg)
Volume (μl/mg)	0.84 ± 0.06	1.20 ± 0.04	0.65 ± 0.04

Light Scatter Response Induced by Various Ionophores - Light scatter responses induced by mitochondrial volume effecting ionophores were examined in figure 3. For this experiment, the AVP induced light scatter response was first examined as a control to demonstrate that the cells were capable of responding. After most of the recovery phase nigericin was added to a final concentration of 5μg/ml, which induced a rapid sustained increase in right angle scattered light. At the point

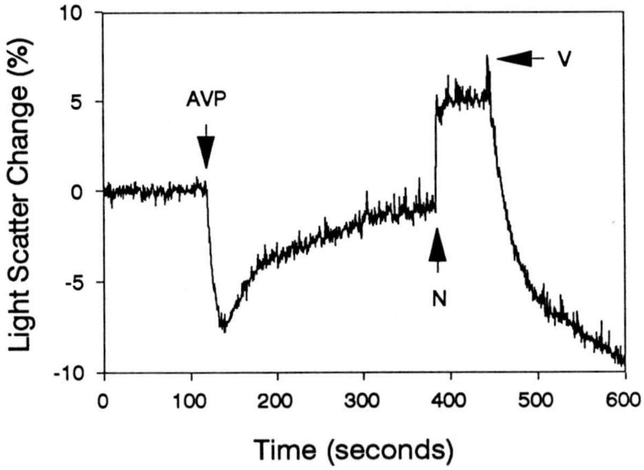

Figure 3 - Effects of mitochondria active ionophores on the right angle light scatter response (AVP - vasopressin; N - 5μg/ml nigericin; V - 1μg/ml valinomycin)

indicated in the figure 1μg/ml of valinomycin was added, inducing a rapid decrease in right angle light scatter.

Comparison of Responses in the Forward and 90° Angle Scattered Light - While it is certain that nigericin and valinomycin induce mitochondrial volume changes, it is also possible that changes in cell volume could also be induced by these ionophores. To determine whether ionophore induced cell volume changes contribute to the light scatter responses forward and right angle light scatter was examined simultaneously by flow cytometry. Forward angle light scatter is strongly dependent on cell volume, while right angle light scatter depends greatly on refractile scatter from internal structures (Salzman et al., 1990). Figure 4A shows the calcium response induced by AVP. Simultaneously with this response, the light scatter responses in the forward and right angle directions were measured, as shown in figure 4B. There was a large response in the right angle direction, with only a very small response in the forward direction. This would suggest that the ionophore induced right angle responses measured in the fluorometer were primarily due to mitochondrial volume changes.

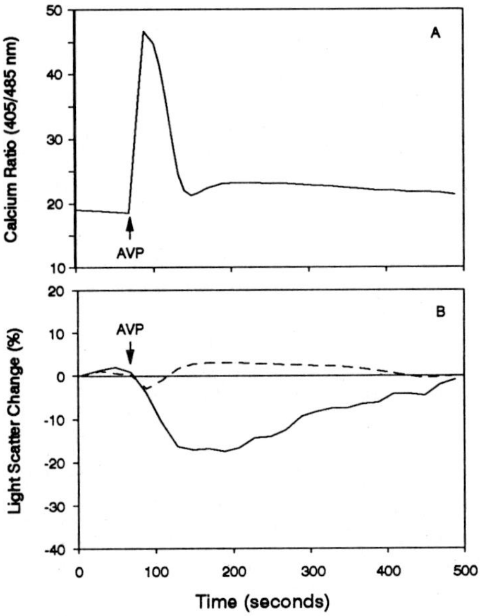

Figure 4 - AVP induced light scatter responses at different angles.

(A) Calcium response
(B) --- Forward angle light scatter (FALS)
—— Right angle light scatter (RALS)

Comparison of Forward and 90° Angle Light Scatter Responses Induced by Ionophores or Direct Osmotic Shrinkage - To confirm that the ionophore induced light scatter responses are not due to changes in the cell volume, responses were examined by flow cytometry. In figure 5A, 5μg/ml nigericin was added where indicated causing a rapid increase in right angle light scatter. Valinomycin (1μg/ml) induced a pronounced decrease in right angle light scatter. There was no change in forward light scatter with either addition. To show that cell volume can affect light scatter, 10x PBS was added to a fresh suspension of cells to a final osmolarity of 3x isotonic (approximately 900 mOsm). This induced a profound decrease in forward light scatter shown in figure 5B, consistent with a cell volume change, while not significantly affecting right angle scattered light.

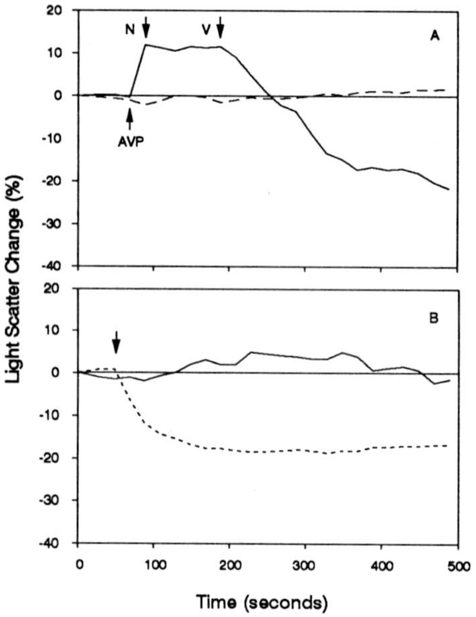

Figure 5 - Light scatter changes induced by ionophores and osmotic changes (--- FALS; —— RALS).

(A) 5μg/ml Nigericin (N) and 1μg/ml Valinomycin (V) induced responses.
(B) Response induced by 3X increase in osmolarity.

Participation of the V1 Receptor in the Light Scatter Response - To determine whether the AVP induced light scatter responses are due to receptor mediated cell activation or are possibly due to some non-specific affect of the hormone, responses were examined in the presence of the vasopressin antagonist SKF 100273. Figures 6A and 6B show the AVP induced calcium response and corresponding right angle light scatter response. If the cells are incubated with 1μM SKF 100273 before the addition of AVP, both responses are eliminated, as shown in the figure. These data indicate that the AVP induced light scatter response is a receptor mediated event. Separate experiment demonstrated that the response was also dose dependent (data not shown).

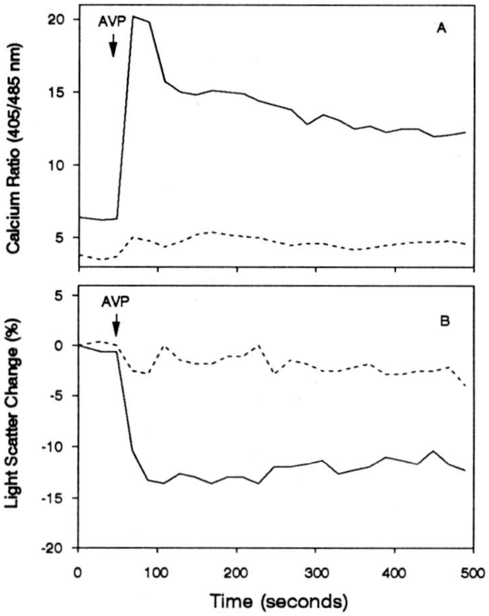

Figure 6 - Effect of vasopressin antagonist (SKF 100273) on calcium and light scatter responses.

(A) Calcium response (—— no SKF 100273; --- 1μM SKF 100273)
(B) Right angle light scatter response

Extracellular Calcium and the Light Scatter Response - Since calcium has been implicated as a regulator of mitochondrial metabolism through its effects on cell volume (McCormack and Denton, 1990; Halestrap, 1989), the light scatter response was examined in the presence and absence of extracellular calcium. Figures 7A and 7B clearly show that receptor mediated influx of calcium is not required for the light scatter response to occur. The response did appear to be reduced somewhat, possibly related to the reduced resting and peak calcium levels in the absence of extracellular calcium.

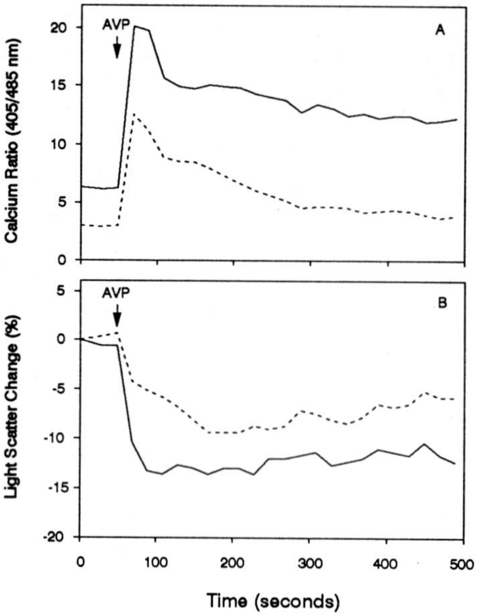

Figure 7 - AVP induced calcium and light scatter responses in the presence and absence of extra-cellular calcium.

(A) Calcium response (—— no EGTA; --- 5mM EGTA)
(B) Right angle light scatter response

Discussion

The data presented indicate that AVP induces the characteristic receptor mediated calcium response in this smooth muscle cell line. Concomitant with the calcium response is a light scatter response which is dose dependent and antagonist inhibitable. Consequently, the light scatter response is a receptor mediated cell activation event. This response has been demonstrated to be consistent with a change in the volume of an intracellular structure. The ionophore effects clearly implicate mitochondria as the internal structures responsible for the light scatter responses. Further, it has been confirmed that part of the light scatter response is directly dependent on mitochondrial respiratory metabolism. It is concluded that the AVP induced light scatter response may be consistent with a receptor dependent mitochondrial volume change which could play a role in cellular energy regulation.

References

Balaban RS and Blum JJ (1982) Hormone-induced Changes in NADH Fluorescence and O_2 Consumption of Rat Hepatocytes. Am. J. Physiol. 242 C172-C177

Berk BC, Brock TA, Gimbrone MA and Alexander RW (1987) Early Agonist-mediated Ionic Events in Cultured Vascular Smooth Muscle Cells. J. Biol. Chem. 262(11) 5065-5072

Grynkiewicz G, Poenie M and Tsien RY (1985) A New Generation of Ca^{2+} Indicators with Greatly Improved Fluorescence Properties. J. Biol. Chem. 6(25) 3440-3450

Halestrap AP (1989) The Regulation of the Matrix Volume of Mammalian Mitochondria In Vivo and In Vitro and Its Role in the Control of Mitochondrial Metabolism. Biochim. Biophys. Acta 973 355-382

Jensen BD, Gunter KK and Gunter TE (1986) The Efficiencies of the Component Steps of Oxidative Phosphorylation. Arch. Biochem. Biophys. 248(1) 305-323

Kimes BW and Brandt BL (1976) Characterization of Two Putative Smooth Muscle Cell Lines from Rat Thoracic Aorta. Exp. Cell Res. 98 349-366

McCormack JG and Denton RM (1990) Role of $Ca2^{+}$ in Mitochondria. Ann. Rev. Physiol. 52 451-466

Monck JR, Reynolds EE, Thomas AP and Williamson JR (1988) Novel Kinetics of Single Cell Ca^{2+} Transients in Stimulated Hepatocytes and A10 Cells Measured using Fura-2 and Fluorescent Videomicroscopy. J. Biol. Chem. 263(10) 4569-4575

Musgrove E, Rugg C and Hedley D (1986) Flow Cytometric Measurement of Cytoplasmic pH: A Critical Evaluation of Available Fluorochromes. Cytometry 7 347-355

Rutter GA and Denton RM (1988) Regulation of NAD^{+}-linked Isocitrate Dehydrogenase and 2-oxoglutarate Dehydrogenase by Ca^{2+} Ions within Toluene Permeabilized Rat Heart Mitochondria. Interactions with Regulation by Adenine Nucleotides and $NADH/NAD^{+}$ ratios. Biochem. J. 252 181-189

Salzman GC, Singham SB, Johnston RG and Bohren CF (1990) Light Scattering and Cytometry. In Flow Cytometry and Sorting (eds. Melamed M, Lindmo T and Mendelsohn ML) Wiley-Liss, New York, NY, pp.81-107

Thomas JA, Buchsbaum RN, Zimniak A and Racker E (1979) Intracellular pH measurement in Ehrlich Ascites Tumor Cells Utilizing Spectroscopic Probes Generated in situ. Biochem. 18 2210-2218

Tsien RY, Rink TJ and Poenie M (1985) Measurement of Cytosolic Free Ca^{2+} in Individual Small Cells using Fluorescence Microscopy with Dual Excitation Wavelengths. Cell Calcium 6 145-157

Studies of Changes in Cytoplasmic pH and Membrane Potential Using Flow Cytometry

Ronald E. Worthington† and Jean-Pierre Aubry*
†Health Sciences Department, Monsanto Corporate Research, St. Louis, MO USA,
and * Glaxo Institute for Molecular Biology, Geneva, Switzerland

Cell activation means the changes in functional and metabolic activity that occur when specific signalling pathways, either alone or in combination, respond to engagement of signal-transducing molecules. These changes include ion fluxes that may result in net alterations of cytoplasmic ion concentration and thereby affect enzyme activation or inhibition, or that may not actually change ion concentration, but may nonetheless cause physical or chemical changes, such as membrane depolarization. Modifications in assembly of the cytoskeleton, secretion of new bioactive metabolites, and expression of stored or masked molecules frequently ensue.

In this article two specific aspects of cell activation will be discussed in the context of their experimental measurement: flux in the ions and solutes that may alter cytoplasmic pH, and membrane depolarization. These discussions are restricted to cells that are not considered excitable in the sense of neuromuscular physiology, and they are not intended to comprise a broad overview. The intent of this article is to provide a few examples of fluorometric measurement of pH and membrane potential changes using reagents that are likely to work with a single argon laser flow cytometer. The purpose here is also to convey three messages: (1) that pathways of signalling, and in particularly those regulating ion flux, do not operate in isolation from each other; (2) that experimental techniques to measure changes in ion flux may profoundly influence general cell metabolism and thus confound interpretation of results; and (3) that significant interactions among pathways do indeed occur.

Methods for studying cell pH changes by fluorescence cytometry

All cytometry to quantitate kinetic changes in fluorescence should be performed using a ratioing technique (Grynkiewicz, 1985). This is because rationing is the only approach that is relatively insensitive to experimentally uncontrollable factors such as dye concentration, optical parameters such as refractive index and cell thickness, and photobleaching, all of which are subject to change during the course of observation. In addition to the obvious requirements of sensitivity and high signal to noise ratio, another important attribute of a pH sensing dye is that the pK_a fall within the range of pH values expected, preferably at midrange. Most of the dyes developed so far for measurement of cytoplasmic pH are of the excitation ratio variety and so are not suitable for single laser cytometry. Several dyes that are acceptable for use with single Ar laser cytometry are listed in Table 1. BCECF can

NATO ASI Series, Vol. H 67
Flow Cytometry
Edited by A. Jacquemin-Sablon

be used in a pseudo-ratioing method as documented by Musgrove et al. (1986). Although the literature using BCECF is the most voluminous, the newer SNAFL-calcein indicator has particularly high potential because it is apparently better retained in the cytoplasm, and the fluorescence signal is much brighter than the parent SNAFL compound. A few questions have arisen regarding the stability and shelf life of the calcein derivatives of the pH sensitive indicators, but these are probably minor technical issues that will be solved by the manufacturer, Molecular Probes.

In general, cells are loaded in a minimal physiological buffer using the lowest dye concentration, e.g., 1-10 µM and shortest time, e.g., 15-60 minutes, that still results in a good fluorescence signal. As mentioned below, problems with aldehyde toxicity may be alleviated by addition of 1 mM pyruvate to the staining buffer. Once staining of the cells has been verified, it is necessary to confirm that the dye correctly responds to changes in cytoplasmic pH. This can be tested be adding 10-50 mM NH_4Cl, upon which an alkalinization signal should be generated. If it is possible to wash out the NH_4Cl, there should be a rapid cytoplasmic acidification followed by an amiloride-sensitive return to near baseline pH. Similar treatment with 0.1 - 1 mM propionic acid induces a rapid acidification that should be compensated by activation of the Na^+-H^+ antiporter in most cells. A failure of these methods to change fluorescence in the expected direction could mean that the cells are not adequately loaded, or that they have loaded and then transported the free acid into vesicles whose pH is relatively independent from that of the cytoplasm. Failure of cells to respond to the NH_4Cl washout could mean that the loading conditions have resulted in metabolic damage to the cells.

Table 1. Dyes for Single Laser Cytometry of Cytoplasmic pH

Dye	pKa	Excitation/ Emission	True Ratio?	Fluorescence Change upon Alkalinization
BCECF	6.98	488/530÷620	No	increase
SNARF-1	7.30	514/640÷610	yes	increase
SNAFL-calcein	7.00	488/530÷620	yes	decrease

Cells stained with any of the dyes listed in Table 1 can be calibrated using the nigericin clamping technique introduced by Thomas et al. (1979). This method involves elevating extracellular K^+ to a concentration close to $[K^+]_i$, then collapsing the residual membrane potential with the ionophore nigericin, which equilibrates both K^+ and protons across the plasma membrane. Then by grace of the Nernst and Henderson-Hasselbalch equations, which define the relationship between internal and external pH (pH_i and pH_o, respectively) under equilibrium conditions for proton flux, it is possible clamp the pH_i to the pH_o, and correlate fluorescence with the experimentally defined pH_o:

$$pH_i = pH_o + (F \bullet V_m) \div (2.3 \bullet R \bullet T)$$

where F is the Faraday constant, V_m is membrane voltage, R is the gas constant and T is absolute temperature.

An example calibration experiment performed in DAUDI (a Burkitt's lymphoma) cells using a Coulter ELITE® cytometer is shown in Figure 1, where it is evident that the fluorescence response is linear over the narrow range of pH_i expected in these and most other cells. The data in Figure 1 represent single emission fluorescence at 530 nm, but ratio data in the same experiment (not shown) were superimposable.

It is also possible to study the pH of cytoplasmic compartments using cationic dyes such as rhodamine 123 for mitochondria (Johnson et al., 1980) and NERF congeners for endocytic vessicles (Haugland, 1991). These probes may be useful in studying the effects of drugs on mitochondrial physiology and in studies of ligand processing following phagocytosis of receptor-ligand internalization.

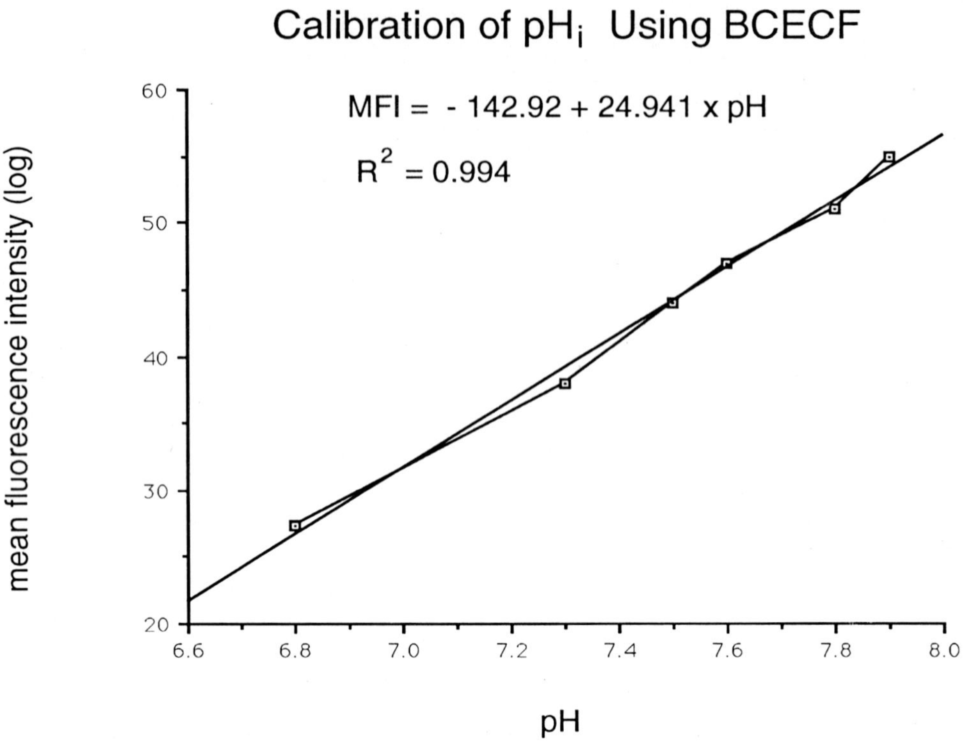

Figure 1

There is controversy concerning both the significance and meaning of cytoplasmic pH change. While the activity of isolated cytosolic enzymes frequently shows a pH dependence, it is difficult to propose a model where, in the highly complex milieu of the cytosol, pH change would work in the same direction for all the necessary enzymes to elicit a concerted functional response. Indeed, some experts have suggested that pH of the cytoplasm cannot deviate much from its norm without inducing pathological changes in the cell (Thomas, 1989). Yet there is no doubt that some of the pH-influencing solute transporters, the best characterized of which is the Na^+/H^+ antiporter, do increase or decrease their activity in response to cell activation.

Why should changes in acid/base pumping activity occur upon cell activation, if not to cause a deliberate change in cytosolic pH? Ganz et al. (1989) suggest that antiporter activity is regulated in anticipation of increased acid/base production expected to occur when cells are metabolically activated. This group also showed that vasopressin-induced alkalinization of mesangial cells depends entirely on the absence of bicarbonate anion in the cell bath, and that when replete with HCO_3^- , experiments with BCECF actually show cytosolic acidification, not alkalinization. While their hypothesis is compelling, others continue to report that in other cells, platelets for example, the alkalinization observed with thrombin activation occurs in the presence or absence of bicarbonate, again using BCECF as the pH reporter (Sage, 1990) These contrasts and contradictions must encourage the prospective student of cytoplasmic pH dynamics to carefully scrutinize the experimental design and buffer composition and pay very close attention to the interpretation of fluorescence data obtained with pH sensitive probes. At the very least, every experiment where the external buffer is nominally bicarbonate-free must be performed in the presence of bicarbonate.

Another problem, that is usually overlooked by eager investigators just entering into studies of cytoplasmic fluorescence with hydrolyzable probes, is the effect that processing of the probe and its metabolites may have on cell metabolism itself. Five sites in the acetoxymethylester of BCECF must be hydrolysed to give the fully functional free acid. Many ostensibly “non-invasive” fluorescent probes are modified with acetoxyester residues to enhance their membrane permeability. The metabolism of BCECF-AM in this example probably is representative for all such derivatives, although the stoichiometry of the byproducts will depend on the specific dye. It is clear that BCECF-AM must be hydrolyzed at 7 sites to yield the active acid. In this process, 2 moles of formaldehyde, 2 moles of acetic acid , and 4 moles of proton are liberated for each mole of BCECF-AM fully hydrolyzed. It is obvious that these metabolites would pose a considerable metabolic burden on the cell. The liberation of four acid equivalents itself forces one to consider the direct effects that loading with BCECF-AM must have on pH homeostasis. It may be less obvious that these reactants can have profound effects on fundamental metabolism, such as ATP production from glucose. For example, the formaldehyde generated by hydrolysis of the AM esters can significantly shift the NADH to NAD ratio, and thereby alter the redox state of the cell. This has been well documented by Tiffert et al. (1984) for calcium-sensitive acetoxymethylester derivatives in red blood cells.These potential toxicities must be kept in mind for all the AM-derived dyes.

Experiments to estimate cytoplasmic pH must involve a calibration procedure to ensure reproducibility and to establish the quantitative relationship between changes of fluorescence and intracellular pH. If the dye has been optimized for the cells and the cytometer is functioning flawlessly, then the procedure is relatively straightforward. After loading with dye, the cells are washed and resuspended in a well buffered, high K^+ medium such as that described in Table 2. The critical ingredient is the antibiotic nigericin, which equilibrates both K^+ and H^+ across the membrane, thereby depolarizing the cell completely and making internal pH approximately equal to the experimentally determined external pH. The relationship between mean fluorescence, recorded either as linear or log-amplified single parameter, or as a ratio of two emissions, and pH should be essentially linear over the range of pH 7.0 to 7.8, for the dyes listed in Table 1. The measurement of the unknown sample must then lie within the range of the curve for meaningful interpolation.

Table 2. Buffers for Calibration of pH_i †

Solute	Primary Buffer	Correction Buffer
KCl	140 mM	140 mM
$CaCl_2$	1 mM	1 mM
NaCl	5 mM	5 mM
tris acid	19.04 mM	0
tris base	0.96 mM	20 mM
nigericin	10 μg/ml	10 μg/ml

† primary buffer is adjusted to pH 6.80, then correction buffer is added in five 1/10 volume aliquots to achieve pH_o = 7.3, 7.5, 7.6, 7.8, 7.9

There are several important problems with studies of continuous change using flow cytometry. Some vendors offer solutions that involve kinetic measurement of fluorescence response over time. However, extremely important issues such as sample temperature, delivery of agonist and/or inhibitor, time delay due to sample transit in the fluidics, and others frequently are not all resolved together in any given system. Frankly, it may be better to begin with a stirred-cuvette spectrofluorometer system to obtain an estimate of the time and condition-dependence of the fluorescence response, and then translate conditions to the flow cytometer for cell-by-cell analysis.

Methods for studying cell depolarization by fluorescence cytometry

All vertebrate cells have a voltage across their membranes which is subject to change. The potential is the result of the distribution of large and small molecular weight ions across membranes that show an enormous range in permeability to those ions, from essentially zero to continuous but low-level flux (leakage). In excitable cells, these changes are used to transmit signals across a network, such as neurons, or to stimulate a high contractile force, such as in myocytes. In non-excitable cells, membrane depolarization usually has nothing to do with transmitting signals across a network or generating strong contraction. Instead, depolarization results from alterations that are occurring in membrane permeability and as a result of other prior events. Some functions, such as the membrane conductance for certain solutes, are in turn influenced by the state of membrane polarization. There is clearly an opportunity for recursive and otherwise complex phenomena to affect membrane potential.

The fact that the causes and effects of membrane depolarization in non-excitable cells are likely to be complicated is illustrated by the cartoons in Fig. 2. These drawings illustrate generically the kinds of channels or transporters that are expected to directly and indirectly influence membrane voltage. In fact, the physiological evidence for the diversity in K^+ channels that is accumulating from major recent experimental innovations, especially patch-clamping, suggests that the situation is far more complex than these cartoons would suggest.

Illustrated are three channels and one transporter, each of which have been shown, in various cells, to influence membrane potential. The prominent system is the transporter that not only pumps two ions against steep chemical gradients, but is also electrogenic: the Na^+/K^+-ATPase. This is a terribly energy-hungry process that can expend a major proportion of a cell's total ATP production. In addition to this system, there exist less well-characterized transport activities that probably are physically present as channels, although this remains to be demonstrated in most cases.

Another channel shown, the ATP-inhibitable K^+ channel, does not consume ATP, but is regulated by cytosolic concentrations of ATP that are physiological, so that conductance by this route is mid-range when ATP levels are normal. But if ATP levels are elevated, such as in pancreatic beta-cells stimulated with the high glucose levels seen in patients with uncontrolled diabetes, then these channels close and may thus promote depolarization of the cell (Boyd III, 1988).

The variety and significance of calcium channels and transport systems is beyond the scope of this discussion, but it is important to stress, within the context of transmembrane voltage, that cytoplasmic calcium concentrations can both regulate and be regulated by membrane potential (Miller, 1992). This is illustrated in the cartoon by the voltage-gated calcium channel, and also the calcium-gated K^+ channel (MacDougall et al., 1988). Part B of this figure illustrates the status of all these channels/transporters in their predicted state after a certain but unknown depolarization threshold has been reached.

Major Ion Channels: Resting, Polarized Membrane

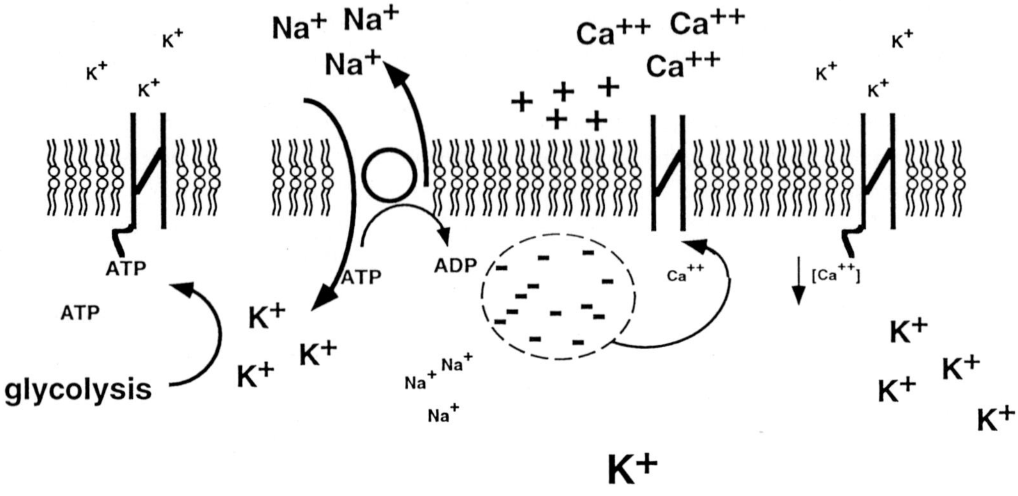

Figure 2a

Major Ion Channels: Depolarized Membrane

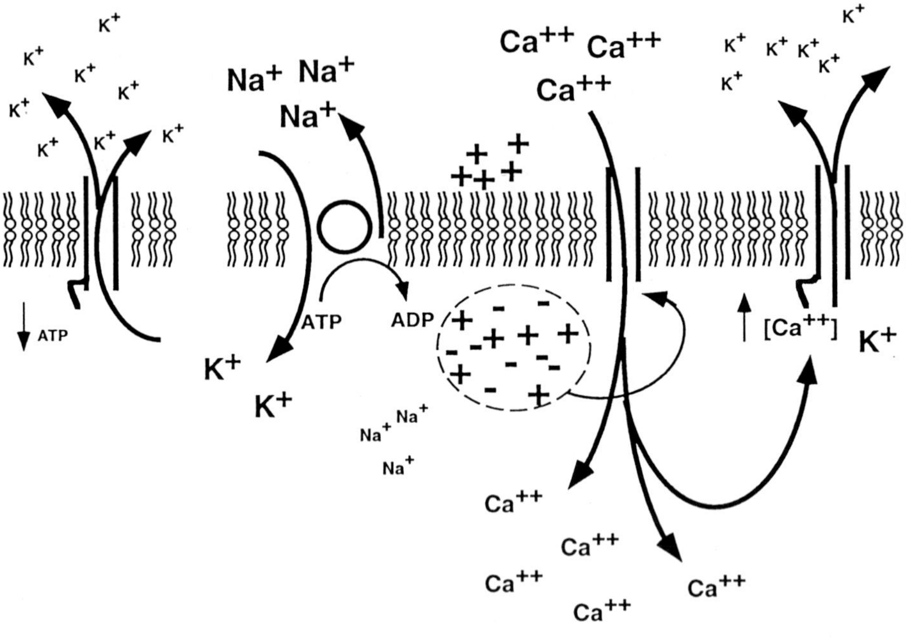

Figure 2b

Table 3. Dyes for Single Laser Cytometry of Membrane Depolarization

Class	Examples	Excitation/ Emission	Charge	Fluorescence Change upon Depolarization
oxonols	DiBAC4(3)	488/520	anionic	increase
carbocyanines	DiOC6(3)	488/510	cationic	decrease
	DiOC2(3)	488/500	cationic	decrease
	DiOC5(3)	488/509	cationic	decrease

The two classes of dyes recommended for use in depolarization studies with a 488 nm excitation line are listed in Table 3. Of these, the oxonol dyes are preferred for three reasons: (1) the fluorescence signal increases with depolarization; (2) the oxonols are anionic and thus do not interact with mitochondrial membranes; and (3) the oxonols are usually less toxic than the cyanine dyes. Because of toxicity and quenching that occur at micromolar concentrations of dye, these are best used in the nanomolar range. In general, 15 minutes at ambient temperature or 37° is sufficient to perform an experiment. The cells must be left in the dye solution, and indeed, the fluidic system of the cytometer must be equilibrated beforehand with buffer containing the same dye concentration.

A good first approach to determine whether the dye is responsive to depolarization is to use the antibiotic gramicidin A. This molecule spontaneously forms a pore in the membrane that for practical purposes renders the membrane infinitely permeable to K^+ and Na^+cations. The channel has a higher equilibrium conductance for K^+, and since the instantaneous electrical gradient favors cation influx, there may be an initial depolarization via K^+ influx with a secondary equilibration of potential based on the movement of Na^+ cations. In any event, gramicidin treatment (10 μg/ml) is probably the most straightforward way to determine whether the staining procedure is working at all.

Another method to verify depolarization responsiveness of the dye is to suddenly increase the extracellular K^+ concentration by addition of a small volume of high strength KCl. Increasing $[K^+]_{out}$ to 50 mM or greater will cause partial depolarization according to the Nernst relationship. However, this method will also increase ionic and osmotic strength significantly, and thus other indirect effects, such as dye efflux, may cause a spurious signal change.

References Cited

Boyd III, A.E. "Sulfonylurea receptors, ion channels, and fruit flies." Diabetes 37 (1988): 847-850.

Ganz, M.B., G. Boyarsky, R.B. Sterzel, and W.F. Boron. "Arginine vasopressin enhances pHi regulation in the presence of bicarbonate by stimulating three acid-base transport systems." Nature 337 (1989): 648-651.

Grynkiewicz, G., M. Poenie, and R. Y. Tsien. "A new generation of Ca++ indicators with greatly improved fluorescence properties." Journal of Biological Chemistry (260 1985): 3440-3450.

Haugland, R.P. (1991) Bioprobes. Volume 13, April. Molecular Probes, Inc. Eugene, Oregon, USA.

Johnson, L.V., Walsh, M.L., Chen, L.B. (1980) Localization of mtiochondria in living cells with rhodamine 123. Proceedings of the National Academy of Sciences (USA) 77:990-998.

MacDougall, S.L., Grinstein, S., Gelfand, E.W. (1988) Activation of Ca^{++} dependent K^{+} channels in human B lymphocytes by anti-immunoglobulin. Journal of Clinical Investigation 81: 449-454.

Miller, Richard J. "Voltage-sensitive Ca^{2+} Channels." The Journal of Biological Chemistry 267 (Jan 25th 1992): 1403-1406.

Musgrove, E., Rugg, C., Hedley, D. (1986) Flow cytometric measurement of cytoplasmic pH: a critical evaluation of available fluorochromes. Cytometry 7:347-355.

Sage, S.O., T.M. Jobson, and T.J. Rink. "Agonist-evoked changes in cytosolic pH and calcium concentration in human platelets: studies in physiological bicarbonate,." Journal of Physiology 420 (1990): 31-45.

Thomas, J.A., R.N. Birschbaum, A. Zimniak, and E. Racker. "Intracellular pH measurements in Ehrlich ascites tumor cells utilizing spectroscopic probes generated in situ." Biochemistry 18 (1979): 2210-2218.

Thomas, R.C. "Bicarbonate and phi response." Nature 337 (1989): 601.

Tiffert, T., J. Garcia-Sancho, and V.L. Lew. "Irreversible ATP depletion caused by low concentrations of formaldehyde and of calcium-chelator esters in intact human red cells." Biochemica et Biophysica Acta 773 (1984): 143-156.

General References

Grinstein S, Dixon SJ (1989) Ion transport, membrane potential, and cytoplasmic pH in lymphocytes: changes during activation. Physiological Reviews 69:417-481.

Haugland, R.P. (1989) Handbook of fluorescent probes and research chemicals. Molecular Probes, Inc. Eugene, Oregon, USA. [Note: new edition available in late 1992]

Justement, L.B., JC Cambier, GT Rijkers, C Fittschen (1990) Use of flow cytometry to study ion transport in lymphocytes and granulocytes. In: Foskett JK & Grinstein S (eds) Noninvasive techniques in cell biology. Modern Cell Biology Vol. 9. Wiley-Liss, New York, 1990.

pH Homeostasis: Mechanisms and Control. D. Häussinger (ed). Academic Press, London, 1988.

Rabinovitch PS and June CH (1990) Measurement of intracellular ionized calcium and membrane potential. In: Melamed MR, Lindmo T, Mendelsohn ML (eds) Flow cytometry and Sorting, 2nd ed., WIley-Liss, New York, 1990.

Roos A, Boron WF (1981). Intracellular pH. Physiological Reviews 61:296-434.

Shapiro, H.M. (1988) Flow cytometric probes of cell activation. In: Practical Flow Cytometry, A. R. Liss, New York, pp. 178-198.

MEASUREMENT OF ENZYME KINETICS BY FLOW CYTOMETRY

Caroline Dive, Paul Workman[1], Babetta L Marrone[2] and James V Watson[3].

CRC Molecular and Cellular Pharmacology Group,
Manchester University School of Biological Sciences,
Oxford Road, Manchester M13 9PT
UK.

INTRODUCTION

Measurement of time-dependent biological events by flow cytometry (FCM) is almost as old as the technology itself, the most widely appreciated examples being measurement of DNA cell cycle kinetics (Gray 1976). In this type of study the time-dependent biological reference frame is long and cell populations are sampled hourly to give the required resolution and information. Here we focus on the use of FCM to study dynamic events within a temporal reference frame of a few seconds to minutes during an enzyme reaction, a technique termed flow cytoenzymology (FCE). The methodology is similar to that currently employed to determine rates of influx or efflux of fluorescent molecules across the cell membrane, binding of ligands to intracellular constituents, and changes in intracellular ions or membrane potential in response to various stimuli.

Enzyme levels can be determined by FCM using antibodies or fluorescently-labelled tight-binding inhibitors (Assaraf et al 1989; Coon & Kaplan 1950). Enzyme activities can be determined

---

[1]CRC Beatson Laboratories, Garscube Estate, Switchback Road, Glasgow B61 1BD, UK

[2]Life Sciences Division: Los Alamos National Laboratory, Los Alamos, New Mexico 87545, USA

[3]MRC Clinical Oncology and Radiotherapeutics Unit, Addenbrookes Hospital, Hills Road, Cambridge CB2 2QQ, UK

NATO ASI Series, Vol. H 67
Flow Cytometry
Edited by A. Jacquemin-Sablon

using fluorogenic substrates where with increasing time after mixing cells with substrate, fluorescent product accumulates within or on the cells and this can be monitored quantitatively. This phenomenon was first described as fluorochromasia by Rotman and Papermaster (1966) and was initially exploited using FCM by Hulett et al in 1969.

FCE has important advantages over conventional biochemical and histochemical assays of enzyme activities. Conventional biochemical methods provide quantitative data, but heterogeneity within biological samples is completely masked by average values. Useful qualitative and structural information regarding the heterogeneous distribution of enzymes within a tissue can be obtained by traditional histochemistry; however, quantitation is often problematic. Furthermore, histochemical procedures are slow and cannot provide dynamic biochemical data. FCE determines the enzyme activity of each intact cell in a sample, and this allows identification and quantitation of cell subpopulations within the sample which differ in enzyme activities. The technique is also rapid and highly sensitive; a detection limit equivalent to 150 molecules of free fluorescein per cell has been achieved (Watson & Walport 1986). In addition, excellent statistical precision is obtained due to the large number of cellular events which can be recorded. Another important advantage of flow cytometry is the capacity for multiparametric analysis on an individual cell basis. Potential exists for measurement of multiple substrates and enzyme activities simultaneously with other biologically and pharmacologically relevant cellular parameters such as DNA content, glutathione and oncoproteins. One disadvantage of FCE with respect to histochemical assays is the loss of structural information in solid tissues due to the requirement for a suspension of single cells; another disadvantage is that the massive amount of data generated takes up massive amounts of storage space! Clearly, in order to measure the dynamics of reactions it is necessary to include time in the data base.

INCORPORATION OF TIME INTO FLOW CYTOMETRIC MEASUREMENT

There are four methods of incorporating time into the flow cytometry data base.

Discontinuous sequential sampling

This is appropriate whenever the dynamic event under study is varying within a time frame of ten minutes to hours. The population is sampled discretely at defined intervals after the initiating event and the measurements are made at defined time intervals. Examples include mixing cells with substrate for enzyme kinetics of 'slow' reactions. After analysis the results obtained at each time point are plotted against the time at which the measurements were recorded.

Continuous interrupted sampling

With this technique, cells are mixed with substrate and the sample is introduced into the instrument as fast as possible, after which the sample runs continuously through the instrument. 'As fast as possible' in this context usually means a 'dead' time of about 20-30 seconds. Thereafter, the data acquisition computer is instructed to record for a defined interval, say 5 seconds, before recommencing data acquisition. The median of the distribution obtained within each 5 second recording interval can be printed out during the period when data are not being collected. The median values are then plotted against time to generate a reaction progress curve (Watson et al 1977; Watson et al 1978).

Continuous time recording

The continuous interrupted sampling technique has one serious defect. The fluorescence from substrate hydrolysis increases during the recording interval; hence the distributions obtained during the recording interval are skewed to the right and the median values are overestimated. This is not a problem for 'slow' reactions but it could introduce a large error in the measurement of 'fast' reactions. Two solutions to this problem where time is introduced directly into the data base were forthcoming. In 1980, Martin and Swartzendruber first used a voltage ramp generator to achieve this, where the potential increased linearly with time and a recording of the voltage was made from the ramp generator each time a cell was analysed in the flow chamber. We incorporated the time stamp from the data acquisition computer directly into the data base (Watson 1987). Hence, events can be timed automatically with 50 msec resolution by the computer clock and this time, in the form of the number of clock 'ticks' (one tick per 50 msec), is recorded for each event analysed.

Reactions in the time frame 1-15 seconds.

Each of the various methods described above suffers from one considerable disadvantage, which is the finite time taken between mixing cells with substrate and recording the first event. This 'dead time' is due to mixing, location of the containing vessel in the instrument, surge pumping the reaction mixture through the sample feed tube into the flow chamber, and restoration of stable flow. There is a clear need to reduce this temporal delay, so as to facilitate the analysis of biochemical events taking place over the time period of a few seconds, not only for FCE but also for measurement of cellular ion flux. We approached this problem in two ways (Habbersett et al 1990, Watson et al 1988). We designed a new sample chamber with a simple rubber septum injection port and mounted it such that the minimum tubing was required to take cells to the analysis point (Habbersett et al 1990). Cells were placed in the chamber to a level above the septum and baseline flow conditions were stabilised. Fluorogenic substrate was rapidly injected through the septum, immediately mixed and boosted under high pressure to the analysis point. The first fluorescence events were measured 1 second after fluorochrome addition. We also developed a system which markedly reduces the temporal reference frame so that this can be varied within the interval 1-20 seconds (Watson et al 1980). This employs computer-controlled precision drive syringe pumps, one for substrate the other for cells, together with variable length tubing between a mixing chamber and analysis point, selected by an array of zero dead-space pinch valves. The different tube lengths at a given pump flow rate give different times between mixing and analysis.

MEASUREMENT OF REACTION KINETICS

Fluorogenic Substrates

In order to define precisely cellular enzyme reaction kinetics by FCE, a fluorogenic substrate is required, ideally with the following characteristics. It must be lipophilic, non fluorescent, non-toxic and specific for the enzyme under study. A schematic for the conversion of such a substrate to product by enzyme action in whole cells is shown in Figure 1.

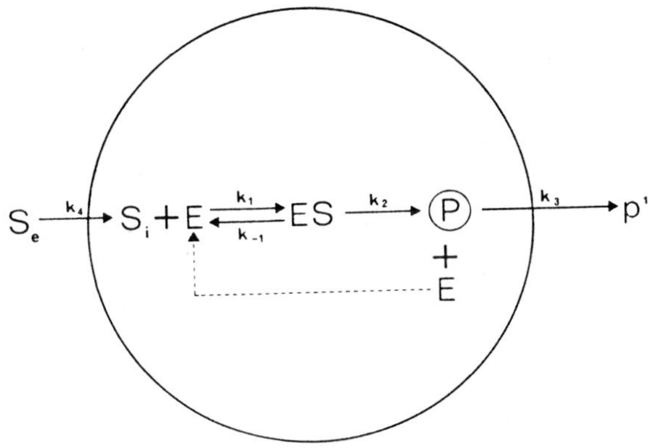

Figure 1 Schematic for the conversion of substrate (S_e) to product (P) by enzyme action in intact cells. Substrate must cross the plasma membrane and interact with the enzyme to generate the fluorescent product which will be detected as it accumulates intracellularly only if it remains inside of associated with the cell. Reproduced from An Introduction to Flow Cytometry by JV Watson with permission of Cambridge Univeristy Press.

Unlike classical biochemical assays, only product which is cell-associated is 'seen' by the flow cytometer. Therefore the reaction product ideally should be highly fluorescent and lipophobic such that it accumulates inside the cell in which it was formed with minimal leakage from that cell. There is now a considerable variety of such substrates based on a number of different fluorophores. These include fluorescein, rhodamine, naphthol, naphthylamine, quinolines, methyl-umbelliferone, resorufins and monochlorobimane for studying esterases, lipases, sulphatases, phosphatases, glucuronidases, transferases, peroxidases, peptidases, transpeptidases, galactosidases, arylamidases, glucosidases, and glutathione S-transferases. In general, the fluorescein and rhodamine based substrates and monochlorobimane are either weakly or non-fluorescent which is ideal. In contrast, the 4-methylumbelliferone conjugates are generally fluorescent but both their excitation and emission wavelengths are much shorter than those of the released product 4-methylumbelliferone and with correct optical design any 'overlap' can be minimized or eliminated. Techniques using the naphthol derivates involve 'trapping' the released product within the cell by coupling with 5-nitrosalicyl-aldehyde which forms an insoluble fluorescent complex with an emission spectrum that is shifted to a longer wavelength.

REACTION KINETICS, PRODUCT EFFLUX KINETICS AND INHIBITION KINETICS

There are now a large number of substrates available for quantitating enzyme reaction kinetics using flow cytometry. Three systems are illustrated here:

a) Measurement of cellular esterases using fluorescein diacetate (FDA). This system is used to exemplify measurement of enzyme inhibition kinetics and of enzyme kinetics in heterogeneous populations of bone marrow cells.

b) Measurement of γ-glutamyl transferase (GGT) using a coumarin-based fluorogenic substrate. This is a membrane associated enzyme which poses particular problems for FCE measurements.

c) Measurement of the activity of the rate limiting enzyme for steroid biosynthesis, cytochrome P-450_{SCC} (cholesterol side chain cleavage), using a specific fluorogenic substrate recently designed and synthesised by Marrone and co-workers in Los Alamos, New Mexico.

a) ESTERASES

Reaction kinetics, product efflux kinetics and inhibition kinetics in EMT6 mouse mammary tumour cells

Non-specific cellular esterases convert FDA to fluorescein whose fluorescence is excited at 488nm and detected between 515-545nm (Figure 2).

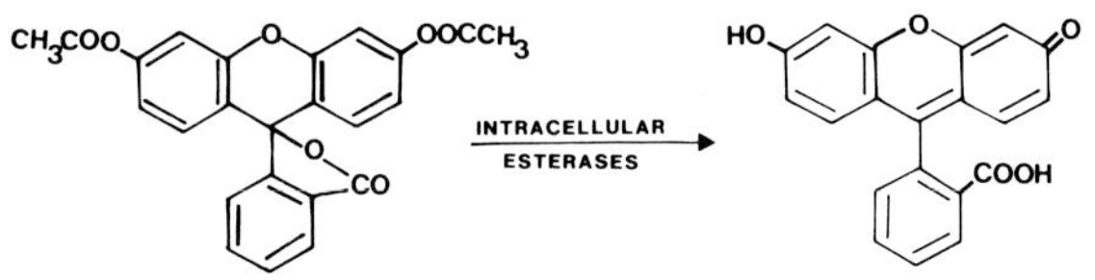

Figure 2 Schematic for the conversion of fluorescein diacetate to fluorescein by non-specific esterases.

The reaction between log phase EMT6 cells (10^6/ml) and FDA, measured over 5 minutes with a flow rate of 166 cells/second can be displayed in several ways as shown in Figure 3 (Dive et al 1987a).

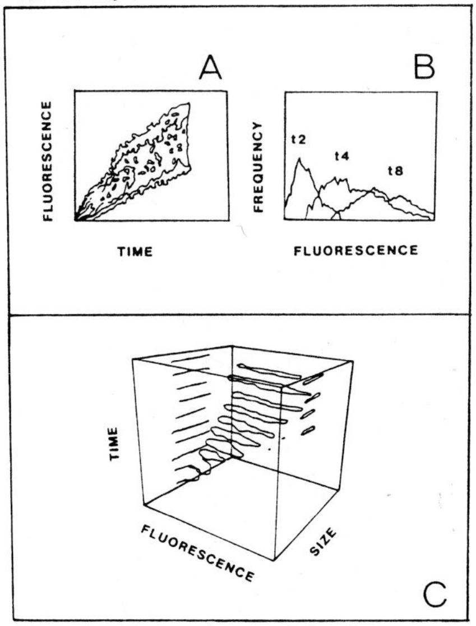

Figure 3 Flow cytometric monitoring of the hydrolysis of FDA by the intracellular esterases of EMT6 mouse mammary tumour cells. Panels A-C illustrate alternative displays of a typical data set for FDA (1μM). **A.** 2-D frequency contour plot of fluorescence versus time. **B.** Sequential fluorescence histograms for the 2nd, 4th and 8th of 8 equal time slices, the overall time being 5 minutes. **C.** 3-D display of fluorescence versus time versus forward scatter (size).

In order to obtain reaction progress curves for substrate concentrations ranging from 0.2-2.0μM, we took the median fluorescence of each of 8 equal time slices throughout the 5 minute reaction. Plotting the reciprocal of the initial velocity versus the reciprocal of substrate concentration (the Lineweaver-Burke Plot) enables calculation of the apparent Michaelis constant for the reaction (Figure 4, Dive et al 1987a).

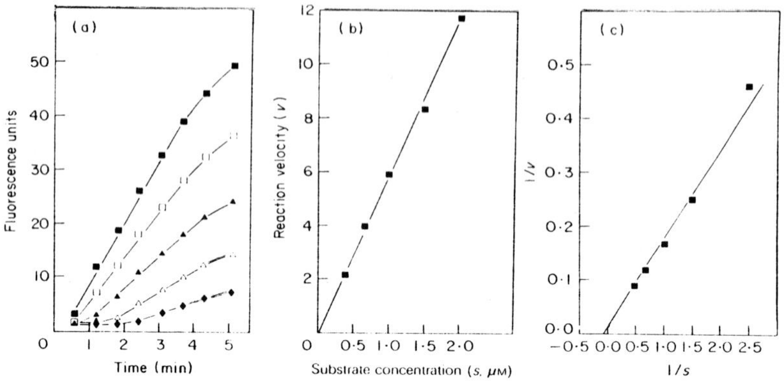

Figure 4 Flow cytometric determination of reaction kinetics for the hydrolysis of FDA by intracellular esterases of EMT6 cells. (a) Enzyme reaction progress curves for varying substrate concentrations: closed diamonds 0.40μM, open triangles 0.67μM, closed triangles 1.00μM, open squares 1.50μM and closed squares 2.00μM. (b) Plot of reaction velocity versus time. (c) Lineweaver Burke Double Reciprocal Plot.

Clearly this apparent Km of 18.6μM reflects the rate of substrate entry into the cell, the rate of reaction of FDA with cellular esterases and the rate of FDA efflux from the cell.

Product Efflux

In order to define the rate of fluorescein efflux from EMT6 cells, they were reacted with FDA for 10 minutes. Subsequently, cells were washed twice in buffer, reintroduced into the flow cytometer and the fluorescence measured over 15-30 minutes (Figure 5). The time in which half the fluorescence is lost is determined as the efflux half life and was determined as 7.6 minutes for 1μM substrate under the conditions of this experiment.

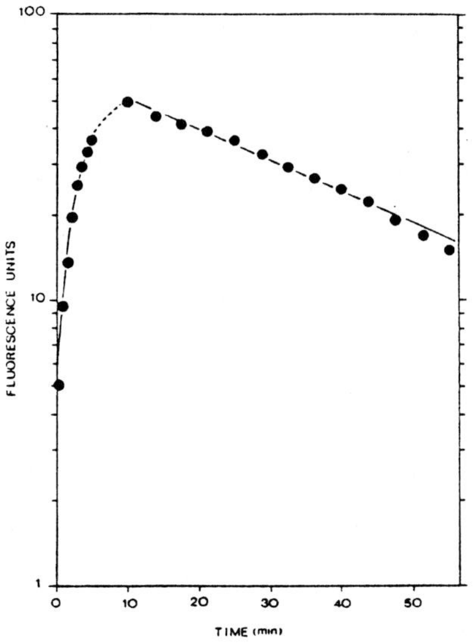

Figure 5 Kinetics of fluorescein efflux from EMT6 cells following a 10 minutes preloading with FDA 1μM.

It is also important to determine whether product lost from one cell enters another. This can be ascertained by analysing the fluorescence in an untreated cell population admixed with a population which has been preloaded with fluorescein and is losing fluorescence with time (see Figure 6).

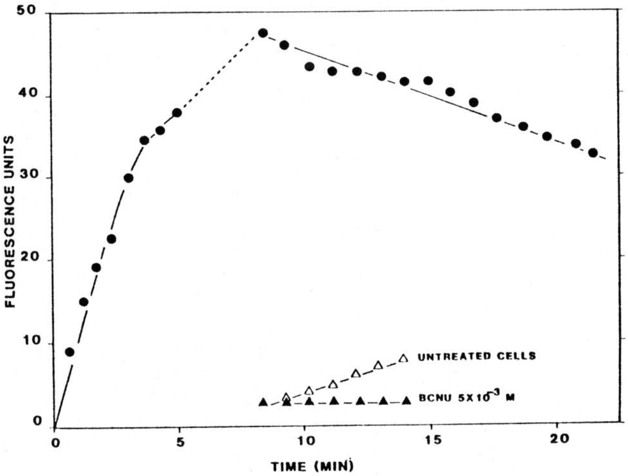

Figure 6 Effect on fluorescein fluorescence when untreated (open triangles) or BCNU-treated ($5x10^{-3}$M for 1h, closed triangles) EMT6 cells ($3x10^5$) were added to an equal number of cells previously reacted with FDA (1μM) for 5 min (closed circles).

The untreated cells accumulated fluorescence with time which could reflect product transfer or transfer of unreacted substrate. To distinguish between these two possibilities a known inhibitor of esterase activity was preincubated with these cells (see below and Figure 6). No increase in fluorescence was then detected in this cell subpopulation suggesting that there was unreacted substrate in the system and product transfer was not contributing to the results obtained.

Inhibitor Kinetics

Chloroethylnitrosoureas (CNUs) are a class of antitumour agent which alkylate DNA. The subsequent cross-linking of DNA has been implicated as the mode of cytotoxicity (Colvin & Brundrett 1981). However, protein carbamoylation may also be involved since organic isocyanates as well as alkylating species are formed when CNUs decompose in aqueous environments. Because of the susceptibility of serine hydrolases to isocyanate inactivation (Brown & Wold 1973), we proposed that flow cytometric measurement of esterase activity might form a basis for the determination of intracellular carbamoylation by CNU-derived isocyanates). This proved to be the case (Dive et al 1987b; Dive et al 1988). EMT6 cells were preincubated for 1h with various concentrations of BCNU before the rate of reaction with 1μM FDA was assessed (Figure 7).

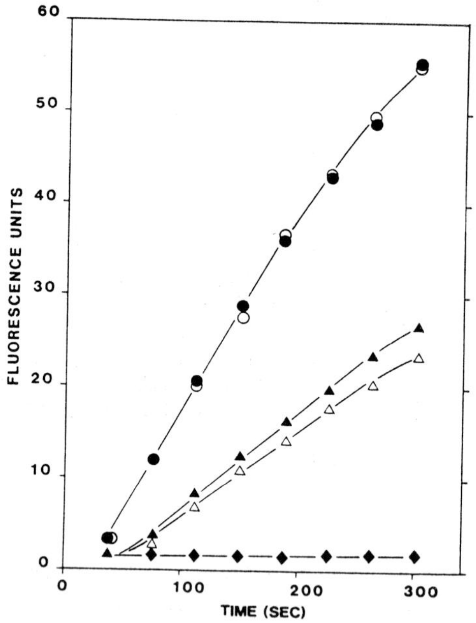

Figure 7 Effect of a 1h pre-exposure to different BCNU concentrations on FDA hydrolysis (1μM) by EMT6 cells. Enzyme reaction progress curves are shown for control (0.5% ethanol, closed circles), 10^{-5}M BCNU (open circles), $2x10^{-4}$M BCNU (closed triangles), $4x10^{-4}$M BCNU (open triangles), and 10^{-3}M BCNU (closed diamonds).

It was then possible to obtain the concentration which causes a 50% reduction in FDA hydrolysis activity, defined as the I_{50} value. Using the I_{50}, it is possible to compare directly the inhibitory effects of a number of CNUs (Figure 8). Chloroethyl- and cyclohexyl- isocyanates, together with BCNU, CCNU and the 4-hydroxy metabolites of CCNU were the more potent esterase inhibitors. Those CNUs which through their chemical structure are able to undergo intermolecular carbamoylation (chlorozotocin, GANU, and the 2-hydroxy metabolites of CCNU) were less potent. Comparison between FCE and fluorimetry of intact and sonicated cells showed that the weak inhibitory potency of chorozotocin and GANU was also due to their reduced ability to cross the plasma membrane since their I_{50} values were about tenfold higher in cells compared with sonicates. Thus enzyme inhibition kinetics by FCE seems capable of assessing not only the ability of such agents to cross the external cell membrane, but also their capacity of inducing damage having entered the cell.

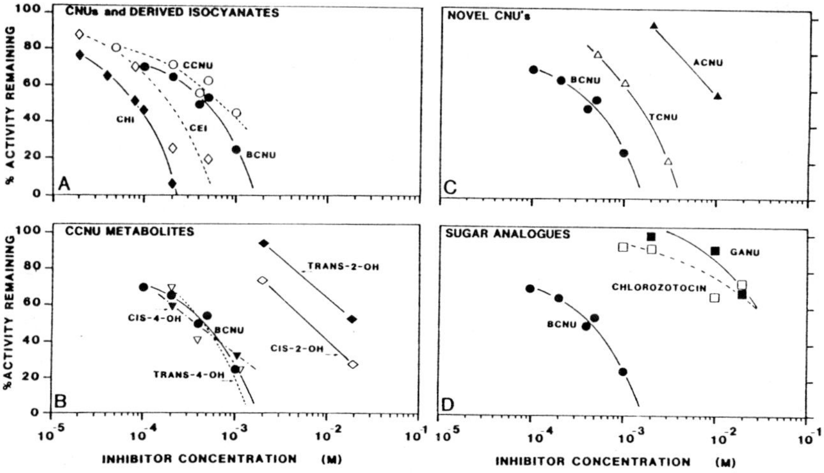

Figure 8 Dose-response curves for the inhibitory effects of chloroethylnitrosoureas, isocyanate metabolites (CHI, cyclohexylisocyanate, CEI chloroethylisocyanate), and CCNU metabolites on the hydrolysis of FDA (1μM) in EMT6 cells. Reproduced from Dive et al, Biochem Pharmacol, 37:3987, 1988 with permission from Pergamon Journals Ltd.

Esterase reaction kinetics in heterogeneous cell populations

In order to ascertain the ability of FCE to distinguish cell subpopulations based on enzyme activities, we studied the conversion of FDA to fluorescein in mouse bone marrow preparations (Dive 1988). Marrow from murine femora was flushed out using RPMI medium, filtered through a 250μm mesh and resuspended at 10^6 cells/ml before addition of 1μM FDA. A light scatter profile revealed 11 subpopulations which were electronically gated (Figure 9A). We assessed the esterase reaction in each of these scatter regions and found a high degree of heterogeneity. For example, scatter region 2 consisted of two subpopulations with high and low enzyme activity. Three subpopulations with different enzyme activities were seen in scatter region 4 (Figure 9 B & C). This study illustrates the powerful ability of FCE to measure enzyme activity in heterogeneous cell populations. Non-specific esterases are exploited in loading many probes of cellular ions such as Indo-1 and BCECF in to cell suspensions. These probes are introduced into cells in a lipophilic acetoxymethyl ester form. The degree of heterogeneity shown above suggests that these probes would be very unevenly loaded into bone marrow cell suspensions and emphasises the important advantage of using ratiometric probes (ie where the fluorescence is ratioed at two wavelengths and the fluorescence of only one varies with ion concentration). Here the problems associated with uneven dye loading are minimised.

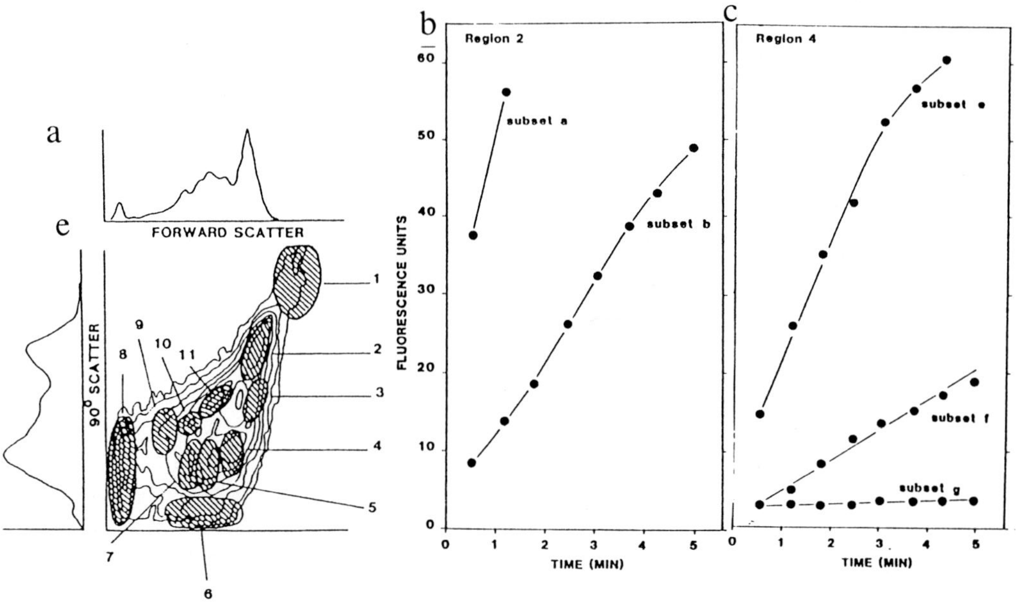

Figure 9 Analysis of FDA hydrolysis in heterogenous cell populations. **A.** 2-parameter frequency contour plot of forward versus 90° light scatter together with scatter histograms for C3H mouse bone marrow cells illustrating the position of 11 gated regions are putative cell subpopulations. **B.** Esterase reaction progress curve for cells in scatter region 2. **C.** Esterase reaction progress curve for cells in scatter region 4.

b) MEASUREMENT OF γ-GLUTAMYL TRANSFERASE (GGT) ACTIVITY, A MEMBRANE BOUND ENZYME.

GGT is of particular interest due to its association with neoplastic transformation and its potential protective role against chemical insult to cancer cells (Curthoys & Hughey 1979). In addition, it provides the technological challenge of membrane-bound enzymes, difficulties having already been identified through FCE studies of membrane-bound phosphatases (Watson et al 1979). The active site of GGT is orientated on the outer surface of the cell membrane (Horiuchi et al 1978), thus any fluorescent product generated at the active site is likely to diffuse away from the cell and be ignored in the analysis. We attempted to develop an assay previously described by Manson et al (1981) to measure GGT levels in transformed (JB1, high GGT activity) and non-transformed (BL8, low GGT activity) liver cell lines using the fluorogenic substrate γ-glutamyl-aminomethylcoumarin (γGAMC). This substrate is converted by GGT to aminomethylcoumarin (AMC) which is excited at 356nm and emits indigo fluorescence in the 420-460nm wavelength range (Figure 10).

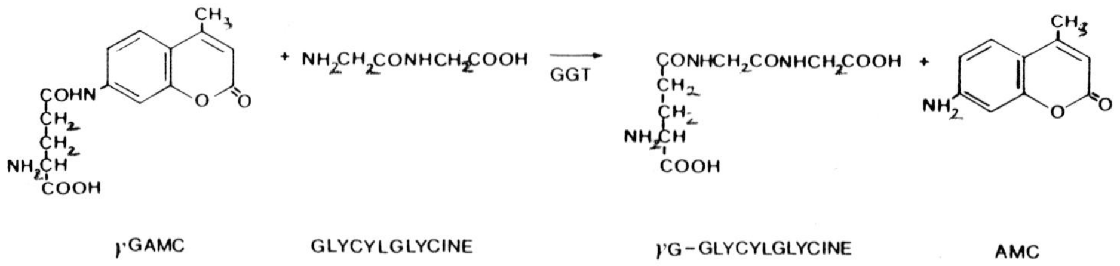

Figure 10 γ-glutamyltransferase (GGT) Enzymology. The reaction scheme for the removal of the γ-glutamyl group from the fluorogenic substrate γGAMC by GGT and its transfer to glycylglycine to yield the fluorescent product AMC.

Efflux experiments show that AMC is indeed rapidly lost from JB1 cells (Figure 11). Furthermore, γGAMC is also fluorescent between 420-460nm and this is problematic especially during the early time points of the reaction.

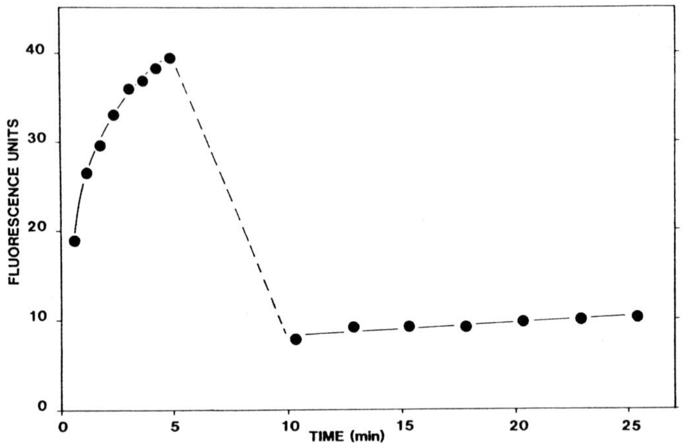

Figure 11 Determination of the rate of loss of AMC from JB1 rat hepatoma cells following a 5 minute reaction with γGAMC. The dotted line indicates the predicted trend in intracellular fluorescence during a period where cells are washed and resuspended in substrate free medium.

Problems associated with substrate fluorescence, a rapid reaction and rapid diffusion of product away from the cell are immediately obvious when the GGT reaction progress curves of JB1 and BL8 cells are compared. The progress curves do not pass through the origin and are non-linear (Figure 12). Rapid diffusion of the product from the membrane surface is likely to be a persistent disadvantage for FCE studies of membrane bound enzymes. In 1979 Watson et al described an FCE assay for plasma membrane phosphatase activity and obtained asymptotic enzyme reaction progress curves. Here the problem of rapid product diffusion was overcome to some extent using novel mathematical analysis to calculate kinetic parameters. This of course does not resolve the problem of interfering substrate fluorescence at early time points. Perhaps an ideal solution to these problems would be to synthesise a non-fluorescent substrate whose product once formed at the membrane immediately binds to a membrane target and then fluoresces? The recent advances in reducing the "dead time" of the reaction (described above) will also aid analysis of this type of reaction.

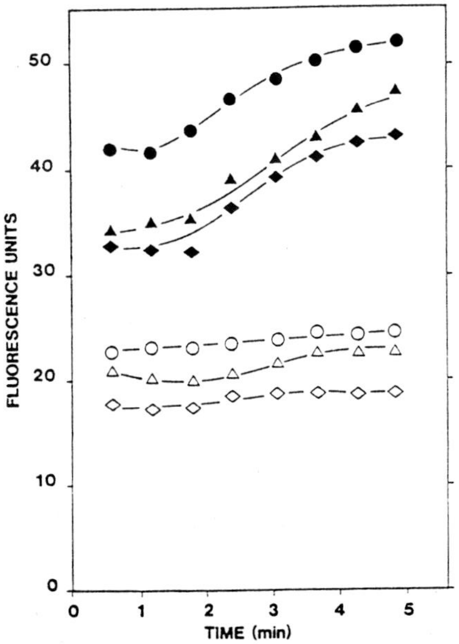

Figure 12 Flow cytometric determination of GGT reaction kinetics in JB1 rat hepatoma cells (closed symbols) and BL8 hepatocytes (open symbols) with varying γGAMC substrate concentrations. Reaction progress curves are shown for γGAMC at 100μM (circles), 50μM (triangles) and 25μM (diamonds).

c) MEASUREMENT OF CYTOCHROME P450$_{SCC}$ ACTIVITY; SINGLE CELL ENDOCRINOLOGY

Cytochrome P450$_{scc}$, the rate limiting enzyme for steroid biosynthesis, is located on the inner mitochondrial membrane of steroidogenic cells and catalyses the conversion of cholesterol to pregnenolone and isocapraldehyde by three sequential oxidations of the cholesterol side chain (Hall 1985). In order to identify steroidogenic cells from heterogeneous populations and further define the signal transduction pathways leading to steroidogenesis, Marrone et al designed and synthesised a mechanism-based fluorogenic probe for P450$_{scc}$ (Simpson et al 1990, Marrone et al 1991). The structure, spectral characteristics and metabolism of the P450 probe are shown below in Figure 13.

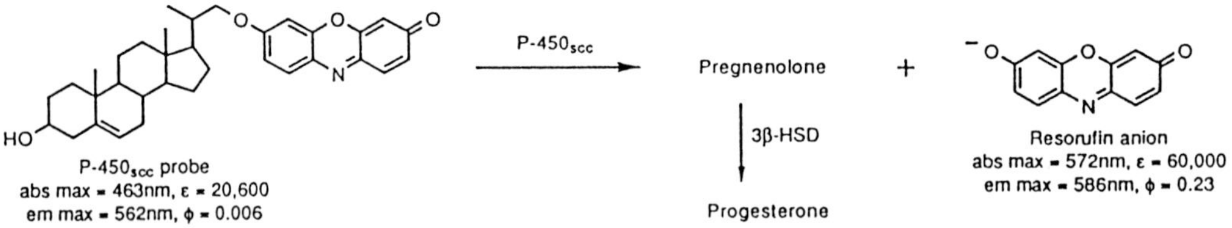

Figure 13 Spectral characteristics and metabolism of the P-450$_{scc}$ fluorogenic substrate probe; E is the extinction coefficient ($M^{-1}cm^{-1}$) and phi is quantum efficiency. Reproduced from Marrone et al, Endocrinology 128:2654 1991 with permission from The Endocrine Society.

Conversion of substrate (green fluorescent) to product by chicken granulosa cells liberates resorufin (red fluorescent) which accumulates with time. Substrate and resorufin fluorescence were measured simultaneously using the Los Alamos Multiparameter flow cytometer with two lasers spatially separated. Substrate fluorescence was excited at 457nm and detected between 495 and 583nm, resorufin fluorescence was excited at 528nm and detected above 583nm. Figure 14 shows enzyme reaction progress curves of substrate and resorufin fluorescence versus time for granulosa cells from chicken follicles of different maturation stages, with cells from the most mature and differentiated follicle (G1) predictably exhibiting the greatest rate of reaction. Further experiments also showed an increased rate of reaction in G1 granulosa cells stimulated with lutenising hormone, and that cultured Chinese Hamster Ovary cells which do not express $P450_{scc}$ do not perform this reaction (Marrone 1991).

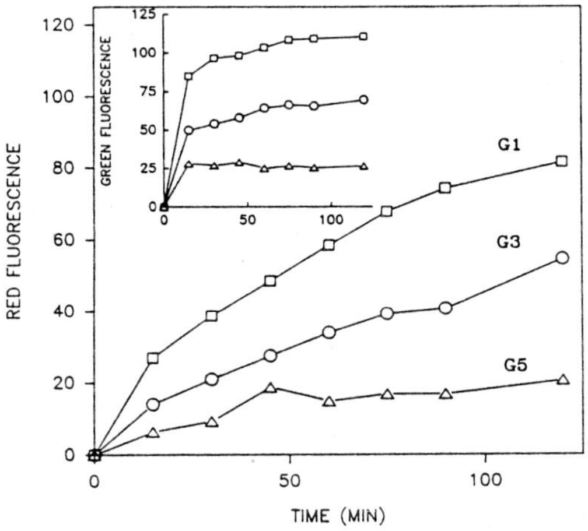

Figure 14 Flow cytometric determination of $P450-_{scc}$ enzyme reaction kinetics in hen granulosa cells (10^6/ml) from the largest (G1), third largest (G3) and fifth largest follicle. Resorufin product fluorescence (red fluorescence) is plotted versus time and the inset shows the mean channel number of green fluorescence from the substrate $P450_{scc}$ probe (1uM) as it is taken up by the same cells over time. Note that resorufin continues to accumulate after the substrate fluorescence reaches a plateaus in the same cells. Reproduced from Marrone et al, Endocrinology 128:2654 1991 with permission from the Endocrine Society.

CONCLUDING REMARKS

"Measurement of dynamic cellular events in intact cells under physiological or near physiological conditions, irrespective of the assay system, is arguably one of the most powerful techniques in cell biology. Physiological and pathological processes are not static but are continuously varying.

Observing and measuring an event in isolation gives us little or no insight into what gave rise to the event or what the future consequences might be" (Watson 1991). The technology for measuring dynamic events in viable cells by FCM is continually improving. The advancing ability to determine enzyme activities in heterogeneous populations of living cells is predicted to provide important insights into understanding cell biology. The analytical and data handling difficulties are largely overcome and the way forward lies in the conception and synthesis of more discriminating designer fluorogenic substrates, exemplified by the $P450_{scc}$ probe described above.

REFERENCES

Assaraf YG, Seamer LC, Schimke RT (1989) Characterisation by flow cytometry of fluorescein-methotrexate transport in Chinese hamster ovary cells. Cytometry 10:50-55

Brown WE, Wold F (1973) Alkyl isocyanates as active site -specific reagents for serine proteases. Reaction properties. Biochem 12:828-834

Colvin M, Brundrett R (1981) Chemical decomposition of choroethylnitrosoureas. In: Prestakyo AW, Baker LH, Crooke ST, Carter SK, Schein PS (eds) Nitrosoureas. Academic Press, New York

Coons AH, Kaplan MH (1950) Localisation of antigen in tissue cells.II. Improvements in a method for the detection of antigen by means of fluorescent antibody. J Exptl Med 91: 1-13

Curthoys NP, Hughey RP (1979) Characterisation and physiological function of rat renal gamma-glutamyl transpeptidase. Enzyme 24:383-403

Dive C, Workman P, Watson JV (1987a) Novel dynamic flow cytoenzymological determination of intracellular esterase inhibition by BCNU and related isocyanates. Biochem Pharmacol 36:3731-3738

Dive C, Workman P, Watson JV (1987b) Improved methodology for intracellular enzyme reaction and inhibition kinetics by flow cytometry. Cytometry 8:552-561

Dive C (1988) Flow Cytoenzymology with special reference to cancer chemotherapy. PhD Thesis, Council for National Academic Awards, UK

Dive C, Workman P, Watson JV (1988) Inhibition of intracellular esterases by antitumour chloroethylnitrosoureas: measurement by flow cytometry and correlation with molecular carbamoylation activity. Biochem Pharmacol 37:3987-3993

Gray JW (1976) Cell cycle analysis of perturbed cell populations; computer simulations of sequential DNA distribution. Cell Tissue Kinet 9:499-510

Habbersett R, Dive C, Marrone BL (1990) An improved injection sample chamber for rapid cellular dynamic measurements by flow cytometry. Cytometry Suppl 4:429

Hall PF (1985) Role of cytochromes P-450 in the biosynthesis of steroid hormones. Vit Horm 42:315-368

Horiuchi S, Inoue M, Morino Y (1978) Gamma-glutamyl transpeptidase: sidedness of its active site on renal brush border membranes. Eur J Biochem 87:429-437

Hulett HR, Bonner WA, Barrett J, Herzenberg LA (1969) Cell sorting: Automated separation of mammalian cells as a function of intracellular fluorescence. Science 166:747-749

Manson MM, Legg RF, Watson JV, Green JA, Neal GE (1981) An examination of the relative resistances to p-phenylene diamine mustard to gamma-glutamyl transpeptidase negative and positive cell lines. Carcinogenesis 2:661-670

Marrone BL, Simpson DJ, Yoshida TM, Unkefer CJ, Whaley TW, Buican TN (1991) Single cell endocrinology: Analysis of P-450_{scc} activity by fluorescence detection methods. Endocrinology 128:2654-2656

Martin JC, Schwartzendruber DE (1980) Time: a new parameter for kineticmeasurement in flow cytometry. Science 207:199-200

Rotman B, Papermaster BW (1966) Membrane properties of living mammalian cells as studied by enzymatic hydrolysis of fluorogenic esters. Proc Natl Acad Sci USA 55:134-141

Simpson DJ, Unkefer CJ, Whaley TW, Marrone BL (1991) A mechanism-based fluorogenic probe for the cytochrome P-450 cholesterol side chain cleavage enzyme. J Organic Chem 56:5391-5396

Watson JV, Chambers SH, Workman P, Horsnell, TS (1977) A flow cyometric method for measuring enzyme reaction kinetics in populations of single cells. FEBS Lett 81:179-192

Watson JV, Workman P, Chambers SH (1978) Differences in esterase activity between exponential and plateau growth phases of EMT6/M/CC cells monitored by flow cytofluorimetry. Br J Cancer 37:397-402

Watson JV, Workman P, Chambers SH (1979) An assay for plasma membrane phosphatase activity in populations of individual cells. Biochem Pharmacol 28:821-827

Watson JV, Walport MJ (1986) Molecular calibration in flow cytometry with subattogram detection limit. J Immunol Methods 77:321-330

Watson JV (1987) Time, a quality control parameter in flow cytometry. Cytometry 8:646-649

Watson JV, Cox H, Hellon C, Workman P, Dive C (1988) Time-resolved flow cytometric measurement of dynamic events in the interval 1-20 seconds in whole cells. Cytometry Suppl 2:93a

Watson JV (1991) Introduction to Flow Cytometry, Ist edn, Cambridge University Press, Cambridge, UK

ENDOCYTOSIS ANALYSIS BY FLOW CYTOMETRY : STUDY OF THE ENDOCYTOSIS OF FLUORESCEINYLATED NEOGLYCOPROTEINS VIA MEMBRANE LECTINS OF MACROPHAGES.

P. Midoux, A.C. Roche and M. Monsigny
Département des Glycoconjugués et Lectines Endogènes
Centre de Biophysique Moléculaire, CNRS
1, rue Haute
45071 ORLEANS cedex02
France

Introduction

Twenty years ago, G. Ashwell, A. Morell and their coworkers showed that parenchymal cells of the liver are able to bind and to internalize glycoconjugates bearing galactosyl residues in a terminal non reducing position (Ashwell and Morell, 1982). Since, endogenous lectins have been characterized at the surface of a number of normal cells (hepatocytes, macrophages, monocytes, lymphocytes and endothelial cells) as well as tumor cells (Monsigny *et al.*, 1983 and 1988).

Synthetic glycoproteins called neoglycoproteins recognized by a given membrane lectin, are easily prepared by substitution of serum albumin or gluconoylated poly-**L**-lysine, with activated sugar derivatives. These glycoconjugates can be labeled by fluorescent probes or substituted with selected drugs and used to target either cytotoxic drugs (Monsigny *et al.*, 1984a), toxins (Roche *et al.*, 1983), immunomodulators (Monsigny *et al.*, 1984b; Roche *et al.*, 1985; Derrien *et al.*, 1989; Petit *et al.*, 1990), antiviral drugs (Fiume *et al.*, 1986; Midoux *et al.*, 1990). Recently, mannosylated streptavidin was shown to increase the cellular uptake by macrophages of a biotinylated oligonucleotide (Bonfils *et al.*, 1992).

Macrophages play a key role in immunological functions and may have antitumoral activities after activation by immunomodulators. Furthermore, ma-

NATO ASI Series, Vol. H 67
Flow Cytometry
Edited by A. Jacquemin-Sablon

crophages and monocytes are infected by various viruses such as herpes virus and HIV virus. Numerous works are done to target to macrophages immunomodulators and antiviral drugs.

Alveolar and peritoneal macrophages from various origins are known to express a membrane lectin that recognizes mannose, *N*-acetyl-β-**D**-glucosamine and fucose bearing neoglycoproteins (Stahl *et al.*, 1978 and 1980; Shepherd *et al.*, 1981; Stahl and Gordon, 1982). Monocytes and bone marrow macrophages do not express the mannose receptor, but express it after maturation (Shepherd *et al.*, 1982). Human monocytes were found to express membrane lectins which recognize and internalize mannose-6-phosphate bearing glycoconjugates (Roche *et al*, 1985).

Glycoconjugates which are recognized by a specific membrane lectin on the surface of the cells, are internalized by the cells by a receptor-mediated endocytosis process. During endocytosis, glycoconjugates are brought into the cells *via* endocytotic vesicles derived from the plasma membrane. The lumen of the endocytotic vesicles is rapidly acidified allowing the dissociation of the ligands from their receptors. The fate of internalized material markedly differs from one to another ligand, a part of internalized ligands is usually delivered to lysosomes where proteolysis occurs (Wileman *et al.*, 1985; Rodman *et al.*, 1990).

In order to determine the efficiency of glycoconjugates as carriers for drug targeting to a given cell, the quantity of endocytosed glycoconjugates, the kinetic of the internalization and the intracellular localization of the glycoconjugates have to be known : A therapeutic agent to be active must reach its target cell and more precisely the right compartments of the target cell which are generally the cytosol and /or the nucleus.

Endocytosis process are usually studied by using radiolabeled ligands, nevertheless flow cytometry can also be used. In this report, we describe a flow cytometry method allowing the quantitative determination of the binding, the uptake and the degradation by macrophages of fluoresceinylated neoglycoproteins . The method is based on the fluorescence properties of fluorescein - the dye most widely employed for labeling of protein ligands because of its high fluorescing power - and on some properties of cell compartments: i) the lumen of endosomes and lysosomes is acidic; ii) the fluorescein fluorescence depends upon the pH of the solution and iii) the fluorescence of the ligand depends upon the number of fluorescein molecules linked to a protein (Monsigny *et al.*, 1984a; Midoux *et al.*, 1986 and 1987).

Neoglycoproteins

Neoglycoproteins are obtained by adding activated sugar, oligosaccharide, or glycopeptide to a protein such as serum albumin (Kieda *et al.*, 1977; Schrevel *et al.*, 1979). Neoglycoproteins are prepared by using glycosylphenylisothiocyanate (Figure 1). This type of neoglycoprotein is well adapted to prepare fluoresceinylated derivatives because the phenylthiocarbamyl group does not impair the emitted fluorescence (Buss and Goldstein, 1968; MacBroom *et al.*, 1972; Roche *et al.*, 1983; Monsigny *et al.*, 1984a). A *p*-nitrophenylglycoside is reduced in the presence of palladium on charcoal; the aromatic amine is transformed into phenylisothiocyanate upon treatment with thiophosgene; and finally the activated sugar is added to a solution of protein at pH 8.5.

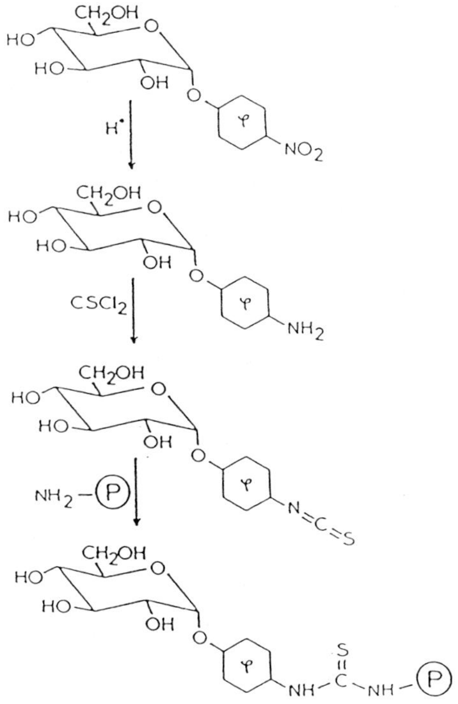

Figure 1 Scheme of synthesis of neoglycoproteins.

Recently, we developped glycosylated polymer as specific biodegradable carriers of either immunomodulators (Derrien *et al.*, 1989) or antiviral drugs (Midoux *et al.*, 1990). These polymers are prepared by sequentially substituting poly(**L**-lysine) with sugar derivative such as hydroxysuccinimidyl α-**D**-mannopyranosylphenylacetate ester and then with either FITC or a biological response-modifier such as an active ester of *N*-acetyl muramyl dipeptide (MDP) or an antiviral drug such as 9-(2-phosphonylmethoxyethyl)adenine (PMEA) and finally with δ-gluconolactone to give a neutral polymer.

Neoglycoproteins such as glycosylated serum albumin, containing about 25 ± 5 sugar residues, or glycosylated polymers, containing about 45 ± 5 sugar residues, are quite useful to study the specificity and the properties of membrane lectins and to target drugs because their apparent binding constant is much higher than that of the related free sugars or glycosides (Townsend and Stahl, 1981; Derrien *et al.*, 1989).

Standardization of the flow cytometer and quantification of the cell-associated fluorescence

The use of a flow cytofluorometer allows a careful analysis of cell-associated fluoresceinylated ligands in both a qualitative and a quantitative ways. The amount of cell associated fluoresceinylated ligand can be assessed when the flow cytofluorometer and the specific fluorescence intensity of a unit amount of fluoresceinylated ligand have been standardizated. This can be done by using fluoresceinylated beads. Fluorescent beads can be easily prepared by mixing a derivative of fluorescein (FTC-ADE) (Figure 2) and commercially available polystyrene sulfonate beads (Dionex DC-4A, Dionex, Sunnyvale, CA, USA).

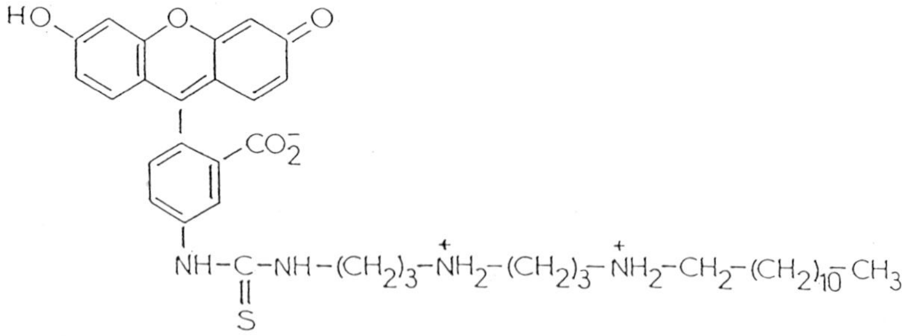

Figure 2 : Structure of 1-(fluoresceinylthioureido)-4,8-diaza-eicosane (FTC-ADE)

The 1-(fluoresceinylthioureido)-4,8-diaza-eicosane (FTC-ADE) is easily synthetized and binds strongly polystyrene sulfonate beads (Monsigny *et al.*, 1984a). Because there is a linear relationship between bead fluorescence intensity (measured with the flow cytofluorometer) and the fluorescence intensity of beads (measured with spectrofluorometer) (Figure 3), the cell fluorescence intensity can be directly converted into quantity of cell associated fluoresceinylated ligand.

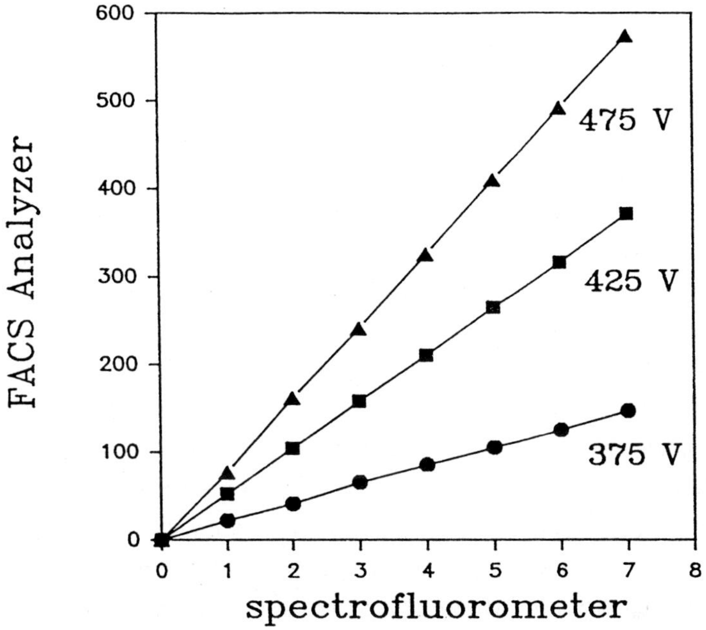

Figure 3 : Relation between the fluorescence intensities of fluoresceinylated polystyrene sulfonate beads measured on the FACS Analyzer with according to the voltage of the FACS photomultiplier and on a spectrofluorometer.

Quantitative measurements

The mean amount (Q) of a fluoresceinylated ligand associated to a single cell is determined by using the following equation :

$$Q = [F/F_n].[d/b] \quad (I)$$

and expressed as pg of fluoresceinylated ligand per cell. F is the cell fluorescence intensity of one labeled cell incubated at 4°C minus the cell autofluorescence, F_n is the fluorescence intensity of 1 μg fluoresceinylated ligand in 1 ml PBS measured with a spectrofluorometer, b is the bead fluorescence intensity of one labeled bead minus the bead autofluorescence and d is the fluorescence intensity of 10^6 fluoresceinylated beads in 1 ml PBS measured with a spectrofluorometer. F_n and d are expressed relatively to the fluorescence intensity of a solution of quinine sulfate (2.55 x 10^{-7} M in 0.1 N H_2SO_4; excitation wavelength : 345 nm; emission wavelength : 460 nm).

pH sensitivity of fluorescein fluorescence

Fluorescein isothiocyanate (FITC isomer I) is usually bound to a ligand by a thiocarbamyl arm bound to the ϵ-amino group of a lysine residue. The fluorescein fluorescence is very sensitive to the pH of the surrounding medium (Martin and Lindqvist, 1975). The fluorescence intensity of fluorescein free or bound to a ligand decreases when the pH decreases and is very low at pH 5 which is the pH of endocytotic vesicles. This property is one of the basis of the flow cytofluorometric method we developed to analyse endocytosis.

The surrounding pH of fluorescein can be measured by monitoring the ratio of the fluorescence emission intensity at 520 nm upon excitation at 450 nm and 490 nm (Tycko and Maxfield, 1982). In this type of measurements, the fluorescein-ligand interaction has to be taken into account. For instance, the pH dependence of the ratios (I_{490}/I_{450}) of fluorescein free and bound to ϵ-aminocaproic acid are quite different (Figure 4A). In addition, the pH dependence of the ratios of fluoresceinylated proteins and fluoresceinylated proteins upon proteolytic digestion are also quite different (Figures 4B-D). The ratios of a fluoresceinylated neoglycoprotein and of a fluoresceinylated IgG are smaller than that obtained upon proteolytic digestion by pronase (Figures 4B and 4C). Conversely, the ratios of native and digested fluoresceinylated poly(L-lysine) are identical (Figure 4D). That indicates that amino acide groups closed to fluorescein residues are involved in the pH dependence of the fluoresceinylated proteins. The ratios of digested proteins and of native and digested poly(**L**-lysine) are closed to that of FITC bound to ϵ-aminocaproic acid.

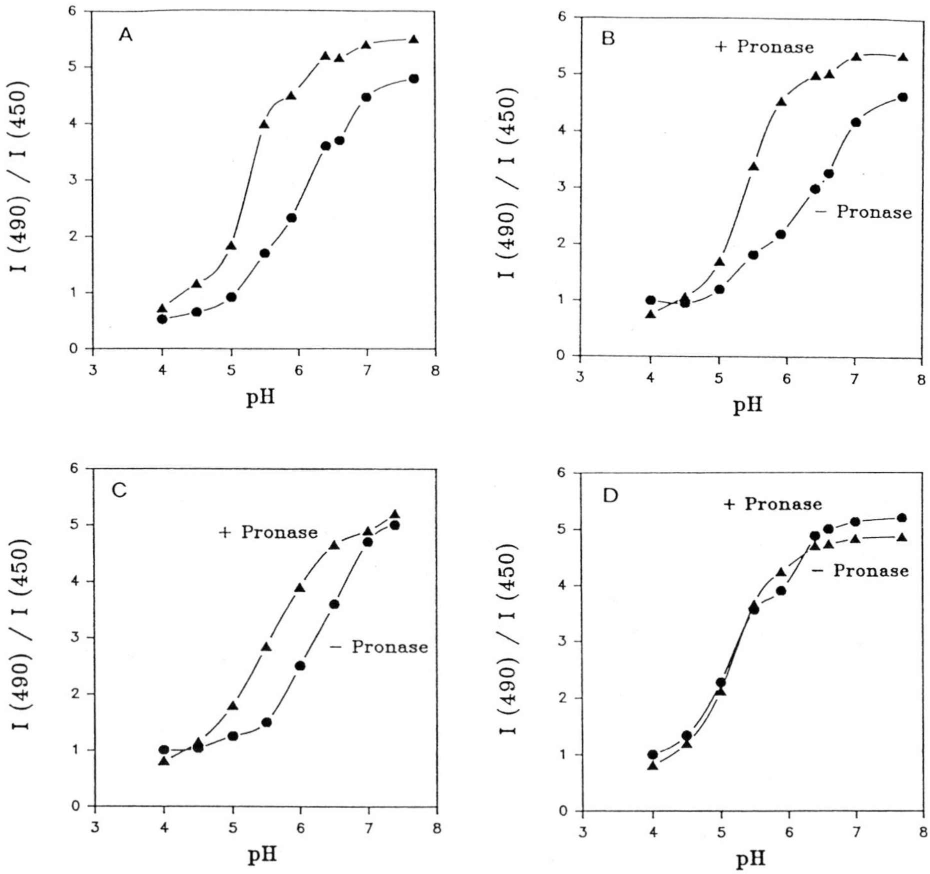

Figures 4 : pH dependence of the ratio (I_{490}/I_{450}) of (A) fluorescein (●) and FITC bound to ε-amino caproic acid (▲) ; fluoresceinylated (B) mannosylated serum albumin ; (C) IgG and (D) mannosylated and gluconoylated poly(L-lysine) before (●) and after (▲) pronase digestion.

Flow cytometry analysis of the endocytosis of neoglycoproteins

A - Internalization

Proteins, glycoproteins, hormones, toxins and viruses are recognized by specific cell surface receptors and are internalized into cells by receptor-mediated endocytosis. During this process, ligands are internalized into endocytotic vesicles derived from plasma membrane. Although the exact pathway followed by the ligands is not known, the key step is a rapid acidification of endocytotic vesicles, allowing the dissociation of the ligands from receptors. The fate of internalized material markedly differs from one to another ligand, but when internalized, ligand-receptor complexes are dissociated, ligands are usually delivered to lysosomes where proteolysis occurs. Because the fluorescein fluorescence is sensitive to pH, the fluorescence intensity of a fluoresceinylated ligand upon internalization by receptor-mediated endocytosis is usually low because of the acidic pH of the endocytotic vesicles. In order to evidence the capacity of cells to internalize a ligand such as a neoglycoprotein, the cells are incubated at 37°C with the fluoresceinylated neoglycoprotein and upon removing the unbound ligand, cells are post incubated at 4°C in the presence of monensin. Monensin is a proton/sodium ionophore used to rapidly raises the pH of the endocytotic vesicles and to allow the measurements of the fluorescence intensity of endocytosed ligand (Maxfield, 1982). So, an enhancement of the cell fluorescence intensity upon a post incubation with monensin will evidence an endocytosis of the ligand (Figure 5).

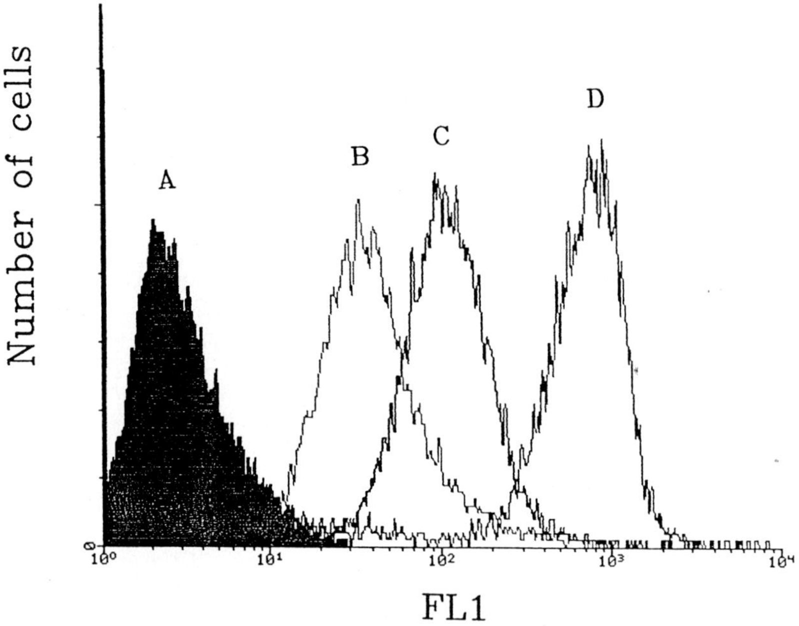

Figure 5 : Flow cytometry analysis of the internalization of fluoresceinylated and mannosylated bovine serum albumin (F-,Man-BSA) by macrophages. (A) cell autofluorescence; (B) cell fluorescence intensity of macrophages incubated at 4°C with F-,Man-BSA; (C) cell fluorescence intensity of macrophages incubated at 37°C with F-,Man-BSA; (D) cell fluorescence intensity of macrophages incubated at 37°C, washed and further incubated at 4°C for 30 minutes in the presence of 50 μM monensin.

B - Intracellular degradation

During receptor-mediated endocytosis, ligands are usually delivered into lysosomes where proteolysis occurs. It may be important to determine the extent of the degradation of the ligands after endocytosis by cells. Intracellular degradation process can be followed by flow cytofluorometry by using fluorescein fluorescence properties of fluoresceinylated ligands.

The extinction coefficient of FITC upon binding to a macromolecular ligand is smaller than that of free FITC. Consequently, the determination of the degree of labeling (F/P) ratio of a fluoresceinylated ligand must be done by using the absorbance at 495 nm after a proteolytic digestion with pronase. Under these conditions, the extinction coefficient of FITC can be used to calculate the F/P ratio (Midoux *et al.*, 1987).

The fluorescence emission of bound FITC is quenched compare to free fluorescein and the extent of the quenching is dependent upon the number of fluorescein residues bound per ligand molecule (Figures 6).

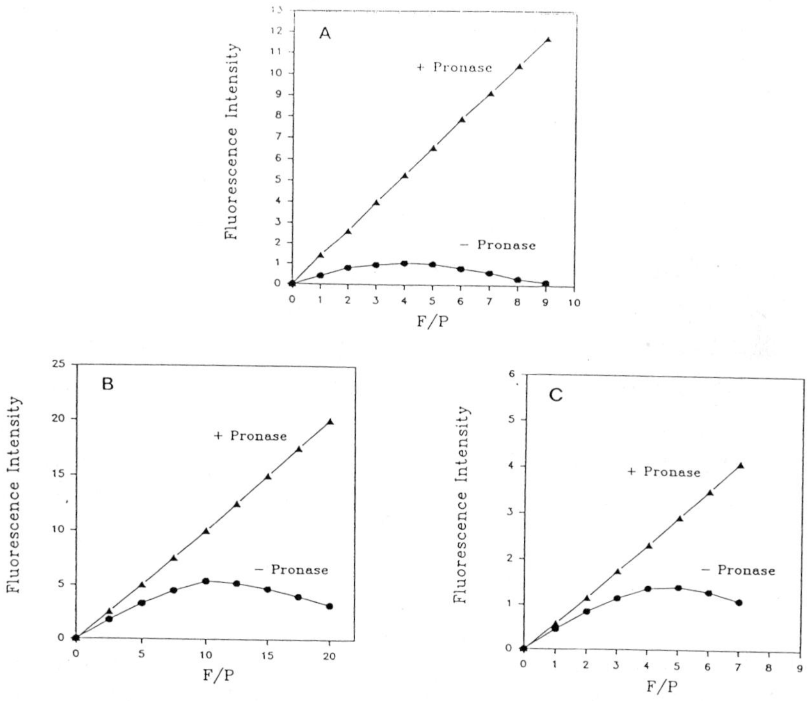

Figures 6 : Fluorescence properties of fluoresceinylated conjugates. (A) neoglycoprotein; (B) poly(**L**-lysine; (C) IgG. (●) fluorescence intensity of fluoresceinylated conjugates in PBS. (▲) fluorescence intensity of fluoresceinylated conjugates in PBS after proteolytic digestion by pronase. F/P is the average number of fluorescein residues bound per macromolecule determined after pronase digestion.

FITC bound to either neoglycoprotein, poly(L-lysine) or IgG is quenched and the quenching increases with the fluorescein/protein ratio (F/P) due to enhancement of energy transfert between fluorescein residues. Upon proteolytic digestion, the fluorescence intensity increases and reaches the intensity of free fluorescein or fluorescein bound to small peptide fragment. In addition, the fluorescence quenching seems also to depend with the nature of the ligand. For a same fluorescein/protein ratio, the fluorescence quenching is greater with BSA than with poly(L-lysine) probably because BSA has a globular conformation wheras poly(L-lysine) has a linear conformation in which the fluorescein residues are far from one another reducing the energy transfert (Figures 6A and 6B): the fluorescence intensity of 4 FITC molecules bound to neoglycoprotein is identical to that of 13 FITC molecules bound to poly(L-lysine). For a same fluorescein/protein ratio, the fluorescence quenching is greater with a neoglycoprotein than with an IgG, probably because the molecular size of BSA is smaller than that of IgG in which the fluorescein residues are far from one another reducing also the energy transfert (Figures 6A and 6C).

a) Time course of the internalization

The time course of the internalization and specially of the intracellular digestion of an internalized fluoresceinylated ligand can be assesed by flow cytometry (Figure 7). Mouse peritoneal macrophages and 2C1112 hybridoma cells (a macrophage cell line) were incubated at 4°C for 1 h with fluoresceinylated mannosylated bovine serum albumin (F-,Man-BSA). The cells were washed to remove unbound neoglycoprotein and then, the cells were incubated at 37°C to allow endocytosis. The cell fluorescence intensity was measured upon time after a post incubation with monensin. The fluorescence intensity of macrophages increased during the first 30 minutes and then decreased. Because fluorescence emission of fluorescein bound to neoglycoprotein is quenched compared to that of free fluorescein, the enhancement of the cell fluorescence intensity compare to that measured after 1 h incubation at 4°C, evidences that the neoglycoprotein is digested after internalization. Consequently, macrophages rapidly internalize mannosylated BSA which is digested after 30 minutes incubation at 37°C and digested fluorescent products are secreted in the medium (fluorescence intensity increased in the medium when the cell fluorescence intensity decreased) (Pimpaneau *et al.*, 1989). Conversely, mannosylated

BSA is slowly digested by 2C1112 hybridoma cells (maximum obtained after 2 h incubation at 37°C) and the digested products were also slowly secreted in the medium.

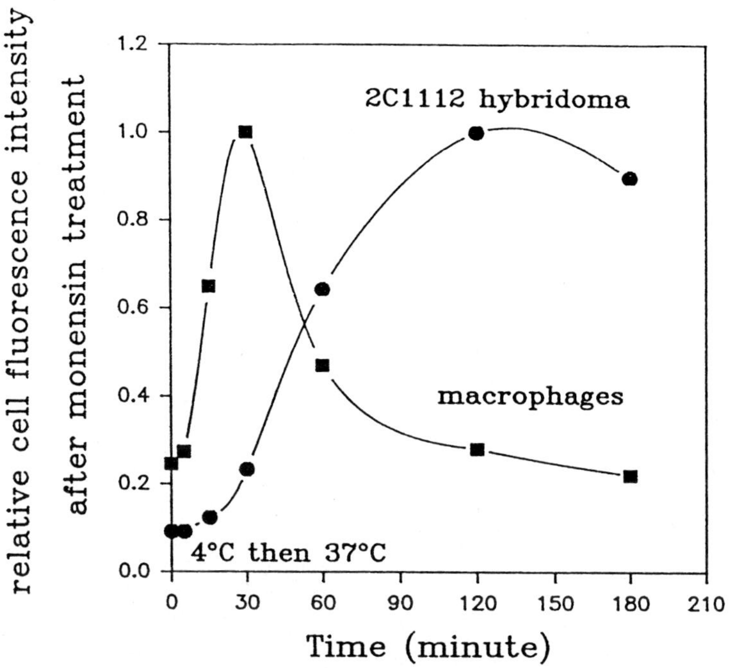

Figure 7 : Kinetics of the endocytosis of fluoresceinylated and mannosylated bovine serum albumin by mouse peritoneal macrophages and 2C1112 hybridoma cells (adapted from Pimpaneau *et al.*, 1989).

b) Quantitative measurements

A quantitative determination of the degradation of a fluoresceinylated ligand can be assessed by using two fluoresceinylated ligands containing either a small or a large number of fluorescein residues. Indeed as shown in figures 6 and Table I, a macromolecular ligand containing a small number of fluorescein residues will give a small enhancement of the fluorescence intensity upon proteolytic digestion while the same macromolecule containing a large number of fluorescein residues will give a greater enhancement of the fluorescence intensity.

Table I : Fluorescence properties of fluoresceinylated and 6-phosphomannosylated bovine serum albumin.

	F_n	F_{np}
$F_{2.2}$-,M6P-BSA	0.21	0.9
$F_{6.6}$-,M6P-BSA	0.28	2.7

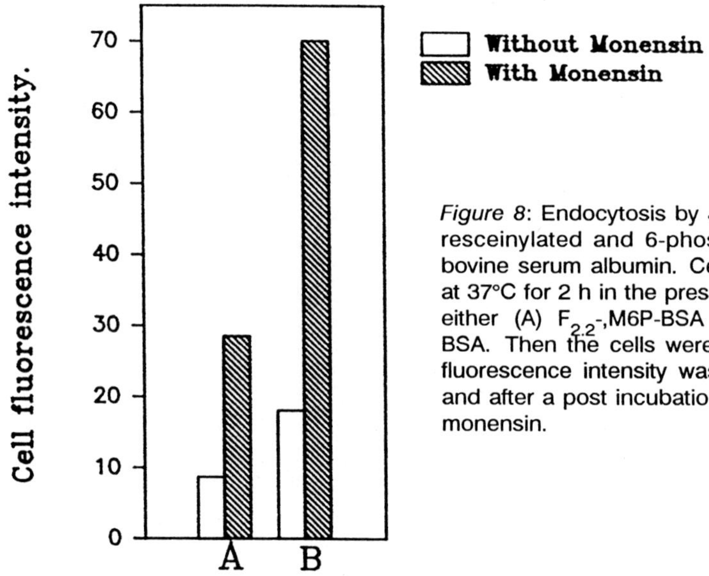

Figure 8: Endocytosis by J774E cells of fluoresceinylated and 6-phosphomannosylated bovine serum albumin. Cells were incubated at 37°C for 2 h in the presence of 25 μg/ml of either (A) $F_{2.2}$-,M6P-BSA or (B) $F_{6.6}$-,M6P-BSA. Then the cells were washed. The cell fluorescence intensity was measured before and after a post incubation at 4°C with 50 μM monensin.

As shown in figure 8, the cell fluorescence intensity (measured upon a post incubation with monensin) of J774E cells (a macrophage cell line) (Diment *et al.*,1987) incubated at 37°C with M6P-BSA substituted with 6.6 fluorescein residues per BSA molecule was greater than that of cells incubated with M6P-BSA substituted with 2.2 fluorescein residues per BSA molecule. When cells are incubated at 37°C, endocytosis process and proteolytic digestion occur

and consequently, the cell fluorescence intensity F(M) (measured upon a post incubation with monensin) of a single cell incubated at 37°C is the sum of two fluorescence in-
ensities coming from a quantity of undegraded fluoresceinylated ligand and a quantity of degraded fluoresceinylated ligand, according to :

$$F(M) = [(Q_n \,.\, F_n) + (Q_{np} \,.\, F_{np})] \,.\, b/d \quad (II)$$

where F_n and F_{np} are the fluorescence intensities of 1 μg/ml undegraded and degraded fluoresceinylated ligand measured with a spectrofluorometer, respectively. Q_n and Q_{np} are the mean amounts of undegraded and degraded fluoresceinylated ligand expressed in pg per cell, respectively. b and d are as defined for equation (I).

The mean amounts Q_n and Q_{np} of undegraded and degraded fluoresceinylated and 6-phosphomannosylated bovine serum albumin associated to J774E cells can be determined by using a 6-phophomannosylated BSA substituted with either a low or a high number of fluorescein residues per protein molecule and indexed 1 and 2, respectively. Q_n and Q_{np} are determined by resolving the followed linear equation system :

$$F(M)_1 = [(Q_n \,.\, F_{n1}) + (Q_{np} \,.\, F_{np1})] \,.\, b/d$$
$$F(M)_2 = [(Q_n \,.\, F_{n2}) + (Q_{np} \,.\, F_{np2})] \,.\, b/d$$

After 2 h incubation in the presence of 25 μg/ml of fluoresceinylated M6P-BSA, J774E cells contained 34.5% and 65.4% of undegraded and degraded neoglycoproteins, respectively.

References

Ashwell G, Morell J (1982) Carbohydrate specific receptors of the liver. Ann. Rev. Biochem. 51: 531-554

Bonfils E, Mendès C, Roche AC, Monsigny M, Midoux P (1992) Uptake by macrophages of a biotinylated oligo-α-deoxythymidylate by using mannosylated streptavidin. Bioconjugate Chem. (in the press)

Buss DH, Goldstein IJ (1968) Protein-carbohydrate interaction. XIV. Carbohydrates containing groups for the alkylation of proteins. J. Chem. Soc. Perkin 12: 1457-1461

Derrien D, Midoux P, Petit C, Nègre E, Mayer R, Monsigny M, Roche AC (1989) Muramyl Dipeptide Bound to Poly-L-Lysine Substituted with Mannose and Gluconoyl Residues as Macrophage Activators. Glycoconjugate J. 6:241-255

Diment S, Leech MS, Stahl PD (1987) Generation of macrophage variants with 5-azacytidine : selection for mannose receptor expression. J. Leukocyte Biol. 42 : 485-490

Fiume L, Bassi B, Busi C, Mattioli A, Spinosa G (1986) Drug targeting in antiviral chemotherapy. A chemical stable conjugate of 9-beta-**D**-arabinofuranosyl-adenine 5'-monophosphate with lactosaminated albumin accomplishes a selective delivery of drug to liver cells. Biochem. Pharmacol. 35: 967-972

Kieda C, Delmotte F, Monsigny M (1977) Preparation and properties of glycosylated cytochemical markers. FEBS Letters 76: 257-261

MacBroom CR, Samanen CH, Goldstein IJ (1972) Carbohydrate antigens : coupling of carbohydrates to proteins by diazonium and phenylisothiocyanate reactions. Meth. Enzymol. 28: 212-219

Martin MM, Lindqvist L (1975) The pH dependence of fluorescein fluorescence. J. Luminesc. 10: 381-390

Maxfield FR (1982) Weak bases and ionophores rapidly and reversibly raise the pH of endocytic vesicles in cultured mouse fibroblasts. J. Cell Biol. 95: 676-681

Midoux P, Nègre E, Roche AC, Mayer R, Monsigny M, Balzarini J, De Clercq E, Mayer E, Ghaffar A, Gangemi JD (1990) Drug targeting: anti-HSV-1 activity of mannosylated polymer-bound 9-(2-phosphonylmethoxyethyl)adenine. Biochem. Biophys. Res. Commun. 167: 1044-1049

Midoux P, Roche AC, Monsigny M (1986) Degradation of endocytosed materials estimated by flow cytofluorometry using two neoglycoproteins containing different numbers of fluorescein molecules. Biol. Cell 58: 221-226

Midoux P, Roche AC, Monsigny M (1987) Quantitation of the binding, uptake and degradation of fluoresceinylated neoglycoproteins by flow cytometry. Cytometry 8: 327-334

Monsigny M, Kieda C, Roche AC (1983) Membrane glycoproteins, glycolipids and membrane lectins as recognition signals in normal and malignant cells. Biol. Cell 47: 95-110

Monsigny M, Roche AC, Bailly P (1984b) Tumoricidal activation of murine alveolar macrophages by muramyl dipeptide substituted mannosylated serum albumin. Biochem. Biophys. Res. Commun. 121: 579-584

Monsigny M, Roche AC, Kieda C, Midoux P, Obrenovitch A (1988) Characterization and biological implications of membrane lectins in tumor, lymphoïd and myeloïd cells. Biochimie 70: 1633-1649

Monsigny M, Roche AC, Midoux P (1984a) Uptake of neoglycoproteins via membrane lectin (s) of L 1210 cells evidenced by quantitative flow cytofluorometry and drug targeting. Biol. Cell 251: 187-196

Petit C, Monsigny M, Roche AC (1990) Macrophages activation by muramyldipeptide bound to neoglycoproteins and glycosylated polymers : cytotoxique factor production. J. Biol. Response Modifiers 9: 33-43

Pimpaneau V, Midoux P, Durand G, De Baetselier P, Monsigny M, Roche AC (1989) Endocytosis of alpha1-acid glycoprotein variants and of neoglycoproteins containing mannose derivatives by a mouse hybridoma cell line (2C11-12). Comparison with mouse peritoneal macrophages. Glycoconjugate J. 6: 561-574

Roche AC, Bailly P, Monsigny M (1985) Macrophage activation by MDP bound to neoglycoproteins: metastasis eradication in mice. Invasion and Metastasis 5: 218-232

Roche AC, Barzilay M, Midoux P, Junqua S, Sharon N, Monsigny M (1983) Sugar-specific endocytosis of glycoproteins by Lewis Lung carcinoma cells. J. Cell. Biochem. 22: 131-140

Roche AC, Midoux P, Bouchard P, Monsigny M (1985) Membrane lectins on human monocytes. Maturation-dependent modulation of 6-phosphomannose and mannose receptors. FEBS Letters 193 : 63-68

Rodman JS, Mercer RW, Stahl PD (1990) Endocytosis and transcytosis. Current Opinion in Cell Biol. 2: 664-672

Schrevel J, Kieda C, Caigneaux E, Gros D, Delmotte F, Monsigny M (1979) Visualization of cell surface carbohydrates by a general two-steps lectin technique : lectins and glycosylated cytochemical markers. Biol. Cell 36: 259-266

Shepherd VL, Campbell TJ, Senior RM, Stahl PD (1982) Characterization of the mannose/fucose receptor on human mononuclear phagocytes. J. Reticoendothel. 32 : 423-431

Shepherd VL, Lee YC, Schlesinger PH, Stahl PD (1981) L-fucose terminated glycoconjugates are recognized by pinocytosis receptors on macrophages. Proc. Natl. Acad. Sci. USA 78 : 1019-1022

Stahl PD, Gordon S (1982) Expression of a Mannosyl-Fucosyl receptor for endocytosis on cultured primary macrophages and their hybrids. J. Cell. Biol. 93 : 49-56

Stahl PD, Rodman JS, Miller MJ, Schlesinger PH (1978) Evidence for receptor mediated binding of glycoproteins, glycoconjugates and lysosomal glycosidases by alveolar macrophages. Proc. Natl. Acad. Sci. USA 75 : 1399-1403

Stahl PD, Schlesinger P, Sigardson E, Rodman JS, Lee YC (1980) Receptor-mediated pinocytosis of Mannose glycoconjugates by macrophages : characterization and evidence for receptor recycling. Cell 19 : 207-215

Townsend R, Stahl P (1981) Isolation and characterization of a mannose / N-acetyl-glucosamine / fucose binding protein from rat liver. Biochem. J. 194: 209-214

Tycko B, Maxfield FR (1982) Rapid acidification of endocytic vesicles containing alpha 2-macroglobulin. Cell 28: 643-651

Wileman T, Harding C, Stahl P (1985) Receptor-mediated endocytosis. Biochem. J. 232: 1-14

ANALYSIS OF INTRACELLULAR ANTIGENS BY FLOW CYTOMETRY: METHODS FOR CELL PERMEABILIZATION TO ANTIBODIES

H. Giloh
Division of Interdepartmental Equipment
The Hebrew University-Hadassah Medical School
Jerusalem
Israel (*)

A. INTRODUCTION

Intracellular molecules have been detected for many years using antibodies and other specific ligands conjugated to fluorescent markers. Cells were prepared by established methods of histochemistry and cytochemistry and were analyzed by fluorescence microscopy. By microscopy not only the presence of an intracellular antigen can be identified but also its localization within the cell. Quantitative analysis, however, suffers from the limitations of microscopy in general, namely (1) the relatively small number of cells of a large population that are analyzed, (2) the slowness and subjectivity of data collection and (3) the restricted sensitivity of the human eye to low-intensity fluorescence.

To overcome the limitations, photomultiplier tube-based detection systems and sensitive video cameras have nowadays been introduced into fluorescence microscopes and increase the sensitivity of light detection by a number of orders of magnitude. However, the only technique by which data from thousands of cells can be collected at a high sampling rate, with high sensitivity and objectivity is flow cytometry. Moreover, the isolation of subpopulations of cells that exhibit interesting intracellular antigenic characteristics has been made possible by the sorting option of the flow cytometer. Since in the flow cytometer tracing of detailed cell

(*) Mailing address: P.O.Box 1172, Jerusalem, Israel 91010

NATO ASI Series, Vol. H 67
Flow Cytometry
Edited by A. Jacquemin-Sablon

morphology is lost, fluorescence microscopy and flow cytometry are complementary for the analysis of antigens in cells and tissues.

In preparing cells for the analysis of intracellular antigens by flow cytometry, the basic problems remain unchanged compared to microscopy. Cells must be treated so as to overcome the membrane barriers to antibody molecules and thereby to allow the antibodies to reach their respective antigens in the cell interior. At the same time, the antigen must not leak out and must be preserved in its natural, antigenic conformation. Cell morphology should also be conserved as much as possible to minimize changes in the light scatter signals that are detected in the flow cytometer and are cell-specific and characteristic.

Since the early years of flow cytometry, attempts were made to analyze intracellular antigens by this novel technique (Horan et al., 1975; Zeile, 1980). Previously established histochemical and cytochemical methods for cell permeabilization were adapted and new methods were worked out that permeabilize the plasma membrane but preserve the cell as an entity that can be analyzed and sorted in a flow cytometer (reviewed in short by Jacobberger et al., 1986; Turner et al., 1989). To date, more than fifty intracellular antigens that occur in nature have been labeled by fluorescent antibodies and analyzed by flow cytometry. In addition, a complex procedure has been worked out to label newly synthesized DNA with bromodeoxyuridine (BrdUrd) and to use anti-BrdUrd antibodies and fluorescence to estimate the fraction of cells in S-phase in a given cell population (Gratzner, 1982; Dolbeare et al., 1990).

Published results show that in order to succeed in binding antibodies to intracellular antigens no single protocol for cell preparation is generally applicable. Therefore, for every new intracellular antigen that has not been analyzed previously, the appropriate procedure for cell permeabilization has to be found by trial and error.

Experience shows that some intracellular antigens can be labeled by a variety of fixation and permeabilization procedures, while for the labeling of others a unique combination of reagents has to be found.

The aim of the present review is to assist the reader to conduct the trial-and-error search systematically. A survey of successfully analyzed antigens is presented and the cell permeabilization procedures that were employed are listed. From the assembled data some general conclusions are drawn that suggest a logical start for the search, as well as priorities for the ongoing steps, to shorten the search as much as possible.

B. CELL PERMEABILIZATION BY ORGANIC SOLVENTS, DETERGENTS AND RELATED REAGENTS

1. General aspects

Antibodies can diffuse from the incubation medium into the cell interior when the plasma membrane is gently permeabilized without causing cell lysis. Most often this is accomplished by partial solubilization of the plasma membrane by adding a detergent to the medium. To avoid leakage of antigen, cells are frequently fixed prior to, or simultaneously with membrane permeabilization. Fixation, however, may induce conformational changes in the antigenic site so as to abolish binding of the specific antibody. Also, extensive cross-linking of cellular proteins by the fixative ("overfixation") may obstruct the access of antibody to its antigen and thereby decrease antigen-specific fluorescence. Moreover, overfixation may interfere with the washing-out of excess, unbound antibody and, hence, may increase non-specific background fluorescence. Therefore, the degree of both permeabilization and fixation must be carefully controlled. The extent of cell fixation and membrane permeabilization needed for each cell system will depend on the cell type, the nature of the antigen and its intracellular location (plasma membrane-bound, diffuse cytoplasmic, confined to the cytoskeleton, the nucleus or other cell organelles or granules).

2. Reagents in use for cell permeabilization and fixation.

2.1 Acetone and alcohols

Acetone, ethanol and methanol (concentration range: 50-100%) fix and simultaneously permeabilize cells by dehydration, lipid extraction and protein precipitation (Bancroft, 1967). From many light and electron microscopic studies it is evident that even if the cells are not lysed, the plasma membrane and internal membranes become highly damaged. Cells fixed by organic solvents tend to shrink (Banckroft, 1967), on the other hand, their refractive index is increased (Shapiro, 1988). Since the major factors that affect forward light scattering are (a) cell size and (b) the difference in refractive index between the cell and the suspending medium (Shapiro, 1988), and since the effects of the organic solvents on light scattering are cell type-specific, the distinction between subpopulations of cells that normally diverge in light scatter intensities may be lost (e.g. lymphocytes and monocytes).

It is also well-known that acetone, ethanol and methanol cause loss of antigenicity of many cell surface antigens, moreover they may cause redistribution of some antigens (see Bauer, 1990). On the other hand, cytoskeleton antigens and a variety of nuclear antigens seem to withstand acetone and alcohol fixation (Table 1).

The most serious problem encountered with fixation by organic solvents is cell clumping which impairs subsequent flow cytometric analysis. Clumping can, however, be minimized by preparing a concentrated cell suspension in isotonic medium to which acetone or alcohol are added dropwise with constant mixing (Rabinovitch, 1986; Morris, Mathews, 1989; Baisch, Gerdes, 1990). A variation on this theme is careful layering of the organic solvent on top of the concentrated cell suspension and mixing the layers gradually by swirling the tube (Jacobberger et al., 1986; Darzynkiewicz et al., 1986; Schimenti, Jacobberger, 1992). Inclusion of a low concentration of Tween-20 or TX-100 in the cell suspension after the fixation step and performing all cell incubations at 0-4 deg may also help to avoid the formation of cell aggregates.

Fixation and permeabilization by acetone, ethanol and methanol are extensive and proceed relatively rapidly. Fine control of the extent of the reactions is therefore difficult. The control of fixation by lowering the temperature (room temperature to -70 deg C.) has not been investigated in a systematic manner, but does not seem to be very effective (Zipf et al., 1984). However, the extent of fixation and permeabilization is to some degree dependent on the concentration of the organic solvent (Schimenti, Jacobberger, 1992). In Schimenti's experimental system, methanol at a concentration of 80-90% was preferable to acetone and ethanol at any concentration, since with methanol maximal immunofluorescence and minimal cell aggregation were obtained. Below a concentration of 50% the organic solvents are usually inefficient as fixatives.

2.2 Aldehydes and other fixatives.

Paraformaldehyde (PFA) fixes cells by cross-linking of proteins (Puchtler, Meloan, 1985). Fixation must, however, usually be combined with plasma membrane permeabilization, e.g. cell freezing and thawing (Rabbitts et al., 1985; Bauer et al., 1986a; Lincoln, Bauer, 1989; Giaretti et al., 1990) or a short incubation of the cells in the presence of a detergent (see next section). The extent of fixation by PFA can easily be controlled by varying the aldehyde concentration in the range of 0.01-4% and the duration of incubation of the cells with the fixative from a few minutes up to hours (Mann et al., 1987, and see Table 1). Also, the reaction rate is highly temperature-dependent and may be very slow at 0-4 degrees C.

(Mann et al., 1987). Therefore, the accurate control of temperature of PFA fixation may be of critical importance.

Formalin solutions are less recommended than fresh or shortly stored PFA solutions because of the effects of alcohol added as a preservative. In addition, upon storage, various condensation and decomposition products form and may interfere in an unpredictable manner with controlled fixation.

Glutaraldehyde possesses two reactive aldehyde groups and is more efficient in cross-linking of proteins than PFA. Therefore, there is an increased danger of overfixation and the applied concentrations of glutaraldehyde are usually small and range from 0.1-0.2%. Moreover, at a concentration higher than 0.25%, glutaraldehyde usually causes an unacceptably high background of nonspecific fluorescence in the cell. Glutaraldehyde is therefore used only infrequently in immunofluorescence.

Fixation by a mixture of PFA (4%) and glutaraldehyde (0.05%) has been described (Reeve et al., 1986). Surprisingly, in this system, the fixative glutaraldehyde also helped to permeabilize the cells so that no detergent treatment was needed. A similar effect of glutaraldehyde addition to PFA has recently be described for the permeabilization of malaria parasites within red blood cells (Pattanapanyasat et al., 1992). Moreover, while PFA fixation of the cells changes the forward light scattering on flow cytometric analysis, the change is prevented when glutaraldehyde is added for cell fixation.

Periodate, lysine and PFA, Nakane's mixture (McLean, Nakane, 1974), is a modification of the aldehyde fixation. This mixture cross-links cell carbohydrate moieties rather than proteins. In some instances the replacement of PFA by Nakane's mixture was the final solution for successful immunofluorescence analysis of an intracellular antigen by flow cytometry (Turner et al., 1988; Drach et al., 1989).

Picric acid is used in histochemistry as a rapidly tissue-penetrating fixative (Bancroft, 1967). It has been successfully applied in combination with PFA to analyze the content of hormones in pituitary cells by flow cytometry (Hatfield, Hymer, 1985) and of tyrosine hydroxylase in nerve cell synaptosomes (Wolf et al., 1989). The PFA/picric acid combination may help to minimize the loss of relatively small-molecular weight substances from the permeabilized cell. Unreacted picric acid stains the cells yellow and should be efficiently washed out.

2.3 Detergents for cell permeabilization

(a) Strong membrane solubilizers

Triton-X100 (TX-100) and the closely related Nonidet-P40 (NP-40) are the most commonly applied detergents for cell permeabilization. They are non-ionic, efficient, general membrane solubilizers. Forward light scattering is greatly reduced when cells are permeabilized by these detergents and analyzed in the flow cytometer . This phenomenon is expected because cell membranes become highly permeable to solutes, including proteins, that may decrease the refractive index of the cell. To prevent extreme solubilization of the membranes and total cell lysis, TX-100 and NP-40 are used at relatively low concentrations (0.01-0.5%), are applied in the cold and for relatively short times only (3-10 min, see Table 1).

(b) Mild membrane solubilizers

Saponin (a mixture of plant glycosides) and digitonin (a defined member of the saponin group) are detergents of more selective and more limited action and have frequently been applied with success in recent years . Saponin and digitonin react specifically with cholesterol and other beta-hydroxysterols (see Fiskum et al., 1980). Since cholesterol is a relatively minor component of cellular membranes and, moreover, is more abundant in the plasma membrane than in intracellular membranes, saponin (0.1-0.3%) and digitonin (10-100ug/ml) permeabilize mainly the plasma membrane and do so without solubilizing it totally (Fiskum et al., 1980). Obviously, intracellular molecules leak out, but when saponin or digitonin are applied at the above mentioned concentrations, 50-60 percent of the protein is retained in the cell (Weigel et al., 1983). In line with reduced protein leakage, compared to permeabilization by TX-100 or NP-40, is the finding that forward light scattering is decreased only slightly. Thus, for example, lymphocytes and monocytes are still separable on the basis of their respective light scatter intensities (Jacob et al., 1991). Even when applied at relatively high concentrations, saponin and digitonin do not lyse the cells but enhance the loss of cellular protein.

Permeabilization by saponin and digitonin enables the passage through the plasma membrane of antibodies and proteins up to a molecular weight of about 200,000 daltons (Fiskum et al., 1980; Weigel et al., 1983). Digitonin has been shown by electron microscopy to give rise to only small morphological changes in the plasma membrane and hardly affects the nucleus, mitochondria and other cell organelles (Fiskum et al., 1980). This notion is corroborated by the finding that to analyze antigens that are located within cytoplasmic granules, within the mitochondria or the nucleus, cell permeabilization by saponin may not suffice and may have to be replaced or supplemented by treatment with a strong solubilizing detergent, e.g. TX-100 (Goldenthal et al., 1985).

Lysolecithin is another natural compound with detergent action that is also applied for gentle cell permeabilization (Schroff et al., 1984 and see Table 1). Lysolecithin is thought to replace phospholipids in the cellular membranes (Weltzien, 1979), but the mechanism of cell permeabilization is not clear. The danger of cell lysis is greater with lysolecithin than with saponin and digitonin that interact only with cholesterol. However, cells can be mildly and reproducibly permeabilized when lysolecithin is applied under stringently controlled conditions. Lysolecithin permeabilizes cells without lysing them only if applied in a restricted concentration range (Weltzien, 1979; Schroff, 1984; Miller, 1985; Dent, 1989), a constant ratio of lysolecithin to the number of cells and carefully controlled temperature (Weltzien, 1979). Lysolecithin treament may be stopped by the addition of nondelipidated serum albumin (final concentration: 1%) which completely inhibits the effects of lysolecithin (Schroff et al., 1984).

Recently, lysolecithin permeabilization was optimized by applying the reagent at pH 4.5 (Labalette-Houache et al., 1991). At the acid pH, the effective lysolecithin concentration for cell permeabilization to antibodies could be decreased by a factor of ten (5ug/ml). Under these conditions, the cells were significantly stabilized, as evident from complete conservation of the forward light scattering pattern in spite of cell permeabilization. Prior to the restoration of neutral pH, the cells were shortly washed in the presence of PFA to further stabilize them and to prevent cell lysis (Labalette-Houache et al., 1991).

<u>Tween-20</u> is quite inefficient in extracting proteins from cell membranes (Hjerten, Johansson, 1972) and in solubilizing phospholipid liposomes (Womack et al., 1983). Consequently, Tween-20 is frequently included (0.1-0.2%) in the rinsing solutions that are used to eliminate unreacted antibody from the cells without causing loss of antigen (see Table 1). Recently, however, Tween-20 has been introduced as a very mild plasma membrane permeabilizer in combination with a relatively low concentration of PFA (Schmid et al., 1991). Another research group used Tween-20 (0.5%) to permeabilize Indian muntjac cells to anti-kinetochore antibody that was used as a probe to label the chromosomes (Levy et al., 1991).

<u>Additional detergents</u> have been tested over the years for their membrane solubilization characteristics. Five chemically pure compounds, commercially produced for applications in biochemistry, were found highly effective in solubilizing egg

phospholipid liposomes and, at the same concentration, did not inactivate three soluble hydrolytic enzymes that were tested, namely esterase, aryl sulfatase and alkaline phosphatase (Womack et al., 1983). Among these compounds, n-octylglucoside has already been introduced for immunofluorescence analysis of intracellular antigens by flow cytometry (Andersson et al., 1989; Hallden et al., 1989). The success with n-octylglucoside as a mild cell permeabilization reagent that preserves antigenicity warrants testing intracellular antigen labeling using the other four detergents (CHAPS (*), CHAPSO, Zwittergent 310 and Zwittergent 312) that have similar lipid solubilization characteristics (Womack et al., 1983).

Finally, regarding the entire group of mild detergents, it should be noted that these detergents can be applied to cells without prior cell fixation (see Table 1). Under well-controlled conditions and in specifically designed incubation media, cells can withstand permeabilization and retain some of their vital functions for 30 min or more (Schulz, 1990; Ahnert-Hilger et al., 1989a; 1989b). Therefore, not only does antibody diffuse into the cell in detergent-permeabilized cells, but it is possible to investigate the biological function of an intracellular antigen by attempting to inactivate the antigen by the antibody (Miller et al., 1985). Similar experiments have been conducted by other means of cell permeabilization, e.g. by liposome fusion and electroporation (see below).

2.4 Cell permeabilization by freezing and thawing and by cell incubation with protease

Two cell treatments that may allow antibody penetration into living cells have not sufficiently been investigated,

(*)Abbreviations: CHAPS, 3-[(3-cholamidopropyl) dimethylammonio]-1-[2-hydroxy-1-propane] sulfonate; CHAPSO, 3-[(3-cholamidopropyl) dimethylammonio]-1-propane sulfonate.

namely the freezing of cells to -70 deg and rethawing them thereafter and the incubation of cells with proteases. Although these two treatments are rather different from each other, their common denominator is that (a) they are routinely used with cultured cells and (b) treated cells stay alive.

<u>(a) Freezing and thawing.</u> Cells are frozen for storage and are thawed prior to their use in further experimentation. Freezing and thawing as part of the treatment of cells that makes them permeable to antibodies is documented in a number of reports cited in Table 1 (Rabbitts et al., 1985; Giaretti et al., 1990; Larsen et al., 1991). The contribution of freezing and thawing to the final extent of cell permeabilization is, however, difficult to assess. What stands out is the fact that in the instances cited above, an intracellular antigen could be labeled without using any additional cell permeabilization procedure. Note that PFA fixation by itself seldom causes cell permeabilization to antibodies.

<u>(b) Protease effects.</u> A somewhat similar situation holds following the incubation of cells with proteases to detach adherent cells from their substrate, or to dissociate pieces of tissue into a single cell suspension. Protease treatment has, however, been specifically worked out as a procedure for cell permeabilization to proteins (Lemons et al., 1988). A hyperpermeable state is induced in fibroblasts following cell incubation with trypsin, pronase or collagenase. The molecular weight of radioisotope-labeled proteins that could penetrate into the protease-treated cells ranged from 6000 (insulin) to 600,000 (thyroglobulin). Thus, antibodies can be expected to enter the cells also. Actually, that may have occurred in at least two experimental systems that are cited in Table 1. In one case there was no need for additional permeabilization of the cells (Nelson et al., 1987) while in the second case, incubation with a very low concentration of detergent was sufficient for antibody introduction into protease-treated cells (Zeile, 1980). Hence, in these cases, cell permeabilization to antibody can be ascribed, at least in part, to protease treatment.

In conclusion, one should be aware of the possible permeabilization effects of freezing and thawing and of protease treatment and of the unique option to exploit these mild treatments to permeabilize living cells to antibodies.

C. INTRODUCTION OF ANTIBODIES INTO CELLS BY OTHER MEANS

A series of additional methods have been worked out to introduce antibodies into cells without fixation and without the application of detergent. The methods include (a) microinjection, (b) fusion of cells with antibody-loaded liposomes, (c) biological means, e.g. Streptolysin O, a bacterial pore-forming protein, (d) pinocytosis combined with osmotic shock and (e) electroporation. Since by all these methods the majority of cells remains alive, the biological effects of the specific antibody on cell metabolism and cell physiology can be investigated.

(a) Microinjection of antibodies into cells (Mercer et al., 1982), although being a most efficient antibody-introducing method, does not lend itself to mass treatment and flow cytometric analysis.

(b) Liposome-cell fusion using liposomes loaded with antibody is another efficient method for introducing antibodies into cells (Thompson, Gross, 1988). Following liposome fusion with *Drosophila* Schneider 2 cells, the presence and the intracellular location of incorporated antibody was visualized by post-fixing the cells and labeling with a FITC-conjugated second antibody. Lupus antibodies against small nuclear ribonucleoprotein were shown to localize in the nucleus, rabbit-derived anti-tubulin antibodies were present predominantly in the cytoplasm, whereas lupus anti-ribosomal antibodies localized both in the nucleolus and in the cytoplasm. Also, it was possible to disturb two normal cellular processes, namely protein synthesis and mitosis, by cell fusion with liposomes that contained anti-ribosomal and anti-tubulin antibodies respectively (Thompson, Gross, 1988).

Interestingly, in contrast to earlier findings in human fibroblasts that only F(ab')2 fragments, but not unfragmented antibodies, are able to traverse the nuclear envelope and enter the nucleus (Einck, Bustin, 1984), in the cultured Drosophila cells and under the experimental conditions of Thompson and Gross, lupus IgG antibodies bound to their nuclear antigens.

(c) Streptolysin O (SLO) from beta-hemolytic streptococci is one of the so-called pore-forming proteins that are cytolytic to mammalian cells. From the monomeric protein, multimeric pores are apparently assembled in the plasma membrane (Ojcius, Young, 1991). The size of the pores that are formed by SLO are large enough for nucleic acid and proteins to diffuse into the cells (Buckingham, Duncan, 1983; Ahnert-Hilger et al., 1989a). A procedure was worked out to permeabilize PC12 pheochromocytoma cells by the rapid formation of pores following a temperature shift. The cells retained their exocytotic capacity. By this procedure, antibodies to cytoplasmic vesicles were incorporated into the cells (Ahnert-Hilger et al., 1989b). Although the experimental results were analyzed by immunofluorescence microscopy only, the method could obviously be combined with flow cytometry.

(d) Pinocytosis in a hypertonic environment, followed by hypotonic shock is another route for antibody introduction into cells (Okada, Rechsteiner, 1982). Four rounds of this procedure could be applied to cells without apparent loss of viability. By using this elegant experimental approach, several rapid cellular insulin effects were inhibited on introducing anti-insulin receptor antibody into the cytoplasm of hepatocytes, adipocytes and a number of cultured cell lines (Morgan, Roth, 1987).

Flow cytometric analysis was used to investigate the effect of incorporation of anti-asparagine synthetase antibodies into L5178Y murine lymphoma cells that grow independently of an exogenous supply of asparagine (Chakrabarti et al., 1989a). Unfortunately, in this experimental system, a considerable, although transitory disturbance of protein and DNA synthesis was caused by the osmotic changes themselves. Nonetheless, a partial dependence on exogenously supplied asparagine was noted after the incorporation of antibody.

(e) Electroporation is by now a technique widely used for gene transfection with good efficiency in many cell systems. Chakrabarti and his coworkers have optimized the conditions of electroporation to retain a high percentage of viable cells and to achieve antibody incorporation into the majority of them (Chakrabarti et al., 1989b). Protein and DNA synthesis were not inhibited in He-La, HT-5 and L5178Y cells that were subjected to electroporation in the presence of antibody to asparagine synthetase. After electroporation, growth of the otherwise asparagine-independent L5178Y cells became clearly dependent on an exogenous supply of this amino acid. The successful introduction of antibodies into mouse melanoma cells using electroporation has also been reported (Berglund, Starkey, 1991).

D. GUIDELINES FOR THE ANALYSIS OF NEW INTRACELLULAR ANTIGENS

1. Conditions for cell permeabilization that allow labeling and analysis of intracellular antigens: a summary and conclusions

Table 1 presents a list of intracellular antigens that have been labeled by fluorescent antibodies and have been analyzed by flow cytometry. As evident from Table 1, the conditions for cell permeabilization/fixation that were needed for successful labeling of intracellular antigens cover an extremely broad range. Conditions range from permeabilization by very low concentrations of mild detergents, without any cell fixation, to complete dehydration by absolute alcohol and hydrocarbons, as is routine for the preparation of biological samples for embedding in paraffin and sectioning.

The aim of the survey presented in Table 1 was to find out whether one can recognize any general principles that relate these widely varying conditions to the characteristics of the antigen. To this end, in Table 1 , the major details of the cell preparation procedure are noted next to each antigen. Details include the fixative, the method of cell permeabilization, the duration and temperatures of the reactions. In addition, the antigens are grouped according to their location in subcellular compartments. Within each group of antigens, functionally related antigens are listed together, e. g. intracytoplasmic immunoglobulins, enzymes, proliferation-associated antigens, etc. Finally, the list of the subgroups of antigens is arranged according to increasing harshness of the fixation and permeabilization conditions. It should be noted that published permeabilization conditions frequently represent the outcome of numerous attempts by the research groups to achieve specific labeling of the antigen. The final, published conditions, can therefore be assumed to represent optimal or near-optimal conditions for cell permeabilization.

When the data assembled in Table 1 are examined according to the above scheme, it becomes evident that a number of general conclusions can be drawn.

(a) <u>Antigens that are located in proximity to the plasma membrane and so-called "soluble" cytoplasmic antigens.</u> For the labeling of these antigens, very mild conditions of cell permeabilization are usually sufficient and obligatory. Cells are not fixed at all or are fixed by PFA, mostly at a relatively low concentration and/or short duration. Fixation by acetone or alcohol is exceptional.

(b) <u>Cytoskeleton-bound and viral antigens.</u> In striking contrast with plasma membrane-bound and "soluble" cytoplasmic antigens, for the analysis of cytoskeleton-bound antigens cells are almost exclusively fixed extensively, by using acetone or alcohol, or PFA at a high concentration (4%). Intracellular viral antigens are also labeled following a similar treatment.

(c) Antigens that are located in cytoplasmic organelles or granules and nuclear antigens. This group of antigens constitutes a mixed group with respect to the fixation/permeabilization conditions. Conditions range from very mild (e. g. for T-cell cytoplasmic granules) to fixation/permeabilization by absolute alcohols.

(d) A small number of antigens can be labeled by a variety of fixation/permeabilization procedures. They are accessible to antibodies after minimal cell treatment and, on the other hand, are also stable to the harshest fixation conditions (e. g. the Ki-67 and c-myc proteins).

Unfortunately therefore, when an antibody to a previously not analyzed intracellular antigen is raised, it is still impossible to prescribe an exact cell permeabilization procedure by which the binding of the antibody to its specific antigen can be predicted to succeed. Thus a search for an appropriate procedure for cell permeabilization must be conducted, based on trial and error. The general conclusions above suggest that a logical and systematic approach to the trial-and-error search may increase the chance to find, and relatively rapidly, conditions to label the new intracellular antigen with the antibody. The more is known about the location of the antigen within the cell, the more logical an approach to its labeling by antibody can be taken.

2. Practical hints for the labeling of new intracellular antigens

2.1 General recommendations

Antibody penetration into permeabilized cells is restricted by the limited size of the pores that were created in cell membranes by the permeabilizing treatment. Therefore, preferentially use an antibody of high affinity and small molecular weight (F(ab')2 or IgG, not IgM) and that is conjugated to a small molecular weight fluorescent probe (FITC, not phycoerythrin). Start the search with the highest possible sensitivity by using a concentrated antibody solution. But, remember that incubation with any antibody at high concentration may result in non-specific binding. Therefore, once a fluorescence signal is obtained, it is imperative to test serial dilutions of all the antibodies that are used in the procedure, to assure specificity and to maximize the signal-to-noise ratio. Also, run appropriate negative controls in parallel, i.e. cell incubation with antibody that was adsorbed with the antigen, or incubation with an isotype, irrelevant antibody.

Single cell suspensions can always be analyzed in a fluorescence microscope. Moreover, only a tiny amount of cells and antibody is needed. Since multi-well microscope slides are commercially available, the screening of cells prepared or labeled in a number of different ways is relatively rapid and economic and is not necessarily dependent on the availability of a flow cytometer. For the microscopic analysis of adherent cells, grow them directly on cover slips or on multi-welled slides specially designed for that purpose.

2.2 Specific guidelines for the trial-and-error search

(a) Literature check. When attempting to label a new intracellular antigen, first screen the literature to establish if a related antigen has already been labeled and analyzed. A good starting point for finding an appropriate cell permeabilization method to analyze a new antigen is to use the same cell preparation method as described in the literature. In the very first experiment, try also some milder and some harsher fixation/permeabilization conditions in parallel to those described in the literature, since your cells, epitope and reagents may differ somewhat from those described in the published report.

(b) Fixation/permeabilization by alcohol and acetone. In spite of the harshness of these reagents, this permeabilization method should be tested with any new antigen at a very early stage of the search. It is the simplest and most rapid method and 40-50% of the antigens listed in Table 1 were labeled by using it. Moreover, with antibodies to nuclear, organelle-bound and cytoskeleton-associated antigens, the chances of immediate success are 60-70%.

All that is needed is a concentrated cell suspension, the careful addition of the organic solvents and careful mixing to avoid cell clumping (see section B.2.1). The method is frequently adequate or even optimal for the labeling of an antigen and the chances for success are high, especially if the antigen is bound to intracellular membranes or is confined to a membrane-enclosed subcellular compartment. Test the following concentrations in parallel: methanol 90%, 80% and 67%, ethanol 90%, 80%, 67% and 50% (Schimenti, Jacobberger, 1992), acetone 90%. Two lengths of incubation should be tested, first a short one (4-5 min, see Hayden et al., 1988) and, only if results are negative, try a longer incubation. Because of the general feeling that the temperature of the reaction with organic solvents is not critical, confine yourself to a single temperature (recommended: -20 deg or ice-cold).

(c) Protease treatment. This treatment is another relatively simple procedure (Lemons et al., 1988) that should be tested before one starts the more laborious experiments of titrating aldehyde fixatives and detergents (see below). That cells become hyperpermeable to proteins after detachment from culture dishes by trypsin, or partial proteolysis of a piece of tissue, may be a quite general phenomenon. Conceivably, mainly cytoplasmic "soluble" and membrane antigens may become accessible to antibodies by this treatment. Adding a mild detergent following protease treatment, or first fixing the cells very gently, may further facilitate antibody access to its antigen.

(d) Permeabilization by detergents without fixation or with very mild PFA fixation. This method has a high chance of success with antigens in the cell periphery, close to the plasma membrane and "soluble" cytoplasmic antigens. Testing this method may involve a larger number of variables than in (b) since it may be necessary to test a number of different detergents. To exclude as many variables as possible, again stay at a constant temperature (ice-cold), at least in the beginning. Similarly, in the beginning, restrict yourself to two lengths of duration of the reaction (in the range of 3-10 min, see Table 1). Thus, in a single run, the concentration of the detergent can be tested in its entire concentration range. If the results are negative, it will be profitable to test additional detergents.

As to the choice of detergent, start with the milder ones. Use the purest reagents and preparations available to minimize the danger of effects of accompanying impurities. Lysolecithin and n-octylglucoside may be the purest ones available. The purest preparations of digitonin are designated 80% pure (Calbiochem). Saponin is a mixture, thus the activity of the reagent may vary from one supplier to another and even between batches. TX-100 has also been used succesfully at a very low concentration (0.01%) without cell fixation (Zeile, 1980). To titrate the concentration of detergent, start with the lowest concentrations cited in the literature (see section B.2.3 or Table 1) and titrate upwards.

At some stage in your search, e.g. once you start getting a fluorescence signal, repeat that experiment and add PFA at a low concentration , e.g. 0.01%, to the incubation medium together with the detergent (Anderson et al., 1989), to find out whether some fixation may be helpful to optimize the procedure.

(e) PFA fixation combined with detergent permeabilization. If up to this stage all the attempts to label the antigen were unsuccessful, fixation by PFA at various concentrations, combined with permeabilization by a detergent, should be tested. It is evident from Table 1 that the concentration range of PFA is wide and, in different antigen systems, 0.01% to 4% solutions of PFA were used for optimal conditions of fixation. However, the majority of antigens could be labeled when the cells were fixed with PFA in the 0.25-4% concentration range. Therefore, at the first attempts, try PFA fixation in this range only (i.e. 5-6 experimental points).

Note, however, that the optimal concentration of detergent used subsequently for cell permeabilization will be dependent on the extent of protein cross-linking by the fixative. Therefore, the concentrations of PFA and detergent must be titrated simultaneously in order not to lose any chance of success. This means that with any concentration of PFA one must test several concentrations of the detergent.

Choose the detergent according to the subcellular location of the antigen, if known. Thus, for a cytoskeleton-bound or organelle-confined antigen, there is a higher chance of rapid success if you start with permeabilization by a strong solubilizer (TX-100 0.1-0.4% or NP-40). On the other hand, start with one of the mild type detergents (i.e. saponin, digitonin, lysolecithin or n-octylglucoside), for a plasma membrane-bound or "soluble" antigen (Goldenthal et al., 1985). If nothing is known about the subcellular antigen location, or if the results with one type of detergent are negative, one representative at least of the second detergent type should be tested as well.

Since titrating both PFA and the detergent is necessary, this method may involve a relatively large set of experiments in order to find conditions for antigen labeling. To exclude as many variables as possible while retainig the highest probability of success, fixation by PFA should be carried out at two temperatures because of the marked effect of the temperature on the reaction (Mann et al., 1987). First conduct the PFA titration experiment in the cold (0-4 deg) for 10 min, to fix the cells only partially and relatively mildly. Treatment with detergent following the fixation should be at 0-

4 deg for 3-5 min. Thus, in a single experiment of PFA fixation, it should be possible to test the above recommended PFA concentrations at a single temperature and at two concentrations of detergent at least. If the results are negative, test a wider range of detergent concentrations. If still negative, repeat the titration at 37 deg and prolong the incubation with PFA to 15-20 min, to achieve a harsher fixation (Mann et al., 1987). Test the same concentrations of detergent, or even higher ones than before.

(f) Additional methods. According to the data assembled in Table 1, the probability of success is 80-90% for labeling the antibody by testing all the above methods thoroughly . However, if no labeling was achieved, the following changes can be tried. (1) Strengthen the fixation further by adding glutaraldehyde or picric acid to PFA. (2) Permeabilize more extensively with detergent by prolonging the incubation and/or raising the temperature of the reaction with detergent. On the other extreme, try also the mildest detergent, namely Tween-20. (4) Replace PFA by Nakane's mixture as a fixative, to minimize protein cross-linking. (5) Try freezing and thawing the cells prior to any other treatment. (6) Finally, switching to introduction of antibodies into cells by one of the additional novel, tested methods (electroporation, etc.) is an open possibility.

(g) Optimalization of the procedure. Once specific labeling of the antigen has been achieved, some additional effort must be devoted to optimize the procedure. The concentration of the reagents, the temperature and duration of the reactions of fixation and of permeabilization should all be optimized to get maximal specific and reproducible immunofluorescence. In this context, a controlled postfixation treatment with protease has been shown to increase immunofluorescence of certain antigens, probably by unmasking additional antigenic sites (Battifora, Kopinsky, 1986). On the other hand, shortcuts can be tried, such as mixing the detergent with the fixative (Garner et al., 1984; Bardales et al., 1989a) and washless staining (Larsen et al., 1991).

3. Final remarks

Although the general conclusions and the suggestions for different groups of intracellular antigens may sound logical, they should be considered with reservation, as they are based on a limited number of antigens analyzed to date. Moreover, because of the natural diversity of cells (internal structure and composition) and of antigens, and because of the multiplicity of epitopes on a single antigen, the conclusions should be considered tentative rather than absolutely true. For instance, it is obvious from Table 1 that some cytoplasmic antigens can be labeled after fixation by an organic solvent. On the other hand, some nuclear antigens are accessible to antibody by mild saponin/digitonin or lysolecithin permeabilization without any cell fixation. The latter finding may especially become relevant when F(ab')2 fragments are used, since they appear to traverse the nuclear envelope without the need for more drastic permeabilization (Einck, Bustin, 1984).

As with any new methodology, the results of the flow cytometric analysis of an intracellular antigen should always be validated by additional, independent experimental procedures. Since the flow cytometer cannot distinguish with certainty between an intact cell, a cell aggregate, cell debris and contaminating fluorescent particles, it should be routine to compare the flow cytometric results of the analysis of a new antigen with the analysis of the same cells and antigen by fluorescence microscopy.

By combining all the information available in Table 1 and the conclusions and suggestions presented above, there is a good chance of succeeding with the analysis of a new intracellular antigen in reasonable time and with a reasonable amount of effort. Obviously, the restraints on the fixation and permeabilization procedure are further increased when double labeling, of two antigens or of an intracellular antigen plus DNA, is attempted. The results presented in Table 1 show that intracellular double labeling too is achievable.

Table 1

Intracellular antigens analyzed by flow cytometry and the conditions for cell permeabilization.

ANTIGEN (*)	FIXATION and PERMEABILIZATION (*)	REFERENCE
1. Plasma membrane, cytoplasmic face		
T-cell receptor zeta-chain	PFA 0.01%, 20', 4 deg Digitonin 10ug/ml, 5', 4 deg	Anderson et al., 1989
Ha-ras (DP)	PFA 0.5%, 10', 4 deg TX-100 0.1% (Ethanol 70% for DNA labeling)	Czerniak et al., 1987
p21-ras	NO FIXATION Electroporation	Berglund, Starkey, 1989 Berglund, Starkey, 1991
V-src pp60, P-tyr residues (DP)	Ethanol 70%, 16h, 4 deg	Preis et al., 1988
2. Cytoplasmic, "soluble" antigens		
Cytoplasmic Ig (DP)	NO FIXATION, protease TX-100 0.01%, 10' (Ethanol 70% for DNA labeling)	Zeile, 1980
Cytoplasmic IgM	NO FIXATION Lysolecitin 50ug/ml, 5', 4 deg	Schroff et al., 1984
Cytoplasmic Ig	PFA 2% + saponin 0.25%, lysolecithin 0.1% or TX-100 0.01% 5', 4 deg	Garner et al., 1984
Cytoplasmic Ig	PFA 2% + saponin 0.25%, 5', 4 deg	Bardales et al., 1989a

(*) Abbreviations used: Ag, antigen; AO, acridine orange; BSA, bovine serum albumin; Carbam.-P synth., Carbamoylphosphate synthase; (DA), DNA label: acridine orange; (DB), DNA label: anti-BrdUrd; (DP), DNA label: propidium iodide or other fluorescent labels; deg, degree Centigrade; EBV, Epstein-Barr virus; FA, formaldehyde; Ig, immunoglobulin; PAP, placental alkaline phosphatase; PCNA, proliferating cell nuclear antigen; PEPCK, phosphoenolpyruvate carboxykinase, PFA, paraformaldehyde; PLP, periodate-lysine-paraformaldehyde (McLean, Nakane, 1974); P-tyr, phosphotyrosine; RBC, red blood cells; RT, room temperature; SnRNP, small nuclear ribonucleoprotein; SV40, Simian virus 40; TdT, terminal deoxynucleotide transferase; TNF, tumor necrosis factor.

ANTIGEN	FIXATION and PERMEABILIZATION	REFERENCE
Cytoplasmic, "soluble" antigens (contd.)		
Cytopl. T-cell antigens (DP)	PFA 3.7%, 5', RT NP-40 0.1%, 5', RT	Zocchi et al., 1990
Cytoplasmic u-chains	Ethanol 50% or 70% or ethanol 95% + acetic acid 5% 20 deg, 4 deg or -20 deg	Zipf et al., 1984
gamma-Interferon	PFA 4%, 5', RT n-octylglucoside 0.6%, 5', RT	Andersson et al., 1988
Hormones: GH, LH, Prolactin (DP)	PFA 2% + picric ac. 7.5%, 30', RT TX-100 0.4%, 30', RT	Hatfield, Hymer, 1985
Progesterone receptor (DP)	PFA 0.5%, 5', RT TX-100 0.1%, 3-5'	Graham et al., 1989
PAP, alkaline phosphatase (DP)	NO FIXATION, protease TX-100 0.02%, pH 5.2 for DNA	Nelson et al., 1987
Ornithine decarboxylase	PFA 2% TX-100 0.05%, 15', 4 deg	Robertson et al., 1990
Tyrosine hydroxylase (in synaptosomes)	PFA 2% + picric acid 7.5%, 30' (antibodies overnight)	Wolf, Kapatos, 1989
Adenosine deaminase	PFA 4%, 60', 4 deg (antibodies overnight)	SenGupta et al., 1985
Enolase	PFA 4% + glutarald. 0.05%, pH 9.5, 30'	Reeve et al., 1986
Ribonucleotide reductase (DP)	PFA 2-4%, 15-30', 4-37 deg TX-100 0.2%, 4 deg (one wash) Tween-20 1% in following washes	Mann et al., 1987
Asparagine synthetase	NO FIXATION Osmotic permeabilization	Chakrabarti et al., 1989a
Asparagine synthetase	NO FIXATION Electroporation	Chakrabarti et al., 1989b
Dictyostelium differentiat. Ag	NO FIXATION, protease Digitonin 0.1 mM, 10', 20 deg	Bernstein et al., 1988

ANTIGEN	FIXATION and PERMEABILIZATION	REFERENCE

3. Cytoplasmic organelles, granules and inclusions

T-cell cytopl. granules	PFA 0.01%, 20', 4 deg Digitonin 10ug/ml, 5', 4 deg	Anderson et al., 1990
Myeloperoxidase (DP)	PFA 1%, 30', 4 deg Lysolecithin 5-40ug/ml	Dent et al., 1989
Lactoferrin	Formol acetone, pH 6.6, 5' Saponin 1% 60', 55 deg	Butler et al., 1988
Melanosome matrix protein (DP)	Ethanol 50%, 30'	Maeda et al., 1990
PEPCK, arginase Carbam.-P synth. (mitochond.) (DP)	Ethanol 70%, -20 deg	van Roon et al., 1989

4. Cytoskeleton

Actin	NO FIXATION, protease Digitonin 50ug/ml, 3', 25 deg	Fiskum et al., 1980
Actin	NO FIXATION Electroporation NBD-phallacidin as fluor. marker	Downey et al., 1989
Cytokeratin (tissue sections)	FA or ethanol, total dehydration, embedding, protease on sections	Battifora, Kopinsky, 1986
Epidermis Ags, Intermed. filamts.	Ethanol 70%, long term storage	van Erp, et al., 1989
Intermediate filaments (DP)	Methanol, -20 deg	Oud et al., 1985
Tubulin (DP)	Ethanol 70%	Ronot et al., 1988
Keratin (DB)	Ethanol 70%	Kirkhus, Clausen, 1990

5. Cytoplasmic and surface antigens

Cytopl. u-chains / CD3 (DP)	PFA 0.25%, 60', 4 deg Tween-20 0.2%, 15', 37 deg	Schmid et al., 1991

ANTIGEN	FIXATION and PERMEABILIZATION	REFERENCE
Cytoplasmic and surface antigens (contd.)		
CD3 zeta-chain / Sezani T-cell Ag	Ethanol 50%, 5', 4 deg or lysolecithin 50ug/ml, 5'	Hayden et al., 1988
Complement receptor CR1	PLP 20', 4 deg Saponin 0.04% + ovalbumin 0.1%	Turner et al., 1988
TNF / TNF receptor	PFA 2%, 15', 4 deg Saponin 0.1%, 10', 4 deg	Hofsli et al., 1989
TNF / plasminogen activator (DP)	Acetone:methanol (1:1), 5', 4 deg Lysolecit. 20 ug/ml, 4', 4 deg	Rice et al., 1991
Vimentin / surface markers	PFA 4%, 5-10', RT n-octylglucoside 0.07-0.8%	Hallden et al., 1989
Cytokeratin / Leukocyte Ag	Ethanol 50%, 4 deg	Zarbo et al., 1989

6. Nuclear antigens

p34, p36, p105 (DP)	PFA 0.5%, 10', 4 deg TX-100 0.1%, 3', 4 deg TX-100 0.1% in washes	Clevenger et al., 1985
p34 (DA)	PFA 0.5%, 10', 4 deg TX-100 0.1%, 3' 4 deg	Bauer et al., 1986b
p34 (DP) (in tissue)	Freeze-thaw, PFA 0.5%, 30' TX-100 0.1% permeabil. and washes	Bauer et al., 1986a
PCNA (DB)	PFA 1% + lysolecit. 0.05%, 5', RT then methanol 10', -20	Kurki et al., 1988
PCNA (DB)	Ethanol 70% or methanol, 10', -20 deg, then NP-40 0.5%, 5', 0 deg or TX-100 0.5% extract. buff.	Landberg, Roos, 1991
PCNA (DP)	Ethanol 75%, 12h, 4 deg	Morris, Mathews, 1989
PCNA, p120, p145 (DP)	PFA 1% + lysolecit. 20 ug/ml, 2', RT Methanol 10', -10 deg then NP-40 0.1% 5', 0 deg	Bolton et al., 1992
Ki-67 (DB) (in nuclei)	NO FIXATION, freeze-thaw lysis buffer	Larsen et al., 1991

ANTIGEN	FIXATION and PERMEABILIZATION	REFERENCE
Nuclear antigens (contd.)		
Ki-67 (DP)	PFA, 0.5%, 5', 4 deg TX-100 0.1%, 5', 4 deg	Palutke et al., 1987
Ki-67, PCNA, p34 p105 (DP) (DA)	PFA 0.5%, 15', 4 deg TX-100 0.1%, 5'	Bruno et al., 1991
Ki-67 (DP)	Acetone 80%, >24h, -20 deg	Baisch, Gerdes, 1990
c-myc (DB)	NO FIXATION, freeze-thaw NP-40 0.1-0.2% + BSA 1%	Giaretti et al., 1990
c-myc (DP)	NO FIXATION, freeze-thaw	Rabbitts et al., 1985; Smith et al., 1985
c-myc (DP)	PFA 0.5%, 30', 4 deg TX-100 0.1%, 3', 4 deg	Engelhard et al., 1991
c-myc (DP)	PFA 1%, 30', 4 deg Lysolecithin 5-40ug/ml	Dent et al., 1989
c-myc (DP)	Methanol 30', 4 deg	Roberts et al., 1989
c-myc (DP) (in tissue)	Ethanol 70%, 2h, -20 deg TX-100 0.1%, 5'	Engelhard et al., 1989
c-myc, p105 (DP)	Freeze-thaw, PFA 0.5%, 30',4 deg	Lincoln, Bauer, 1989
c-myc, c-myb, c-fos, Ki-67 (DP)	PFA 1%, 10', 4 deg TX-100 0.1%, 3', 4 deg TX-100 0.1% in washes	Kastan et al., 1989
c-myc (DP)	Archival (FA-fixation, total dehydration, embedding)	Watson et al., 1985
p53 (DP)	NO FIXATION Nuclei isolated in Tween-80, 0.2%	Remvikos et al., 1990
p53 (stability to pepsin) (DP)	Archival (see above), nuclei or cells: PFA, 0.5%, 5'	Morkve, Laerum, 1991
p53 (DP) (DA)	80% Ethanol:acetone (1:1), 14h, 0 deg	Darzynkiewicz et al., 1986
p53 (DP)	Glutaraldehyde 2%, 2h, 4 deg Methanol:acetone (1:1), 5', -20 deg	Shen et al., 1983
TdT	PFA 1%, 15' TX-100 0.1%, 3'	Gore et al., 1990

ANTIGEN	FIXATION and PERMEABILIZATION	REFERENCE
Nuclear antigens (contd.)		
TdT	PFA 2% + saponin 0.25% 5', 4 deg	Bardales et al., 1989b
TdT, Ki-67 (DB)	Cold methanol 30' (NaOH 0.007N, 10-15s for BrdUrd)	Campana et al., 1988
Kinetochore (in cells) (DP)	NO FIXATION, protease Tween-20 0.5%, 30', 4 deg, then hypotonic chromos. isolat. buffer Tween-20 0.5%, 20', 4 deg	Levy et al., 1991
DNA-polymerase alpha	NO FIXATION, protease Lysolecitin 0.05%, 1', 0 deg	Miller et al., 1985
DNA-polymerase alpha (DP)	PFA 1%, 1h, 0 deg TX-100 0.1%	Stokke et al., 1991
Topoisomerase-2 (in nuclei) (DP)	Methanol 50%	Epstein et al., 1989
DNA-methylase, Ki-67 (DP)	Ethanol 70%, 10', 4 deg	Neubauer et al., 1989
7. Nuclear and surface antigens		
PCNA and CD3 or (DB)	PFA 1%, 10', RT, post surface IF NP-40 0.1%, 5', 0 deg	Kurki et al., 1987
DNA-polymerase alpha and CD4 (DP)	PLP or PFA 2-4%, 30' 4 deg	Shibata et al., 1989
Ki-67 and surface markers	PLP 15', -10 deg	Drach et al., 1989
TdT and CD2, CD22	FA 9.2% + acetone, 45%, 2 sec, RT	Slaper-Cortenbach et al., 1988
Ki-67 and T11	Acetone 95%, 10', RT	Schwarting et al., 1986
8. Nuclear and cytoplasmic antigens		
Ki67 and cytoplasmic IgE (DP)	NO FIXATION Lysolecithin 5ug/ml,pH4.5 2', 4 deg, then PFA 0.78%	Labalette-Houache et al., 1991

ANTIGEN	FIXATION and PERMEABILIZATION	REFERENCE
Nuclear and cytoplasmic antigens (contd.)		
Ki-67 and cytoplasmic CD3, keratin	NO FIXATION Saponin 0.3%, 30' (antibody incub.) Saponin 0.1% in all washes	Jacob et al., 1991
Various nuclear, cytoplasmic and surface Ags	PFA 3.7% Saponin 0.1%, 20', RT +/- TX-100 0.1%, 3', RT	Goldenthal et al., 1985
SnRNP, ribosomes, tubulin	NO FIXATION Liposome fusion	Thompson, Gross, 1988

9. Viral antigens

EBV p63	FA 1% + sucrose 5%, 20'	Mann et al., 1985
EBV capsid (VCA)	FA 1% + sucrose 5%, 20'	Fronko et al., 1989
Adenovirus E1A	PFA 2%, 15', RT TX-100 0.2% 5', 4 deg	White et al., 1988
Herpes simplex type 2	Ethanol 70%, 4 deg	Goolsby et al., 1988
SV40 T-Ag	Acetone:methanol (70:30), 20', 4 deg	Horan et al., 1975
SV40 T-Ag	Methanol 90%, 4 deg	Jacobberger et al.,1986
SV40 T-Ag, V-Ag, p53 and DNA	Methanol 90%, -20 deg TX-100 0.002%	Laffin et al., 1989
SV40 T-Ag and DNA	Compared: PFA 0.5-2%, 10', 37 deg or methanol, ethanol 67-90%, or acetone:methanol (7:3), -20 deg	Schimenti, Jacobberger, 1992
Cytomegalovirus alpha-antigen	Methanol 90%, -20 deg	Elmendorf et al., 1988

REFERENCES

Ahnert-Hilger G, Mach W, Foehr KJ, Gratzl M (1989a) Poration by alpha-toxin and streptolysin O: an approach to analyze intracellular processes. In: Turner JR, Tartakoff AM (ed) Methods in cell biology, vol 31 . Academic Press, San Diego New York Berkeley Boston London Sydney Tokyo Toronto, p 63

Ahnert-Hilger G, Bader M-F, Bhakdi S, Gratzl M (1989b) Introduction of macromolecules into bovine adrenal medullary chromatin cell and rat pheochromocytoma cells (PC12) by permeabilization with streptolysin O: inhibitory effect of tetanus toxin on cathecholamine secretion. J Neurochem 52:1751-1758

Anderson P, Blue M-L, O'Brien C, Schlossman SF (1989) Monoclonal antibodies reactive with the T cell receptor zeta chain: production and characterization using a new method. J Immunol 143:1899-1904

Anderson P, Nagler-Anderson C, O'Brien C, Levine H, Watkins S, Slayter HS, Blue M-L, Schlossman SF (1990) A monoclonal antibody reactive with a 15-kDa cytoplasmic granule-associated protein defines a subpopulation of CD8+ T lymphocytes. J Immunol 144:574-582

Andersson U, Hallden G, Persson U, Hed J, Moeller G, DeLey M (1988) Enumeration of IFN-gamma-producing cells by flow cytometry. J Immunol Methods 112:139-142

Baisch H, Gerdes J (1990) Identification of proliferating cells by Ki-67 antibody. In: Darzynkiewicz Z, Crissman HA (eds) Methods in Cell Biology, vol 33. Academic Press, San Diego New York Boston London Sydney Tokyo Toronto, p 217

Bancroft JD (1967) An introduction to histochemical technique, Butterworth, London

Bardales RH, Al-Katib AM, Carrato A, Koziner B (1989a) Detection of intracytoplasmic immunoglobulin by flow cytometry in B-cell malignancies. J Histochem Cytochem 37:83-89

Bardales RH, Carrato A, Fleischer M, Schwartz MK, Koziner B (1989b) Detection of terminal deoxynucleotidyl transferase (TdT) by flow cytometry in leukemic disorders. J Histochem Cytochem 37:509-513

Battifora H, Kopinski M (1986) The influence of protease digestion and duration of fixation on the immunostaining of keratins. A comparison of formalin and ethanol fixation. J Histochem Cytochem 34:1095-1100

Bauer KD (1990) Analysis of proliferation-associated antigens. In: Darzynkiewicz Z, Crissman HA (eds) Methods in Cell Biology, vol 33. Academic Press, San Diego New York Boston London Sydney Tokyo Toronto, p 235

Bauer KD, Clevenger CV, Endow RK, Murad T, Epstein AL, Scarpelli DG (1986a) Simultaneous nuclear antigen and DNA content quantitation using paraffin-embedded colonic tissue and multiparameter flow cytometry. Cancer Res 46:2428-2434

Bauer KD, Clevenger CV, Williams TJ, Epstein AL (1986b) Assessment of cell cycle-associated antigen expression using multiparameter flow cytometry and antibody-acridine orange sequential staining. J Histochem Cytochem 34:245-250

Berglund DL, Starkey JR (1989) Isolation of viable tumor cells following introduction of labeled antibody to an intracellular oncogene product using electroporation. J Immunol Methods 125:79-87

Berglund DL, Starkey JR (1991) Introduction of antibody into viable cells using electroporation. Cytometry 12:64-67

Bernstein RL, Browne LH, Yu SC, Williams KL (1988) Detergent treatment of Dictyostelium discoideum cells allows examination of internal cell type-specific antigens by flow cytometry. Cytometry 9:68-74

Bolton WE, Mikulka WR, Healy CG, Schmittling RG, Kenyon NS (1992) Expression of proliferation associated antigens in the cell cycle of synchronized mammalian cells. Cytometry 13:117-126

Bruno S, Crissman HA, Bauer KD, Darzynkiewicz Z (1991) Changes in cell nuclei during S-phase: progressive chromatin condensation and altered expression of the proliferation-associated nuclear proteins Ki-67, cyclin (PCNA), p105, and p34. Exp Cell Res 196:99-106

Buckingham L, Duncan JL (1983) Approximate dimensions of membrane lesions produced by Streptolysin S and Streptolysin O. Biochim Biophys Acta 729:115-122

Butler TW, Heck LW, Huster WJ, Grossi CE, Barton JC (1988) Assessment of total immunoreactive lactoferrin in hematopoietic cells using flow cytometry. J Immunol Methods 108:159-170

Campana D, Coustan-Smith E, Janossy G (1988) Double and triple staining for studying the proliferative activity of human B and T lymphoid cells. J Immunol Methods 107:79-88

Chakrabarti R, Pfeiffer NE, Wylie DE, Schuster SM (1989a) Incorporation of monoclonal antibodies into cells by osmotic permeabilization. J Biol Chem 264:8214-8221

Chakrabarti R, Wylie DE, Schuster SM (1989b) Transfer of monoclonal antibodies into mammalian cells by electroporation. J Biol Chem 264:15494-15500

Clevenger CV, Bauer KD, Epstein AL (1985) A method for simultaneous nuclear immunofluorescence and DNA content quantitation using monoclonal antibodies and flow cytometry. Cytometry 6:208-214

Czerniak B, Herz F, Wersto RP, Koss LG (1987) Expression of Ha-ras oncogen p21 protein in relation to the cell cycle of cultured human tumor cells. Am J Pathol 126:411-416

Darzynkiewicz Z, Staiano-Goico L, Kunicka JE, DeLeo AB, Old LJ (1986) p53 content in relation to cell growth and proliferation in murine L1210 leukemia and normal lymphocytes. Leuk Res 10:1383-1389

Dent GA, Leglise MC, Pryzwansky KB, Ross DW (1989) Simultaneous paired analysis by flow cytometry of surface markers, cytoplasmic antigens, or oncogene expression with DNA content. Cytometry 10:192-198

Dolbeare F, Kuo W-L, Beisker W, Vanderlaan, Gray JW (1990) Using monoclonal antibodies in bromodeoxyuridine-DNA analysis. In: Darzynkiewicz Z, Crissman HA (eds) Methods in Cell Biology, vol 33. Academic Press, San Diego New York Boston London Sydney Tokyo Toronto, p 207

Downey GP, Chan CK, Grinstein S (1989) Actin assembly in electropermeabilized neutrophils: role of G-proteins. Biochem Biophys Res Commun 164:700-705

Drach J, Gattringer C, Glassl H, Schwarting R, Stein H, Huber H (1989) Simultaneous flow cytometric analysis of surface markers and nuclear Ki-67 antigen in leukemia and lymphoma. Cytometry 10:743-749

Einck L, Bustin M (1984) Functional histone antibody fragments traverse the nuclear envelope. J Cell Biol 98:205-213

Elmendorf S, McSharry J, Laffin J, Fogelman D, Lehman JM (1988) Detection of an early Cytomegalovirus antigen with two-color quantitative flow cytometry. Cytometry 9:254-260

Engelhard HH, Butler AB, Bauer KD (1989) Quantification of the c-myc oncoprotein in human glioblastoma cells and tumor tissue. J Neurosurg 71:224-232

Engelhard HH, Krupka JL, Bauer KD (1991) Simultaneous quantification of c-myc oncoprotein, total cellular protein, and DNA content using multiparameter flow cytometry. Cytometry 12:68-76

Epstein RJ, Smith PJ, Watson JV, Waters C, Bleehen NM (1989) Oestrogen potentiates topoisomerase-2-mediated cytotoxicity in an activated subpopulation of human breast cancer cells: implications for cytotoxic drug resistance in solid tumors. Int J Cancer 44:501-505

Fiskum G, Craig SW, Decker GL, Lehninger AL (1980) The cytoskeleton of digitonin-treated rat hepatocytes. Proc Natl Acad Sci USA 77:3430-3434

Fronko GE, Long W, Wu B, Papadopoulos T, Henderson EE (1989) Relationship between methylation status and expression of an Epstein-Barr virus (EBV) capsid antigen gene. Biochem Biophys Res Commun 159:263-270

Garner JG, Colvin RB, Schooley RT (1984) A flow cytometric technique for quantitation of B-cell activation. J Immunol Methods 67:37-51

Giaretti W, Di Vinci A, Geido E, Marsano B, Minks M, Bruno S (1990) Measurement of c-myc protein content and cell cycle kinetics of normal and spontaneously transformed murine mastocytes by bivariate flow cytometry. Cell Tissue Kinet 23:473-485

Goldenthal KL, Hedman K, Chen JW, August JT, Willingham, MC (1985) Postfixation detergent treatment for immunofluorescence suppresses localization of some integral membrane proteins. J Histochem Cytochem 33:813-820

Goolsby C, Gay H, Docherty JJ, Todd P (1988) Flow cytometric detection of Herpes simplex virus type 2 infected and transformed cells by immunoenzymatic and indirect immunofluorescence staining. Cytometry 9:126-130

Gore SD, Kastan MB, Goodman SN, Civin CI (1990) Detection of minimal residual T cell acute lymphoblastic leukemia by flow cytometry. J Immunol Methods 132:275-286

Graham ML, Dalquist KE, Horwitz KB (1989) Simultaneous measurement of progesteron receptors and DNA indices by flow cytometry: analysis of breast cancer cell mixtures and genetic instability of the T47D line. Cancer Res 49:3943-3949

Gratzner HG (1982) Monoclonal antibody to 5-bromo and 5-iododeoxyuridine: a new reagent for detection of DNA replication. Science 218:474-475

Hallden G, Andersson U, Hed J, Johansson SGO (1989) A new membrane permealization method for the detection of intracellular antigens by flow cytometry. J Immunol Methods 124:103-109

Hatfield JM, Hymer WC (1985) Flow cytometric immunofluorescence of rat anterior pituitary cells. Cytometry 6:137-142

Hayden GE, Walker KZ, Miller JFAP, Wotherspoon JS, Raison RL (1988) Simultaneous cytometric analysis for the expression of cytoplasmic and surface antigens in activated T cells. Cytometry 9:44-51

Hjerten S, Johansson K-E (1972) Selective solubilization with Tween-20 of membrane proteins from Acholeplasma laidlawii. Biochim Biophys Acta 288:312-325

Hofsli E, Bakke O, Nonstad U, Espevik T (1989) A flow cytometric and immunofluorescence microscopic study of tumor necrosis factor production and localization in human monocytes. Cell Immunol 122:405-415

Horan M, Horan PK, Williams CA (1975) Quantitative measurement of SV40 T-antigen production. Exp Cell Res 91:247-252

Jacob MC, Favre M, Bensa JC (1991) Membrane cell permeabilization with saponin and multiparametric analysis by flow cytometry. Cytometry 12:550-558

Jacobberger JW, Fogleman D, Lehman JM (1986) Analysis of intracellular antigens by flow cytometry. Cytometry 7:356-364

Kastan MB, Stone KD, Civin CI (1989) Nuclear oncoprotein expression as a function of lineage, differentiation stage, and proliferative status of normal human hematopoietic cells. Blood 74:1517-1524

Kirkhus B, Clausen OPF (1990) Cell kinetics in mouse epidermis studied by bivariate DNA/bromodeoxyuridine and DNA/keratin flow cytometry. Cytometry 11:253-260

Kurki P, Lotz M, Ogata K, Tan EM (1987) Proliferating cell nuclear antigen (PCNA)/cyclin in activated human T-lymphocytes. J Immunol 138:4114-4120

Kurki P, Ogata K, Tan EM (1988) Monoclonal antibodies to proliferating cell nuclear antigen (PCNA)/cyclin as probes for proliferating cells by immunofluorescence microscopy and flow cytometry. J Immunol Methods 109:49-59

Labalette-Houache M, Torpier G, Capron A, Dessaint JP (1991) Improved permeabilization procedure for flow cytometric detection of internal antigens. Analysis of interleukin-2 production. J Immunol Methods 138:143-153

Laffin J, Fogleman D, Lehman JM (1989) Correlation of DNA content, p53, T-antigen, and V-antigen in Simian virus 40-infected human diploid cells. Cytometry 10:205-213

Landberg G, Roos G (1991) Antibodies to proliferating cell nuclear antigen as S-phase probes in flow cytometric cell cycle analysis. Cancer Res 51:4570-4574

Larsen JK, Christensen IJ, Christiansen J, Mortensen BT (1991) Washless double staining of unfixed nuclei for flow cytometric analysis of DNA and a nuclear antigen (Ki-67 or bromodeoxyuridine). Cytometry 12:429-437

Lemons R, Forster S, Thoene J (1988) Protein microinjection by protease permeabilization of fibroblasts. Anal Biochem 172:219-227

Levy HP, Schultz RA, Ordonez JV, Cohen MM (1991) Anti-kinetochore staining for single laser, bivariate flow sorting of Indian muntjac chromosomes. Cytometry 12:695-700

Lincoln ST, Bauer KD (1989) Limitations in the measurement of c-myc oncoprotein and other nuclear antigens by flow cytometry. Cytometry 10:456-462

Maeda K, Dixon WT, Jerry LM (1990) Characterization of human melanosome-associated antigen recognized by monoclonal antibody HMSA-2. J Invest Dermatol 94:221-226

Mann GJ, Dyne M, Musgrove EA (1987) Immunofluorescent quantification of ribonucleotide reductase M1 subunit and correlation with DNA content by flow cytometry. Cytometry 8:509-517

Mann KP, Staunton D, Thorley-Lawson DA (1985) Epstein-Barr virus-encoded protein found in plasma membranes of transformed cells. J Virol 55:710-720

McLean IW, Nakane PK (1974) Periodate-lysine-paraformaldehyde fixative, a new fixative for immunoelectron microscopy. J Histochem Cytochem 22:1077-1083

Mercer WE, Nelson D, DeLeo AB, Old LJ, Baserga R (1982) Microinjection of monoclonal antibody to protein p53 inhibits serum-induced DNA synthesis in 3T3 cells. Proc Natl Acad Sci USA 79:6309-6312

Miller MR, Ulrich RG, Wang TS-F, Korn D (1985) Monoclonal antibodies against human DNA polymerase-alpha inhibit DNA replication in permeabilized human cells. J Biol Chem 260:134-138

Morgan DO, Roth RA (1987) Acute insulin action requires insulin receptor kinase activity: introduction of an inhibitory monoclonal antibody into mammalian cells blocks the rapid effects of insulin. Proc Natl Acad Sci USA 84:41-45

Morkve O, Laerum OD (1991) Flow cytometric measurment of p53 protein expression and DNA content in paraffin-embedded tissue from bronchial carcinomas. Cytometry 12:438-444

Morris GF, Mathews MB (1989) Regulation of proliferating cell nuclear antigen during the cell cycle. J Biol Chem 264:13856-13864

Nelson KG, Haskill JS, Sloan S, Siegfried JM, Siegal GP, Walton L, Kaufman DG (1987) Flow cytometric analysis of human uterine sarcomas and cell lines. Cancer Res 47:2814-2820

Neubauer A, Serke S, Siegert W, Kroll W, Musch R, Huhn D (1989) A flow cytometric assay for the determination of cell proliferation with a monoclonal antibody directed against DNA-methyltransferase. Br J Haematol 72:492-496

Ojcius DM, Young JD-E (1991) Cytolytic pore-forming proteins and peptides: is there a common structural motif? Trends Biochem Sci 16:225-229

Oud PS, Henderik JBJ, Beck HlM, Veldhuizen JAM, Vooijs G, Herman CJ, Ramaekers FCS (1985) Flow cytometric analysis and sorting of human endometrial cells after immunocytochemical labeling for cytokeratin using a monoclonal antibody. Cytometry 6:159-164

Palutke M, KuKuruga D, Tabaczka P (1987) A flow cytometric method for measuring lymphocyte proliferation directly from tissue culture plates using Ki-67 and propidium iodide. J Immunol Methods 105:97-105

Pattanapanyasat K, Webster HK, Udomsangpetch R, Wanachiwanawin W, Yongvanitchit K (1992) Flow cytometric two-color staining technique for simultaneous determination of human erythrocyte membrane antigen and intracellular malarial DNA. Cytometry 13:182-187

Preis PN, Waldman FM, Frackelton AR, Saya H, Levin VA (1988) Protein-tyrosine kinase activity and pp60 v-src expression in whole cells measured by flow cytometry. Cancer Res 48:4633-4638

Puchtler H, Meloan SN (1985) On the chemistry of formaldehyde fixation and its effects on immunohistological reactions. Histochemistry 82:201-204

Rabbitts PH, Watson JV, Lamond A, Forster A, Stinson MA, Evan G, Fisher W, Atherton E, Shepport R, Rabbitts FH (1985) Metabolism of c-myc gene products: c-myc RNA and protein expression in the cell cycle. EMBO J 4:2009-2015

Rabinovitch PS, Torres RM, Engel D (1986) Simultaneous cell cycle analysis and two-color surface immunofluorescence using 7-amino-actinomycin D and single laser excitation: applications to study of cell activation and the cell cycle of murine LY-1 B cells. J Immunol 136:2769-2775

Reeve JG, Stewart J, Watson JV, Wulfrank D, Twentyman RR, Bleehen NM (1986) Neuron specific enolase expression in carcinoma of the lung. Br J Cancer 53:519-528

Remvikos Y, Laurent-Puig P, Salmon RJ, Frelat G, Dutrillaux B, Thomas G (1990) Simultaneous monitoring of p53 protein and DNA content of colorectal adenocarcinoma by flow cytometry. Int J Cancer 45:450-456

Rice GC, Pennica D, Borree JA, Williams SR (1991) Measurement of transient cDNA expression in mammalian cells using flow cytometric cell analysis and sorting. Cytometry 12:221-233

Roberts P, Jones M, Gale R, Thomas S, Tidman N, Linch D (1989) The c-myc oncogene is regulated independently of differentiation in myeloid cell lines. Leuk Res 13:651-659

Robertson FM, Gilmour SK, Beavis AJ, O'Connel WSM, Conney AH, Huang M-T, Laskin JD, Hietala OA, O'Brien TG (1990) Flow cytometric detection of ornitine decarboxylase activity in epidermal cell subpopulations. Cytometry 11:832-836

Ronot X, Gaillard-Froger B, Hainque B, Adolphe M (1988) In vitro aging of articular chondrocytes identified by analysis of DNA and tubulin content and relationship to cell size and protein content. Cytometry 9:436-440

Schimenti KJ, Jacobberger JW (1992) Fixation of mammalian cells for flow cytometric evaluation of DNA content and nuclear immunofluorescence. Cytometry 13:48-59

Schulz I (1990) Permeabilizing cells: some methods and applications for the study of intracellular processes. In: Fleischer S, Fleischer B (eds) Methods in enzymology, vol 192. Academic Press, San Diego New York Boston London Sydney Tokyo Toronto, p 280

Schmid I, Uittenbogaart CH, Giorgi JV (1991) A gentle fixation and permeabilization method for combined cell surface and intracellular staining with improved precision in DNA quantification. Cytometry 12:279-285

Schroff RW, Bucana CD, KLein RA, Farrell NM, Morgan AC (1984) Detection of intracytoplasmic antigen by flow cytometry. J Immunol Methods 70:167-177

Schwarting R, Gerdes J, Niehus J, Jaeschke l, Stein H (1986) Determination of the growth fraction in cell suspensions by flow cytometry using the monoclonal antibody Ki-67. J Immunol Methods 90:65-70

SenGupta S, Petsche D, Gelfand EW, Chechik BE (1985) A flow cytometric method for the detection of adenosine deaminase in mononuclear cells. J Immunol Methods 80:155-162

Shapiro HM (1988) Practical flow cytometry, 2nd edn. Alan R Liss, New York

Shen D-W, Real FX, DeLeo AB, Old LJ, Marks PA, Rifkin RA (1983) Protein p53 and inducer-mediated erythroleukemia cell commitment to terminal cell division. Proc Natl Acad Sci USA 80:5919-5922

Shibata M, Hirono Y, Takahashi M, Kaneda T (1989) Expression of DNA polymerase alpha and Leu3a molecules in growing and saturated cultures of human leukemic cells: phenotype analysis of proliferating cells by flow cytometry. Jap J Cancer Res 80:1083-1088

Slaper-Cortenbach ICM, Admiraal LG, Kerr JM, van Leeuwen EF, von dem Borne AEGK, Tetteroo PAT (1988) Flow-cytometric detection of terminal deoxynucleotidyl transferase and other intracellular antigens in combination with membrane antigens in acute lymphatic leukemias. Blood 72:1639-1644

Smith PJ, Nakeff A, Watson JV (1985) Flow cytometric detection of changes in the fluorescence emission spectrum of a vital DNA-specific dye in human tumor cells. Exp Cell Res 159:37-46

Stokke P, Erikstein B, Holte H, Funderud F, Steen HB (1991) Cell cycle-specific expression and nuclear binding of DNA polymerase alpha. Molec Cell Biol 11:3384-3389

Thompson WS, Gross RH (1988) Antibodies introduced into living cells with liposomes localize specifically and inhibit specific intracellular processes. Gene Anal Techn 5:73-79

Turner JR, Tartakoff AM, Berger M (1988) Intracellular degradation of the complement C3b/C4b receptor in the absence of ligand. J Biol Chem 263:4914-4920

Turner JR, Tartakoff AM, Berger M (1989) A flow cytometric method for the quantitative analysis of intracellular and surface membrane antigens. In: Tartakoff AM (ed) Methods in Cell Biology, vol 31. Academic Press, San Diego New York Berkeley Boston London Sydney Tokyo Toronto, p 351

van Erp PEJ, Rijzewijk JJ, Boezeman JBM, Leenders J, de Mare S, Schalkwijk J, van de Kerkhof PCM, Ramaekers FCS, Bauer FW (1989) Flow cytometric analysis of epidermal subpopulations from normal and psoriatic skin using monoclonal antibodies against intermediate filaments. Am J Pathol 135:865-870

van Roon MA, Aten JA, van Oven CH, Charles R, Lamers W (1989) The initiation of hepatocyte-specific gene expression within embryonic hepatocytes is a stochastic event. Dev Biol 136:508-516

Watson JV, Sikora K, Evan GI (1985) A simultaneous flow cytometric assay for c-myc oncoprotein and DNA in nuclei from paraffin embedded material. J Immunol Methods 83:179-192

Weigel PH, Ray DA, Oka JA (1983) Quantitation of intracellular membrane-bound enzymes and receptors in digitonin-permeabilized cells. Anal Biochem 133:437-449

Weltzien HU (1979) Cytolytic and membrane perturbing properties of lysophosphatidylcholine. Biochim Biophys Acta 559:259-287

White E, Spector D, Welch W (1988) Differential distribution of adenovirus E1A proteins and colocalization of E1A with the 70-kilodalton cellular heatshock protein in infected cells. J Virol 62:4153-4166

Wolf ME, Zigmond MJ, Kapatos G (1989) Tyrosine hydroxylase content of residual striatal dopamine nerve terminals following 6-hydroxydopamine administration: a flow cytometric study. J Neurochem 53:879-885

Womack MD, Kendall DA, Macdonald RC (1983) Detergent effects on enzyme activity and solubilization of lipid bilayer membranes. Biochim Biophys Acta 733:210-215

Zarbo RJ, Visscher DW, Crissman JD (1989) Two-color multiparametric method for flow cytometric DNA analysis of carcinomas using staining for cytokeratin and leukocyte-common antigen. Anal Quant Cytol Histol 11:391-402

Zeile G (1980) Intracytoplasmic immunofluorescence in multiple myeloma. Cytometry 1:37-41

Zipf TF, Bryant LD, Koskowich GN, MacGregor SE, Chin L, Johnson H (1984) Enumeration of cytoplasmic u immunoglobulin positive acute lymphoblastic leukemia cells by flow cytometry: comparison with fluorescence microscopy. Cytometry 5:610-613

Zocchi MR, Marelli F, Poggi A (1990) Simultaneous cytofluorometric analysis for the expression of cytoplasmic antigens and DNA content in CD3- human thymocytes. Cytometry 11:883-887

EXPRESSION AND ROLE OF SURFACE MARKERS OF IMMUNOCOMPETENT CELLS IN NON LYMPHOID SYSTEMS

S. PLAISANCE, A. ALILECHE, D. HAN, E. RUBINSTEIN, Y. SAHRAOUI, C. JASMIN and B. AZZARONE

INSERM UNITE-268
Hôpital Paul-Brousse
14, Av P.V. COUTURIER
F-94800 VILLEJUIF
FRANCE

SUMMARY

The α and ß chains of the Interleukin 2 receptor (IL2Rα and IL2Rß) were detected at the surface of cultured melanoma and fibroblastic cells by flow cytometry, using monoclonal antibodies (mAbs) directed against the IL2Rα and the IL2Rß. Both cell types bound ^{FITC}-IL2 and this binding was inhibited by an excess of cold ligand and by mAbs recognizing the IL2 binding sites of the α and ß chains. Internalisation studies show that the fibroblastic IL2R/IL2 complex is internalized at 37°C. By Northern Blot analysis we detected the presence of specific transcripts for the IL2Rα and IL2Rß genes.

Finally, the addition of exogenous IL2 specifically modified the surface expression of different antigens involved in the process of immunosurveillance. Indeed, IL2, at concentrations affecting the high affinity IL2R, caused the down regulation of ICAM-1 protein. IL2 also decreased the surface expression of the class I and class II HLA. By contrast, the use of IL2 concentrations which saturate the intermediate affinity IL2Rß, as well as the use of high concentrations of anti IL2Rß mAbs caused the up regulation of the surface expression of the ICAM-1 protein.

ICAM-1 is the natural ligand for the LFA-1 integrin expressed at the surface of lymphoid cells. ICAM-1/LFA-1 interactions favour homotypic and heterotypic cell-cell adhesion. Since human fibroblasts express an LFA-1 like molecule, we propose that in these cells IL2 can modify homotypic and heterotypic interactions acting on the surface expression of ICAM-1 protein. The presence

NATO ASI Series, Vol. H 67
Flow Cytometry
Edited by A. Jacquemin-Sablon

of the IL2R on human tumor cells and fibroblasts as well as the IL2 dependent modulation of ICAM-1, class I and class II HLA antigens may influence the immune response inside reactive tissues and tumors.

INTRODUCTION

Interleukin 2 (IL2) is a T cell derived lymphokine necessary for clonal proliferation of antigen stimulated T lymphocytes. In T cells, it delivers a growth signal through a specific receptor (IL2Rαß) which binds IL2 with high affinity (Kd 10-50 pM) and is composed of at least two distinct IL2 binding molecules: the α chain (IL2Rα) and the ß chain (IL2Rß) displaying low (Kd 10nM) and intermediate affinity (Kd 1-2 nM) respectively (Waldmann, 1991). In vivo infusion of IL2 and lymphokine activated killer cells (LAK) induces tumor regression in 10-20% of patients with melanoma. However this treatment causes several side effects. Since IL2 displays several effects on lymphoid cells including B cells, myelomonocytic cells, NK cells, and on cells of nervous origin (Waldmann, 1991; Saneto et al., 1986; Okamoto et al., 1990); we hypothesized that IL2 could also act directly on melanoma and on other non lymphoid cell systems.

In this respect, we have recently shown the constitutive expression of the α chain of the IL2R in human embryonic lung fibroblasts and in the HT1080 fibrosarcoma cell line (Carloni et al. 1989). By contrast, in the adult normal fibroblasts tested, in two non tumorigenic revertants derived from the HT1080 fibrosarcoma cells and in senescent human embryonic fibroblasts the IL2Rα gene is poorly expressed or not expressed at all (Carloni et al., 1989; Plaisance et al. in press). This identification has been performed by studies showing both the presence of a membrane molecule recognized by anti IL2Rα mAbs and of three transcripts (3.6, 2.8 and 1.7kb) specific for this gene (Carloni et al., 1989).

Moreover ^{125}I-IL2 and ^{FITC}-IL2 binding affinity studies revealed the presence of two IL2 binding peptides of 55 and 70 kDa respectively as well as the existence of both high and low affinity receptors (Plaisance et al. in press). Here we show that also human melanoma cells may express the α and the ß chains of the IL2R and that this receptor binds IL2 and is functional. It is indeed involved in the control of several surface molecules important for immune response.

MATERIAL AND METHODS

CELL LINES: The origin and the characteristics of the human embryonic lung fibroblastic cell line HEL/ICIG-7 and of M14 melanoma cells have been detailed elsewhere (Carloni et al., 1989; Plaisance et al., in press; Cillo et al., 1984). The origin and the characteristics of the murine L929 cells transfected with cDNA coding for the human IL2Rα gene and expressing constitutively this antigen (L-TRANS3 clone) have been detailed elsewhere (Rubin et al., 1987). The origin and the characteristics of the human B cell line SKW6.4 have been reported elsewhere (Dukovich et al., 1987).
Cells were maintained in culture in Minimal Eagle's Medium [MEM] supplemented with 10% foetal calf serum [FCS], 2mM glutamine and Penicillin Streptomycin Neomycin ([PSN] 5-5-10μg/ml, GIBCO BRL France). Cultures were transferred weekly at 1:2 split ratio after trypsin treatment (GIBCO BRL, France). Human recombinant IL2 used for the biological assays was obtained from EuroCetus (The Netherlands) and Roussel-Uclaf (France). Embryonic fibroblast cells were tested between the 12th and the 20th passage.

IMMUNOFLUORESCENCE: Immunofluorescent staining of living cells was performed as previously described (Boucheix et al., 1986). Each cell line was examined in at least three independent assays. The Mean Fluorescence Intensity of labelling [MFI] were determined by flow cytometry (EPICS-C, Coulter, Hialeah FL and ODAM-BRUKER ATC 3000, WISSEMBOURG, France) on cell detached with 5mM EDTA in PBS, washed in PBS and incubated for 30m at + 4°C with monoclonal antibodies [mAbs]. Cells were then washed twice and incubated for 30mn at +4°C with FITC-coupled goat anti-mouse (Fab)'2 antibodies (FITC-GAM, Immunotech, Luminy, France). Negative controls were defined by labelling cells with FITC-GAM alone or with a non-relevant antibody followed by FITC-GAM. The monoclonal antibodies directed against the IL2 binding site of the IL2Rα (IOT14, Immunotech Luminy, France), the IL2 binding site of the IL2Rß (MIKß1; Tsudo et al., 1989), against the Intercellular Adhesion Molecule-1 (ICAM-1, 84H10, Immunotech Luminy, France), against the α chain of the LFA-1 molecule (6B11, Lorenzo Moretta, IST, Genoa, Italy) and against the class I HLA and HLA-DR (IOT2, IOT2B, Immunotech, Luminy, France) were used at 10μg/ml.

GEL ELECTROPHORESIS ANALYSIS OF ^{125}I-IL2 BINDING PROTEINS: Cells (20x10^6 per point) were detached with a rubber policeman and washed in

PBS. Radioiodinated IL2 (5nM, specific activity of about 40 μCi/μg, NEN DuPont, RFA) was then added to the suspension for 1h at + 4°C. Cells were cross-linked with disuccinimidyl suberate (DSS, 2mM), washed twice in Tris-EDTA (10mM-1mM) and lysed in extraction buffer before electrophoretic separation according to a previously described technique (Green et al., 1989).

^{FITC}-IL2 AND ^{125}I-IL2 AFFINITY BINDING STUDIES

- ^{FITC}-IL2 binding analysis was performed by flow cytometry according to a previously described technique (Harel Belan et al., 1989) and analysed as described for mAbs. Preincubations were performed for 1 h at +4°C with human rIL2 (IL2°: 64,5μM) anti IL2Rß mAb (MIKß1, 50μg/ml) and anti-IL2Rα mAb (MAR93, 50μg/ml); ^{FITC}-IL2 (6.4μM; F/P=3) was then added for 45mn at +4°C.
- Scatchard plot analysis of human recombinant ^{125}I-IL2 binding to a suspension of human M14 melanoma cells detached by 5mM EDTA in PBS was performed exactly as described elsewhere (Green et al., 1989).

IMMUNOPRECIPITATION STUDIES

Metabolic cell labelling with ^{35}S-methionine, membrane glycoprotein extraction and immunoprecipitation were performed as previously described (Azzarone et al. 1986). Radiolabelled extracts were made in 0.75% NP-40, 0.5% deoxycholate and 0.05% sodium dodecyl sulphate (SDS) just before use. After adsorption of immune complexes onto Staphylococcus aureus Cowan I strain, 3 washes with 0.5% NP-40, 0.5% DOC and 0.05% SDS in phosphate-buffered saline pH 7.2 were carried out. Five percent 2-mercaptoethanol was present in the loading buffer.

RNA EXTRACTION

The RNA extraction and poly(A) containing mRNA preparation and RNA blotting were done as previously described (Carloni et al., 1989). Detection of IL2Rα gene specific transcripts was performed by hybridization with the complementary DNA (cDNA) probe pIL2R2 specific for this gene (Leonard et al., 1984). Then, the pIL2R2 signal was washed off and the same filters were rehybridized to the murine actin probe PAL-41 as a control for the quantity and quality of polyadenylated RNAs used in these experiments. Detection of IL2Rß transcripts was performed using the cDNA probe pUCß30 (Hatakeyama et al., 1989).

RESULTS

IL2Rα AND IL2Rß CHAINS IN HUMAN FIBROBLASTS AND MELANOMA CELLS: FLOW CYTOMETRY ANALYSIS.

We investigated by flow cytometry on suspensions of fibroblastic and melanoma living cells the staining patterns of anti IL2Rα (left panels) and anti IL2Rß (right panels) mAbs. Analysis of FIGURE 1 shows that human embryonic lung fibroblasts (HEL/ICIG-7; 1A) and the melanoma cell line M14 (1B) coexpress the two chains of the IL2R.

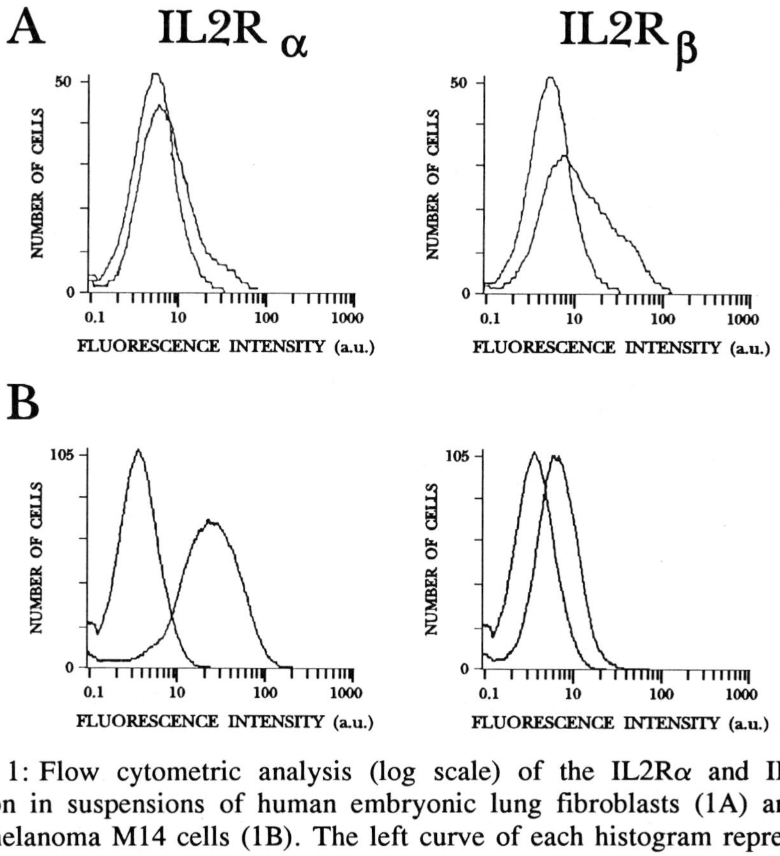

FIGURE 1: Flow cytometric analysis (log scale) of the IL2Rα and IL2Rß expression in suspensions of human embryonic lung fibroblasts (1A) and of human melanoma M14 cells (1B). The left curve of each histogram represents the background binding of FITC-GAM.

ANALYSIS OF ^{FITC}-IL2 BINDING BY FLOW CYTOMETRY.

We then analyzed the capacity of both cell types to bind IL2. Analysis of FIGURE 2 shows that suspensions of HEL/ICIG-7 cells and of M14 cells bind ^{FITC}-IL2 since preincubation with 20 fold molar express of human recombinant IL2 (IL2°: EuroCetus, NL) significantly reduces the intensity of labelling. Moreover, preincubation with anti IL2Rα and/or anti IL2Rß mAbs almost completely abolished the ^{FITC}-IL2 labelling, therefore showing the identity of the IL2 binding peptides.

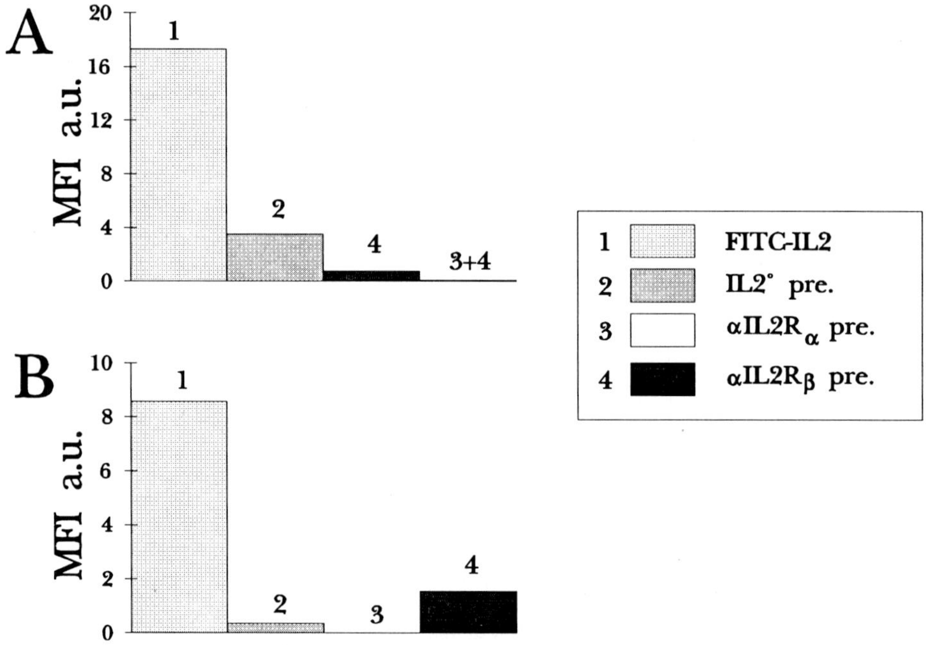

FIGURE 2: Analysis of ^{FITC}-IL2 binding to human embryonic lung fibroblasts (2A) and to M14 melanoma cells (2B). Effect of preincubation with human rIL2 (64.5uM) and anti IL2R mAbs (50μg/ml). Data show the Mean Fluorescence Intensity [MFI] expressed in arbitrary units [a.u.]; negative control value has been subtracted.

ANALYSIS OF THE FIBROBLASTIC IL2 BINDING PEPTIDES BY LIGAND RECEPTOR AFFINITY CROSS-LINKING.

We then used the technique of ligand receptor affinity cross linking to analyze the number and the size of the IL2 binding proteins present on PHA blasts and human fibroblasts. FIGURE 3 displays labelling of cells with ^{125}I-IL2. PHA blasts and human embryonic fibroblasts express two IL2 binding proteins of about 55 and 75 kDa. Preincubation at +37°C for 1h before crosslinking at +4°C strongly decreased the intensity of labelling of ^{125}I-IL2 to both chains, therefore suggesting their internalisation.

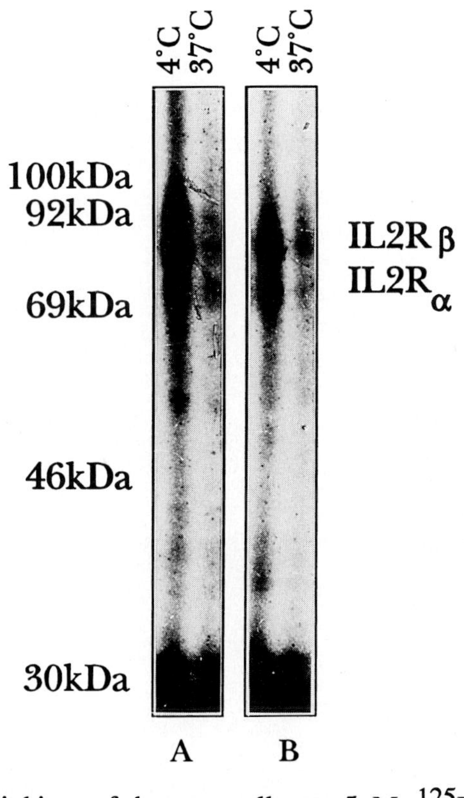

FIGURE 3: Cross-linking of human cells to 5nM ^{125}I-IL2 with DSS and analysis by SDS PAGE. Control PHA blasts (3A) and human embryonic fibroblasts (3B) express two IL2 binding peptides of 75 and 55 kDa. The apparent MW observed is increased by the cross-linked IL2 (15kDa).

NORTHERN BLOT ANALYSIS OF THE TRANSCRIPTS SPECIFIC FOR THE HUMAN IL2Rα AND IL2Rß GENES IN FIBROBLASTIC AND MELANOMA CELLS.

The specificity of immunofluorescence and cross linking experiments was confirmed by the detection in Northern blots of specific transcripts for the IL2Rα and for the IL2Rß genes.

- Total mRNAs from HEL-ICIG7 and M14 cells were hybridized against the IL2Rα cDNA probe pIL2R2. Three classes of transcript (1.7, 2.8 and 3.6 kb) were detected in both cultures (4A).
- Total mRNAs from the HEL/ICIG-7 and M14 cells were hybridized against the human IL2Rß gene probe pUCß30. As positive control we used the SKW6.4 B cell line. Three transcripts of 1.6, 3.2 and 4kb were detected in the three cell lines. In the SKW6.4 B cells line the predominant class of transcripts was represented by the 4kb mRNA, whereas in fibroblastic and melanoma cells the 3.2 and the 1.6kb mRNA were the predominant forms (4B).

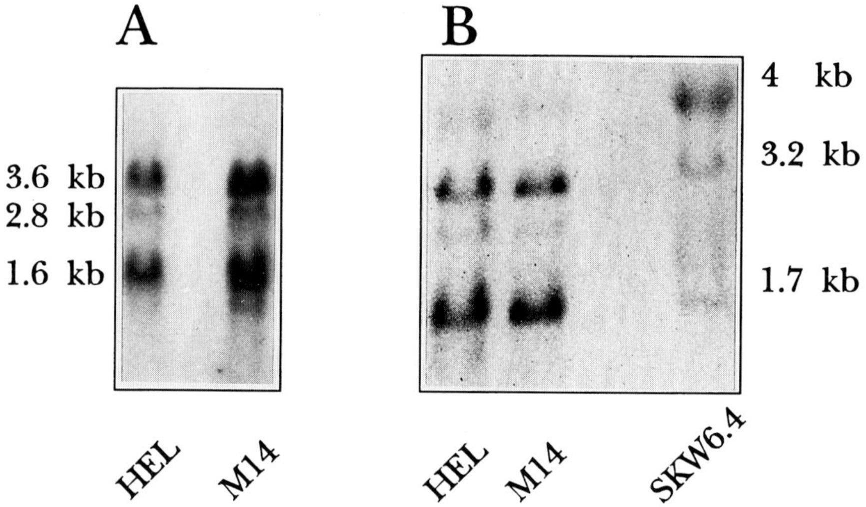

FIGURE 4: Northern blot analysis of the IL2Rα and IL2Rß genes expression. (4A) autoradiograms of blots of total RNAs from human embryonic fibroblasts (HEL/ICIG7) and M14 melanoma cells, hybridized against the pIL2R2 probe. (4B) autoradiograms of blots of total RNAs from human embryonic fibroblasts (HEL/ICIG7), M14 melanoma cells and SKW6.4 B cell line used as positive control, hybridized against the pIL2Rß30 cDNA probe.

IL2 MODULATION OF CLASS I AND CLASS II HLAs EXPRESSION BY FIBROBLASTIC CELLS.

Analysis of FIGURE 5 shows typical experiments illustrating the effects of IL2 and IFN gamma on class I HLA (5A and 5B) and on class II HLA (FIGURE 5C).

- In sparse cultures (5A; 10-20x10^3 cells/cm²), IFN gamma only induces, a 10% increase in the surface levels of class I HLA whereas IL2 (1000IU/ml : 3.6nM) seems to be more effective, causing a 2 fold increase of class I HLA expression. The two cytokines together exhibit a powerful synergistic effect, increasing by 7 fold the surface expression of this molecule.

- In confluent cultures (5B; 1-1.5x10^5 cells/cm²), we observed a quite different pattern. Indeed, IFN gamma alone induces a 5 fold increase of class I HLA level. By contrast, IL2 alone or in combination with IFN gamma strongly inhibits the spontaneous and induced expressions of class I HLA. Concerning the modulation of class II HLA (5C), human embryonic fibroblasts do not spontaneously express HLA-DR molecules. In these cells, however IFN gamma triggers the surface expression of HLA-DR. This induction is almost completely inhibited by simultaneous incubation with IL2.

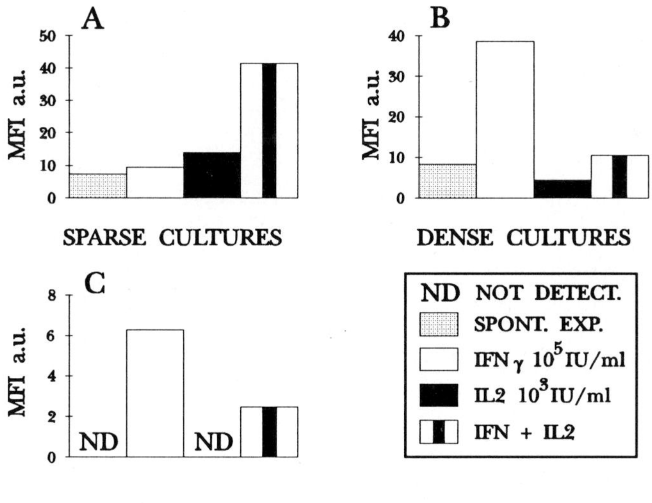

FIGURE 5: Flow cytometry analysis of class I HLA (5A and 5B) and class II HLA-DR (5C) in human embryonic fibroblasts and effect of incubation with human rIL2 and/or human rIFN gamma. The biological effect is achieved after 24h. Data show the mean fluorescence intensity expressed in arbitrary units; negative control value (FITC-GAM) has been subtracted. (ND: Not Detectable)

IL2 MODULATION OF ICAM-1 PROTEIN EXPRESSION BY FIBROBLASTIC CELLS.

FIGURE 6 presents typical experiments illustrating the effect of IL2 on ICAM-1 expression.

- In sparse cultures, where the spontaneous expression of ICAM-1 is weak, IL2 causes within 1 hour a 2 fold increase in the surface levels of ICAM-1 (6A, left panel). A similar effect is obtained, incubating the cells with IFN gamma, whereas a 5 fold increase is obtained by the simultaneous use of these two cytokines (6A, right panel).
- By contrast, in confluent cultures (6B), where the spontaneous expression of ICAM-1 is higher, IL2 causes, within 4 hours, the down modulation of ICAM-1 (50% decrease). The effect is maximal at IL2 concentrations corresponding to the high affinity IL2Rαß heterodimer (10IU/ml : 36pM).

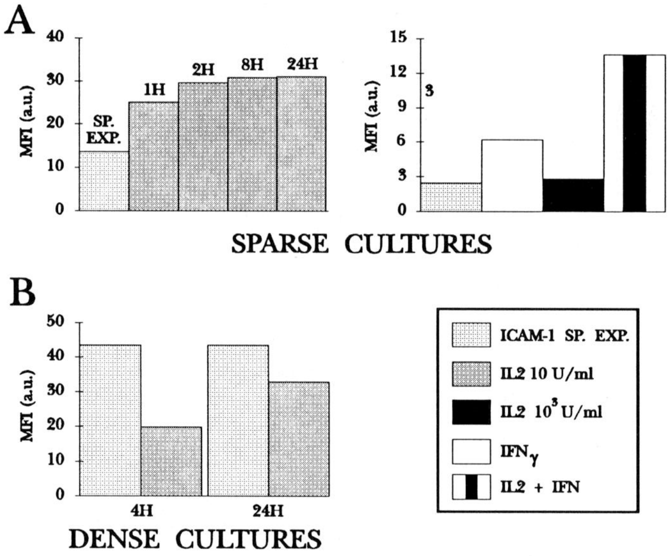

FIGURE 6: Flow cytometry analysis of ICAM-1 protein expression in human embryonic fibroblast sparse cultures (6A) and dense cultures (6B) and effect of incubation with human rIL2 and/or human rIFN gamma. Data show the mean fluorescence intensity expressed in arbitrary units; negative control value (FITC-GAM) has been subtracted.

LFA-1 EXPRESSION ON HUMAN FIBROBLASTS: FLOW CYTOMETRY AND IMMUNOPRECIPITATION.

ICAM-1 is the natural ligand of the LFA-1 molecule (Makgoba et al. 1988). Therefore, we investigated the expression of the two molecules (CD11a and CD18) which form the LFA-1 complex.

In FIGURE 7A, flow cytometry analysis with mAb recognizing the CD11a and the CD18 show the presence of these two antigens at the surface of human embryonic fibroblasts.

The immunofluorescence studies were confirmed by the immunoprecipitation performed with an anti CD11a mAbs (7B). Under reducing condition, we detected on embryonic fibroblasts two bands of 140 and 95 kDa respectively.

As positive control we used PHA blasts, in these cells, we observed a doublet of 180 and 95 kDa. These results suggest that human embryonic fibroblasts may express an LFA-1 like complex at their surface.

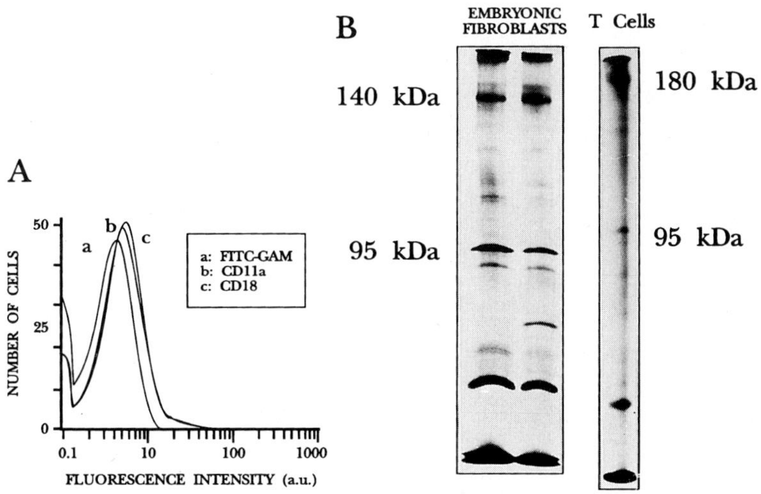

FIGURE 7: Flow cytometry analysis and SDS PAGE electrophoretic analysis after immunoprecipitation of the two chains (CD11a, and CD18) of the LFA-1 integrin in human embryonic fibroblasts and PHA blasts. (7A) the left curve represents the background binding of FITC-GAM. (7B) cultures were labelled with ^{35}S-methionin and cell membrane extracts were immunoprecipitated with the anti-CD11a mAb 6B11. Under reducing conditions PHA blasts exhibit two major bands of 180 and 95 kDa respectively, whereas human embryonic fibroblasts exhibit two bands of 140 and 95 kDa respectively.

IL2 MODULATION OF CLASS I AND CLASS II HLAs EXPRESSION BY MELANOMA CELLS.

Analysis of FIGURE 8 shows typical experiments illustrating the effects of IL2 and IFN gamma on the surface expression on class I HLA (8A) and class II HLA (8B) in human melanoma cells. IFN gamma increases the surface expression of both class I and class II HLA. By contrast, IL2 decreases the expression of these two molecules and inhibits the stimulatory effect of IFN gamma.

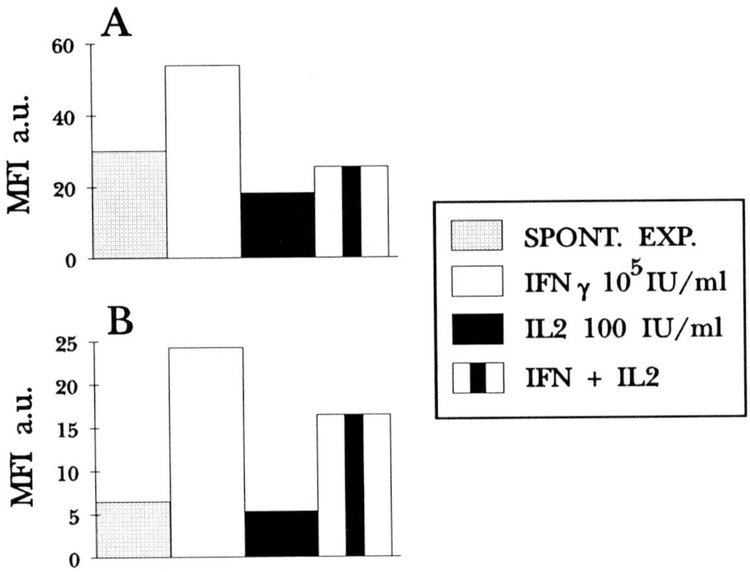

FIGURE 8: Flow cytometry analysis of class I (8A) and class II (8B) HLAs in human M14 melanoma cells, effect of incubation with human rIL2 and/or rIFN gamma. The biological effect is achieved after 24 hours. Data show the mean fluorescence expressed in arbitrary units; negative control value (FITC-GAM) has been subtracted.

IL2 MODULATION OF ICAM-1 PROTEIN EXPRESSION BY MELANOMA CELLS.

- In melanoma cells suspensions (FIGURE 9A, left panel) we observed high levels of ICAM-1 protein. In this culture condition IL2 induces a dose dependent down modulation of ICAM-1 surface expression. The maximal effect (60% decrease) is obtained with 360pM IL2 (100IU/ml). Preincubation with anti IL2Rα or IL2Rß mAbs (FIGURE 9A right panel) partially inhibits the IL2 dependent down modulation of ICAM-1.
- In sparse adherent M14 cultures (100-200x10^3 cells/cm²), we detected low surface levels of ICAM-1. In this culture condition, incubation with IL2 (1000IU/ml : 3.6nM) or with anti IL2Rß (5μg/ml) induces a strong increase (40-80%) of ICAM-1 surface levels.

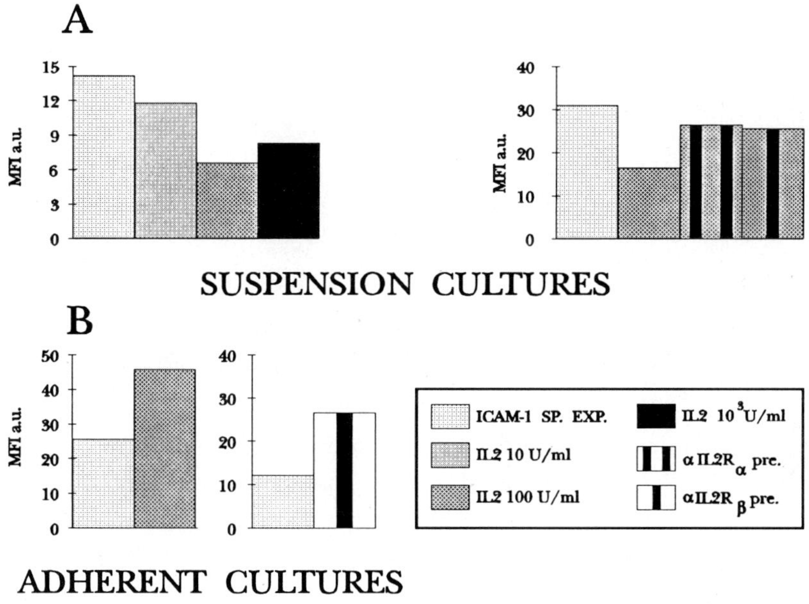

FIGURE 9: Flow cytometric analysis of ICAM-1 protein expression on human melanoma M14 cells, cultured in suspension (9A) or adherent to the plastic surface (9B). Effect of overnight incubation with IL2 and preincubation with anti IL2Rα or IL2Rß mAbs. Data show the mean fluorescence expressed in arbitrary units; negative control value (FITC-GAM) has been subtracted.

DISCUSSION

The IL2R system is a very interesting model since complex inter-molecular interactions are involved between one cytokine (IL2) and at least two receptor subunits (IL2Rα and IL2Rß). Though its functional specificity was though to be restricted to immunocompetent cells (Waldmann, 1991), here we show that human embryonic lung fibroblasts and human melanoma cells express a functional IL2R. Both anti IL2Rα and IL2Rß mAbs block albeit not completely the binding of ^{FITC}-IL2 to either cells. These observations indicate that the IL2R spontaneously expressed on non-lymphoid cells, is able to specifically bind IL2 alone.

In this respect, human fibroblastic and melanoma cells behave differently from murine fibroblasts. Indeed, the 70 kDa peptide expressed by murine fibroblastic cells after transfection of the human IL2Rß cDNA is unable to bind IL2 (Minamoto et al., 1990).

In human fibroblastic and melanoma cells, the association of the two chains occurs since preincubation with IL2 at 37°C strongly decreases ^{125}I-IL2 labelling on both chains at 4°C, suggesting that the IL2Rαß/IL2 complex has been internalized. The specificity of flow cytometry and IL2 binding experiments was further confirmed by the detection, in either cell types of specific transcripts for the IL2Rα gene (3.6, 2.8 and 1.7 kb) and for the IL2Rß gene (4, 3.2 and 1.6 kb). The size of the three transcripts for the IL2Rα gene is identical to that previously described in embryonic fibroblasts and fibrosarcoma cells (Carloni et al., 1989). Concerning the mRNAs specific for the IL2Rß gene, embryonic fibroblasts and melanoma cells exhibit a pattern of transcripts identical to that found in the control B cell line SKW6.4. In the lymphoid B cell line, however, the 4kb species is predominant and the 3.2 and 1.6kb represent minority forms whereas in fibroblastic and in melanoma cells the 3,2 and 1,6kb are the predominant forms.

Since IL2 does not affect the proliferation of human fibroblasts (Carloni et al., 1989) and of melanoma cell lines (not shown) we decided to check whether IL2 could modify other biological properties. So, we chose to analyze the effect of IL2 on the expression of membrane molecules which control the adhesion of cytolytic T cells to target cells (ICAM-1) (Makgoba et al., 1989) or which act as antigen presenting molecules (HLA class 1 and class II HLA) (Krief et al.,

1987). Moreover the ICAM-1 protein is not present on normal melanocytes and its expression appear to correlate with increased metastatic potential on melanoma cells (Johnson et al., 1989). Here, we show that IL2/IL2R complex is internalized and down modulates the surface expression of ICAM-1, class I and class II HLA. IL2 also inhibits the induction of class II HLA by IFN gamma. The down modulation of ICAM-1 depends on the culture conditions and on the concentration of IL2. Indeed, IL2 decreases the surface expression of ICAM-1 only in confluent fibroblastic cells where this expression is maximal (Dustin et al., 1986) and on melanoma cells in suspension (also exhibiting a higher ICAM-1 expression).

IL2 is, to date, the first cytokine to lower the expression of ICAM-1 at the cell surface. This effect is specific and dependent on the high affinity heterodimer since it is partially prevented by preincubation with mAbs recognizing the IL2 binding sites on the α (H93) and ß chains (MIKß1). It has been recently shown that ICAM-1 and IL2Rα are physically linked at the cell surface, one of the possible mechanisms for the ICAM-1 down-modulation could be a cointernalization of the two molecules upon IL2 stimulation as proposed by Burton et al (1990).

Nevertheless, the IL2Rα can only be internalized when part of the IL2Rαß receptor. This model would therefore predict that only the ICAM-1 molecules complexed with the IL2Rαß heterodimers will be internalized. The low number of high affinity receptors detected on fibroblasts and melanoma cells together with the higher effect observed after 48-72h suggest that other mechanisms are involved in the diminution of the ICAM-1 surface expression induced by IL2.

By contrast, IL2 increases the surface expression of ICAM-1 when added to sparse fibroblastic cultures or to sparse adherent melanoma cells, where the expression of this molecule is very low (Dustin et al., 1986). In these culture conditions ICAM-1 up regulation is also observed after overnight incubation with saturating amounts of anti IL2Rß mAb (5μg/ml). These data strongly suggest the specific involvement of the IL2Rß chain in the up regulation of ICAM-1 protein. Interestingly, in human fibroblastic cultures, the cytokine modulation of class I HLA antigens is apparently dependent on the cell density. Indeed, in sparse cultures IFN gamma or IL2 cause a modest increase in the surface levels of this antigens and only simultaneous addition of the two cytokines induces a 7 fold increase in the class I HLA expression. In confluent cultures, we observed a quite different behaviour; indeed, IFN gamma alone is very efficient in inducing a 7 fold increase in the class I HLA surface expression whereas IL2 exhibits a strong inhibitory effect, reducing of about 50% their spontaneous expression, and blocking, almost completely the up regulation produced by IFN gamma.

In summary, human cultured fibroblasts and melanoma cells may simultaneously express the two chains of the IL2R; the addition of IL2 induces the formation of high affinity heterodimers and is directly involved in the control of several surface antigens. The expression of the IL2Rαß in these cell types and the complex modulation of ICAM-1, class I and class II HLA antigens by IL2 may constitute a relevant regulatory mechanism of cell mediated immune response inside reactive tissues and tumors. Changes in the surface levels of ICAM-1 protein may also affect homotypic cell adhesion of embryonic fibroblasts since these cells also express an LFA-1 like molecule, the natural ligand of the ICAM-1 protein.

REFERENCES

Azzarone, B., Eid, P., Malpiece, Y., Andeol, Y., Fauci, A.S., Suarez, H. (1986). Evolution in the structure and distribution of 4F2-antigen from the oncofetal to the adult phenotype of human fibroblasts. Int. J. Cancer. 38:177-181

Boucheix C., Krief, P., Perrot, J., Y., Mirshahi, M., Rosenfeld, C. (1986). Method for rapid detection of membrane antigens by immunofluorescence and its application to screening antibodies. In Methods in Enzymology Immunological Techniques Part I Hybridoma Technology and Monoclonal Antibodies. Eds JJ. Langone, H. Vand Vunakis. Academic Press INC. 121:580-587

Burton J., Goldman, C. K., Rao, P., Moos, M., Waldmann, T. A. (1990). Association of intercellular adhesion molecule 1 with the multichain high affinity interleukine 2 receptor. Proc. Natl. Acad. Sci. USA 87, 7329-7333

Carloni G., Paterson, H., Augery-Bourget, Y., Sahraoui, J., Rubinstein, E., Suarez, H. and Azzarone, B. (1989). N-ras dependent revertant phenotype in human HT1080 fibrosarcoma cells is associated with loss of proliferation within normal tissues and expression of adult membrane antigenic phenotype. Oncogene 4, 873-880

Cillo C., Mach, J.P., Schreyer, M., Carrel, S. (1984). Antigenic heterogeneity of clones and subclones from human melanoma cell lines demonstrated by a panel of monoclonal antibodies and flow microfluorometry analysis. Int. J. Cancer 34: 11-20

Dukovich M., Y. Wano, L. Bich-Thuy, P. Katz, B.R. Cullen, J.H. Kehrl, and W.C. Green. (1987). A second human interleukin-2 binding protein that may be a component of high-affinity interleukin-2 receptor. Nature 327:518

Dustin L.M., Rothlein, R., Bhan, A.K., Dinarello, C.A. and Springer, T.A. (1986). Induction by IL1 and INFg : Tissue distribution, biochemistry, and function of a natural adherence molenule (ICAM-1). Int. J. Immunol. 137:245-254

Green, J. E., Begley, C. G., Wagner, D. K., Waldmann, T. A. and Jay, G. (1989). Trans Activation of Granulocyte-Macrophage Colony-Stimulating Factor and the Interleukin-2 receptor in transgenic mice carrying the human T-lymphotropic virus type 1 tax gene. Molec. and Cell. Biol. 9, 11:4731-4737

Harel Bellan A., Mishal, Z., Willette Brown, J. and Farrar, W. L. (1989). Detection of low and high affinity binding sites with fluoresceinated human recombinant Interleukin-2. J. Immunol. Meth. 119, 127-133

Hatakeyama M., Tsudo, M., Minamoto, S., Kono, T., Doi, T., Miyata, T., Miyasaka, M., and Taniguchi, T. (1989). Interleukin-2 receptor ß chain gene: Generation of three receptor forms by cloned human α and ß chain cDNA's. Science 244, 551-556

Johnson, J. P., Stade, B. G., Holzmann, B., Schwable, W. and Riethmuller, G. (1989). De novo expression of intercellular adhesion molecule 1 in melanoma correlates with increased risk of metastasis. Proc. Natl. Acad. Sci. USA 86, 641-644

Krief P., Boucheix, C., Billard, M., Mishal, Z., Van-Agthoven, A., Fiers, W., Azzarone B. (1987). Modulation of expression of class II histocompatibility antigens by secretion of a cellular inhibitor in K562 leukemic cells. Eur. J. Immunol. 17:1021-1025

Leonard W. J., Depper, J. M., Crabtree, G. R., Rudikoff, S., Pumphrey, J., Robb, R. J., Kronke, M., Svetlik, P. B., Peffer, N. J., Waldmann, T. A. and Greene, W. C.(1984). Molecular cloning and expression of cDNAs for the human interleukin 2-receptor. Nature 311, 626-631

Makgoba M. W., Sanders, M. E., Ginther Luce, G. E., Dustin, M. L., Springer, T. A., Clark, E. A., Mannoni, P., Shaw, S. (1988). ICAM-1 a ligand for LFA-1 dependent adhesion of B, T and myeloid cells. Nature 331, 86-88

Minamoto S., Mori, H., Hatakeyama, M., Kono, T., Doi, T., Ide, T., Uede, T., Taniguchi, T. (1990). Characterisation of the heterodimeric complex of human IL-2 receptor α.ß chains reconstituted in a mouse fibroblast cell line, L929. J. Immunol. 145, 2177-2182

Okamoto Y., Minamoto, S., Shimizu, K., Mogami, and Tanigushi, T. (1990). Interleukin-2 receptor ß chain expressed in an oligodendroglioma line binds interleukin-2 and delivers growth signal. Proc. Natl. Acad. Sci. USA 87, 6584-6588

Plaisance S., Rubinstein, E., Alileche, A., Sahraoui, Y., Krief, P., Augery-Bourget, Y., Jasmin, C., Suarez, H., and Azzarone, B. (in press). Expression of the Interleukin-2 receptor on human fibroblasts and its biological significance. Intern. Immunol.

Rubin L.A., Hoekzema, G.S., Nelson, D.L., Greene, W.C., Jay, G. (1987). Reconstitution of a functional interleukin-2 receptor in a non-lymphoid cell. J. Immunol. 139:2355-2360

Saneto R. P., Altman, A., Knobler, R. L., Johnson, H. M. and DeVellis (1986). Interleukin-2 mediates the inhibition of oligodendrocyte progenitor cell proliferation in vitro. Proc. Natl. Acad. Sci. USA 83, 9221-9225

Tsudo M., Kinamura, F. and Miyasaka, M. (1989). Characterisation of the Interleukin 2 receptor ß chain using three distinct monoclonal antibodies. Proc. Natl. Acad. Sci. USA. 86, 1982-1986

Waldmann T.A. (1991). The Interleukin-2 receptor. The J. of Biol. Chem. 266:5, 2681-2684

CHANGE IN LIPID COMPOSITION AND IN MEMBRANE FLUIDITY OF HUMAN PERIPHERAL BLOOD LYMPHOCYTES UNDERGOING BLASTIC TRANSFORMATION

Mishal Z.*, Anel A.+, Aguilar J.J.+ and Uriel J.+
* CNRS - CTSRC
Laboratoire de Cytométrie, 94800 Villejuif, France

INTRODUCTION

The process of lymphocyte activation which leads to proliferation is involved in the majority of immunological events. Although the early metabolic changes in response to specific antigens or mitogenic lectins are well studied, the mechanism, however, is not yet completely understood. The increase in cell size is the first morphological change that we can observe, because the cell diameter more than doubles. This increase is followed by the enhancement of the synthesis of membrane phospholipids and by the change in their fatty acid composition implicated deacylation-reacylation reaction (Shinitzky M.).

There are few studies on changes in fatty acid composition of human T-lymphocytes during the entire period of blastic transformation but there are several studies on the modification of membrane fluidity measured by fluorescence polarization of DPH and other hydrophobic probes following blastic transformation (Resh K et al).

In these studies the time of activation was less than four hours and the variations of P values were not followed by a parallel study of lipid composition ; cholesterol to phospholipid (CH/PL) molar ratio and ratio of saturated to unsaturated fatty acids.

+ CNRS Institut de Recherches Scientifiques sur le Cancer
Laboratoire de Chimie des Protéines
94801 Villejuif, France

NATO ASI Series, Vol. H 67
Flow Cytometry
Edited by A. Jacquemin-Sablon

Measurements of fluorescence polarization give information about the environment of the fluorescent molecule. This environment can modify the properties of the fluorescence detected from that molecule and hence give information about that environment.This can be done by exciting the fluorescent molecule with polarized light and measuring the extent to which the emitted fluorescence retains the polarization of the light source.The measured polarization value depends upon both the life time of the fluorescent state and the rate of rotation of the fluorescent molecule (Weber G.).

In our experiments, the term fluidity is used to describe the rotational freedom of a lipid soluble molecular probe (DPH) within a membrane bilayer. Determination of absolute fluidity is limited in an anisotropic membrane suspension because of the inability to reproduce accurately the three dimensional structure of the hydrophobic bilayer. Therefore the steady state fluorescence anisotropy is employed to estimate the degree of fluidity after probe incorporation into the bilayer. In the most common case in flow cytometry, the fluorescent sample illuminated by a vertically polarized plane parallel excitation beam lying in a horizontal plane, the emission in this plane is partially vertically polarized.The degree of polarization can be measured by means of plane polarizer and photo detector (PMT). The emission is usually observed at an angle 90° to the excitation. When horizontally polarized excitation is used, the emission at 90° to the excitation is unpolarized because of the symmetry of the system at this angle. This fact can be used in flow cytometry polarization measurements for calibration (Gorvel J.P. et al) by rotating the plane of polarization from vertical to horizontal.This orientation results in unpolarized fluorescence, the measurement of which is electronically adjusted to fluorescence polarization zero.

Fluorescence polarization is one of the most sensitive fluorescence factor to changes in conformation and structure such as the one that occurs in cell membranes and cytoskeleton which are among the initial events following binding of mitogen or antigen to a responsive cell. The molecular changes that accompany lymphocyte activation will be potent factor in changing the fluorescence polarization of cells labelled with the two lipophilic probes : Diphenylhexatriene (DPH) and its Trimethyl-Ammonium derivative (TMA-DPH).

MATERIALS AND METHODS

Purified T lymphocytes (Stubbs et al) were activated by adding 2µg of PHA-M (Sigma, UK) per ml of culture medium. This concentration proved to be optimal for cell activation, as determined from previous experiments.

Cell lipids were extracted with chloroform:methanol (2:1, V/V). The lipid extracts were fractionated in the different lipid classes by thin layer chromatography in silicagel plates (Merck, Germany). In others experiments, the cell lipid extracts were used to prepare the methyl esters of total fatty acids as described (Stubbs et al).

FLUORESCENCE POLARIZATION MEASUREMENTS

The fluorescent hydrocarbon DPH was used is as probe for determining changes in dynamic nature of hydrocarbon regions of the cell membrane bilayer. For labelling, a solution of 2mM DPH in tatrahydrofuran was diluted 1000 fold by injection into vigorously stirred phosphate buffer saline (PBS). The resulting dispersion of 2µM DPH is clear and practically void of fluorescence.

The cells were washed twice with PBS at 25°C and the final pellet was resuspended with DPH dispersion, and incubated for 30 min at 25°C. The fluorescence polarization was measured on a single cell level by FACS 440 (Becton Dickinson, USA) cell sorter. The FACS analyzes cells in aqueous suspension as they pass through a focused argon-ion laser beam (model 2020 Spectra Physics) at 363 nm at an output of 100 mW by simultaneous detection of apparent parallel and perdendicular fluorescence emission intensities. Fluorescence polarization P for each cell was calculated using equation

$$P = (I_{//} - I_{\perp}) / (I_{//} + I_{\perp}).$$

In FACS 440 system a silvered 50:50 mirror is used to direct the fluorescence emission equally to the PMT without materially affecting any fluorescence parameter. Polarizing filters are placed in front of each PMT and oriented to transmit the adequate polarized component. When horizontally polarized excitation is used in the same geometry the emission at 90° with respect to direction of the exciting beam (fig.1) is unpolarized because of the symmetry of the system at this angle. This fact is used to balance the two channels, that means to ensure that the PMTs give equal magnitude signal for the same fluorescence intensity. For this purpose we use half wave retarder (Oriel corp. USA) to rotate the plane the polarization of the exciting laser beam from vertical to horizontal.

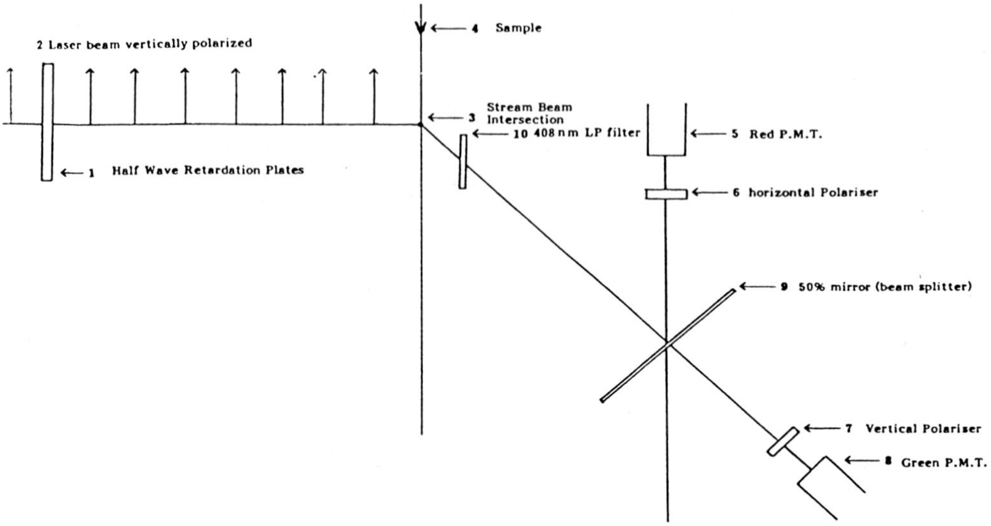

Fig. 1 Fluorescence polarization measurements in flow cytometer

The ratio $I_{//}/I_{\perp}$ is measured and used to calculate P. However the quantity known as emission anisotropy r is of a more (2)
$r = (I_{//} - I_{\perp}) / (2I_{//} + I_{\perp})$ fundamental significance than P.
r represents the proportion of all light emitted to all direction which is polarized, $(2I_{//} + I_{\perp})$ is proportional to the total emission intensity.
We have measured the order parameter S, because the membrane lipid layer is an ordered fluid (Shinitzky M.) where orientation perpendicular to membrane plane is more probable then in plane.
The order parameter S is defined as :

3) $S = \left\{ [1-2r/r_o + 5 \ (r/r_o)^2]^{0,5} - 1 + r/r_o \right\} : r/r_o$

r is defined in equation (2) r_o is the emission anisotropy in the absence of any rotational motion of the probe r_o is about 0.4, S varies between -1/2 and 1.
These limiting cases has the following meaning, when S=1 the molecules are all exactly perpendicular to the membrane plane.
When S= -1/2 they are randomly distributed in the plane. S=0 corresponds to a complete random distribution of the molecular axes.

RESULTS

The total fatty acid composition of T lymphocytes is shown in table I.

TABLE I

Total fatty acid composition of quiescent and PHA-activated human T lymphocytes after 24 hours of culture.

Data are expressed as percents of the total fatty acids and are the mean ± SD of two analyses on cells from three different donors DMA, dimethyl acetal from alkenyl side chains of plasmalogens.

(From A. Anel, J. Naval, B. Gonzalez, JM. Torres, Z. Mishal, J. Uriel and A. Pineiro. Biochem. Biophys. Acta 1044 (1990) 323-331. With permission)

Fatty acid	quiescent	24 h in culture	24 h PH activated
16:0 DMA	1.2 ±0.5	0.9±0.1	0.8±0.1
16:0	27.5±1.7	23.0±1.7	23.7±0.7
16:1 (n-7)	1.7±0.7	1.1±0.4	1.4±0.0
18:0	16.3±1.0	18.4±0.9	18.4±0.5
18:1 (n-9)*	19.8±2.0	16.5±1.0	18.2±0.2
18:2 (n-6)	11.1±2.0	9.9±1.8	9.4±0.2
20:1 (n-9)	0.4±0.1	1.2±0.5	0.4±0.1
20:2 (n-6)	1.5±0.2	2.1±0.3	0.8±0.0
20:3 (n-6)	2.7±0.3	2.1±0.7	2.5±0.3
20:4 (n-6)	10.1±1.5	18.0±1.1	17.4±1.2
20:5 (n-3)	0.4±0.1	1.4±0.1	0.5±0.0
22:4 (n-6)	1.1±0.2	2.0±0.5	1.2±0.2
22:5 (n-3)	N.D	1.1±0.3	2.3±0.1
22:6 (n-3)	0.9±0.0	1.8±0.5	2.2±0.2
saturated/ unsaturated	0.98±0.08	0.73±0.06	0.75±0.07

*To indicate that oleic acid is an 18-carbon fatty acid with one double bond the shorthand 18:1 us used. The number before the colon denotes the number of carbon atomes, and the number following refers to the number of double bonds. To designate on individual fatty acid within a family of structurally related acid, the (n-) nomenclature, is used. Here the position of the first double bond from the methyl end is described.

As shown in table I a culture time of 24 hours was sufficient to modify the fatty acid composition of quiescent cells without affecting their viability (more than 90%). An enrichment in arachidonic acid was observed in both quiescent and activated cells relative to freshly activated cells. This caused a 24% reduction in saturated/unsaturated fatty acid ratio. The proportion of 18:1 (n-9) and 22:5 (n-3) was higher in activated lymphocytes than in quiescent ones. These differences become more significat for a longer time of activation. The amount of cholesterol and phospholipids present in lipid extracts of quiescent and PHA activated human T-lymphocytes, as well as the corresponding cholesterol to phospholipids molar ratios, are shown in table II.

TABLE II
Cholesterol/phospolipid (CH/PL) molar ratio of resting (cultured for 24 h.) and activated human T lymphocytes . Data are the mean ±SD of duplicate determination on cells from three different donors.
(From A. Anel, J. Naval, B. Gonzalez, JM. Torres, Z. Mishal, J. Uriel and A. Pineiro. B.B.A.1044 (1990) 323-331. With permission)

	cholesterol (nmol/10^6 cells)	phospholipids (nmol/10^6 cells)	CH/PL molar ratio
restings T lymphocytes	2.5 ± 0.2	5.3 ± 0.3	0.47 ± 0.03
48 h activated T lymphocytes	3.7 ± 0.2	10.2 ± 0.9	0.36 ±0.01

Resting T cells maintained for 24 h in cultures without PHA were used as control. We observed that the activation process induced a great increase in the amount of phospholipids in T activated lymphocytes. Since the amount of cholesterol increased in a lesser extent the resulting CH/PL molar ratios were reduced.

Fluorescence polarization results

The results of fluorescence polarization measurements are shown in table III.

TABLE III

Values of fluorescence polarization (P), fluorescence emission anisotropy (r) and order parameter (S) of resting and activated T lymphocytes. Data are the mean of duplicate determination on cells from different donors. Values in brackets indicate the number of different donors used in each case. SD was never greater than 15% of the mean.

(From A. Anel, J. Naval, B. Gonzalez, JM. Torres, Z. Mishal, J. Uriel and A. Pineiro. Biochem. Biophys. Acta 1044 (1990) 323-331. With permission)

resting	P	r	S
T-lymphocytes non cultured (2)	0.330	0.247	0.737
24 h in culture (4)	0.283	0.208	0.640
48 h. " " (2)	0.278	0,205	0,629
72 h. " " (2)	0.235	0.170	0.530
PHA activated T-lymphocytes			
30 min (1)	0.330	0.247	0.737
3 h. (1)	0.326	0.244	0.730
6 h. (1)	0.309	0,229	0.693
24 h. (3)	0.221	0.159	0.495
48 h. (3)	0.186	0.132	0.410
72 h. (2)	0.152	0.107	0.326

Short time activation did not provoke any significante change in the order parameter S indicating that in the early steps of activation, no significant changes occured in the packing of the membrane contituents. The culture of T resting cells for 24 h produced a slight decrease of S value. At longer time of culture, the difference between resting and activated cell were significant. The culture of T resting cells in serum-supplemented medium caused a moderate decrease in S value comparatively less than that observed at any time in activated cells.

Discussion

The data of the order parameter S indicates that the activation process provoked an important reduction on membrane packing after periods between 24 h and 72 h . This reduction should be related to the change in the cholesterol to phospholipid molar ratio and fatty acid composition. The ratio of saturated/unsaturated fatty acid did not change significantly during blastic transformation. So the low level of the CH/PL ratio found in activated T-lymphocytes relative to resting cells must be one of the important parameters that provoked the reduction of membrane packing.

The main methodological difference between this actual study and others (Parola et al) is that we used measurements of P by flow cytometry which enabled us to exclude dead cells from the measurements by light scatter gating while by fluorometry it is impossible to do that.

The important percentage of dead cells in mitogen-treated cultures explain that contrary to others, we have found differences of P values between activated and resting T cells.

The significant changes in fatty acid composition occur during activation of human T lymphocytes may contribute to the increase in membrane fluidity. But the major factor must be the decrease in cholesterol to phospholipid molar ratio.

The latter was confirmed in a recent study (Aguilar et al) showing that the cholesterol/phospholipid ratio, was clearly higher in HIV-I virus infected than uninfected T-lymphocytes. This finding fits well with the high cholesterol content and high membrane rigidity of the HIV-virus envelope and suggests that the cholesterol/phospholipid ratio of cell membranes of infected cells tend to approach that of the virus envelope.

REFERENCES

Aguilar JJ., Anel A, Torres JM, Semmel M, and Uriel J (1991) Changes in lipid composition of human peripheral blood lymphocytes infected by HIV. Aids Research and Human Retroviruses, 7, pp 761-765

Anel A, Naval J, Gonzalez B, Torres JM, Mishal Z, Uriel J and Pineiro A (1990) Fatty acid metabolism in human lymphocytes. I. Time-course changes in fatty acid composition and membrane fluidity during blastic transformation of peripheral blood lymphocytes.
Biochim. Biophys. Acta 1044. pp 323-331

Gorvel JP, Mawas C, Maroux S and Mishal Z (1984) Flow cytometry is a new method for the characterization of intestinal plasma membrane. Biochem. J. 221. pp 453-457

Resch K and Ferber E (1987) Structure and role of phospholipids in the lymphocyte plasma membrane. In : Marchalonis JJ (ed.) The Lymphocyte Structure and function, Marcel Dekker Inc. New York, pp171-222

Shinitzky M (1984) The lipid fluidity of cell membranes In : CRC Press Inc. (ed) Physiology of Membrane Fluidity Vol I, Boca Raton Florida, pp 1-57

Stubbs CD, Tsang WM, Belin J, Smith AD and Johnson SM (1980) Incubation of exogenous fatty acids with lymphocytes. Changes in fatty acids composition and effects on the rotational relaxation time of 1,6-Diphenyl-1,3,5-hexatriene. Biochemistry, 19, pp 2756-2762

Weber G (1973) Polarized fluorescence. In : Thaer AA and Sernetz M (ed.) Fluorescence techniques in cell biology, Springer-Verlag, Berlin, pp 5-13

Acknowledgments: We thank Arlette Vervisch for valuable advice and Dominique Laird for preparation of the manuscript.

PART II

Membrane-Ligand Interactions and Cell Identity

CELLULAR PROLIFERATION HISTORY BY FLUORESCENT ANALYSIS

Paul Karl Horan
Sue E. Slezak
Bruce D. Jensen
Zynaxis Cell Science, Inc.
371 Phoenixville Pike
Malvern, PA 19355

A variety of biochemical and cellular assays have been developed to make it possible to distinguish between dividing and non-dividing cells (Darzynkiewicz, Z., 1984, and Darzynkiewicz, Z., Traganos, F., and Kimmel, M., 1986, and Dolbeare, F., Beisker, W., Pallavicini, M.G., Banderlaan, M., and Gray, J.W., 1985, and Crissman, H.A., Steinkamp, J.A., 1987, and Muirhead, K.A., Horan, P.K., and Poste, G., 1985). Most of these assays can only determine the state of proliferative activity at the time the cells are "harvested" and do not permit the observer to determine the activity at times prior to harvesting. For example, the elegant methods developed by Darzynkiewicz, et al. (Darzynkiewicz, Z, 1984, and Darzynkiewicz, Z., Traganos, F. and Kimmel, M., 1986) permit the investigator to distinguish cycling from non-cycling cells at the time of harvest using dual-fluorescence measurements of DNA and RNA levels, but they do not permit evaluation of the proliferative activity of these same cells at times prior to harvesting. In this manuscript, a novel method of labeling cells is presented which does not affect the viability or proliferative activity of cells, but permits retrospective monitoring of their growth for as many as eight cell divisions. The dye, PKH-1 (Zynaxis Cell Science, Inc., 371 Phoenixville Pike, Malvern, PA USA 19355), is applied to the cells and results in uniform staining of the plasma membrane. The cellular growth rate is unaffected by the presence of the dye. Further, the dye does not leak out of the cells and it is not transferred from cell to cell within the same culture. When a cell divides, each resulting daughter cell receives one-half of the total dye present in the parent cells membrane and each fluoresces at one-half the intensity of the parent cell. Thus, by determining the fluorescence of cells at the beginning of the experiment, it is possible to calculate the number of divisions a cell has made at any time by merely determining the resultant fluorescence intensity.

NATO ASI Series, Vol. H 67
Flow Cytometry
Edited by A. Jacquemin-Sablon

A.

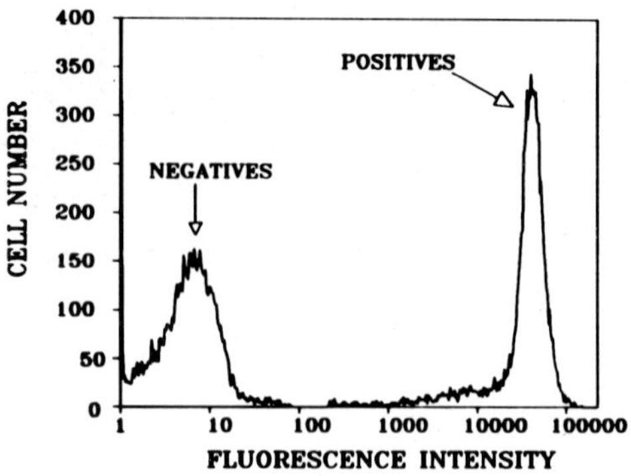

B.

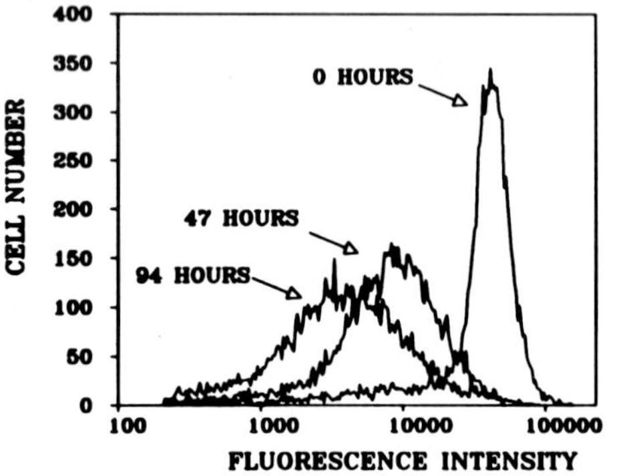

FIGURE 1: Fluorescence intensity distribution for Yac-1 cells labeled with PKH-1. Logarithmically growing Yac-1 cells were washed once in PBS and resuspended in 10^{-5}M PKH-1 solution at a concentration of 10^6 - 10^7 cells/ml for 5 minutes at room temperature. The staining reaction was terminated by adding an equal volume of fetal calf serum and the cells were pelleted at 400xg for 5 minutes, washed (3x) and resuspended in complete media. Fluorescence intensity measurements were achieved using a Coulter EPICS 753 flow cytometer. 200mW of 488nm light was used to excite the dye and the green fluorescence of propidium iodide negative cells was measured using a 525nm band pass interference type filter. Note the large dynamic range between fluorescence positive and fluorescence negative cells. To monitor the growth rate of cells, PKH-1 stained cells were placed into the incubator in complete culture media (DMEM High Glucose + 20% FBS) and analyzed daily using flow cytometric techniques (Muirhead, K.A., Horan, P.K., and Poste, G., 1985). The instrument was aligned daily and intensity settings reproduced using fluorescent microbead standards. From each fluorescence profile the mean fluorescence intensity and standard deviation were determined. Note in the lower panel that as the fluorescence intensity decreased, the standard deviation increases.

From Figure 1 (Upper Panel), it can be seen that the fluorescence intensity of stained Yac-1 cells is extremely bright compared to the unstained controls. The dynamic range of the staining is so large that the fluorescence intensity in this and subsequent figures must be plotted on a semi-logarithmic axis. This is advantageous to the analysis, since cell growth kinetics are exponential in nature. Therefore, a semilog plot of the fluorescence kinetics should be linear during the log phase growth (see Figure 2). Note that when the measurements are performed on subsequent days the fluorescence intensity diminishes but in a uniform manner (Figure 1, Lower Panel). Using the dye, PKH-1, it is possible to follow these cells for six to eight divisions. Additionally, the cells can be fixed in paraformaldehyde (Lanier, L.L. and Warner, N.L., 1981) for storage so that the samples can all be run at the same time at the conclusion of experimentation. In Figure 2, growth rates are compared for PKH-1 stained and unstained Yac-1 cells. When Yac-1 cells are stained at the concentrations indicated in Figure 2, the cellular growth rate is essentially unaffected by the dye. In some instances, stained cells have exhibited a decrease (less than 5%) in their doubling time. However, it should be noted that it is possible to increase the concentration of stain to a point where there is some increase in the doubling time of the stained cells and concentration dependent cytotoxicity may result. These concentration specific effects vary with the cell type being used. For example, tissue cultured cells can tolerate staining concentrations as high as 2-4 x 10^{-5} M but primary spleen cells must be stained at much lower levels.

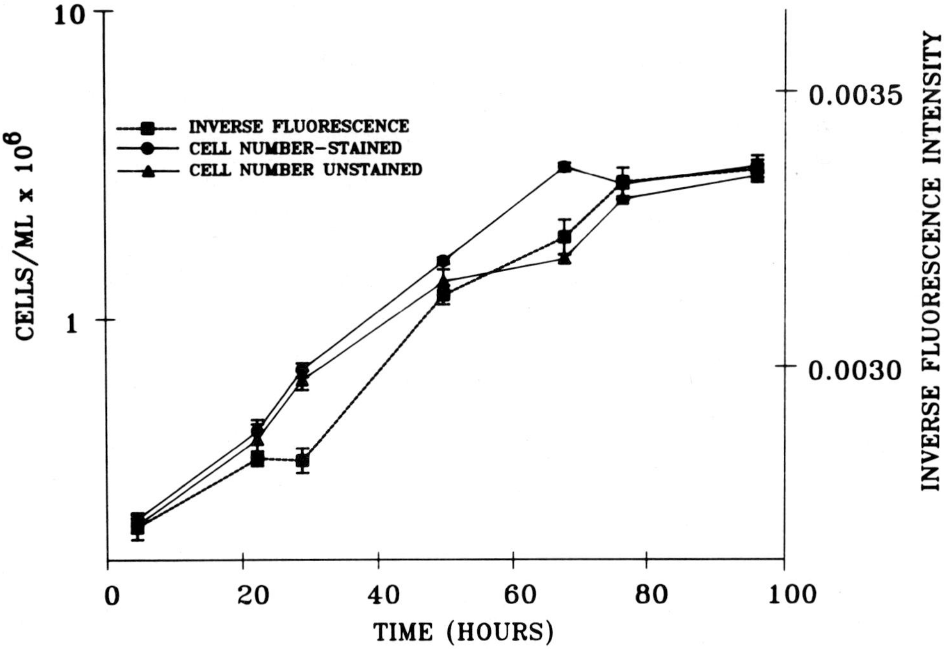

FIGURE 2: Cell number and inverse fluorescence intensity as a function of time. Separate flasks of PKH-1 labeled and non-labeled control cultures (Yac-1) were each set at 2 x 10^5 cells/ml at time zero. Cell counts were obtained using a Coulter ZBI cell counter (Coulter Electronics, Inc. Hialeah, Fl). Fluorescence intensity measurements were determined as described in Figure 1. Mean log fluorescence intensities were determined for each time point and the inverse was calculated and plotted. Standard deviations were calculated and plotted at each time point having been determined on quadruplicate samples.

Figure 2 demonstrates that the drop in fluorescence intensity is directly proportional to the number of divisions of the cell population. This demonstration was achieved by plotting the log fluorescence intensity as a function of time on the y_2-axis and comparing it to the increase in cell number plotted on the y-axis. Note that the growth curve and the plot of the inverse of fluorescence intensity are virtually superimposable. In this logarithmically growing culture, it can, therefore, be concluded that the drop in fluorescence intensity is directly proportional to the number of divisions the population has undergone.

Since the inverse fluorescence curves and cell growth curves are superimposable in Figure 2, it indicates that the dye is tightly bound and does not leak out of the cells. It is also important to insure that no transfer occurs between cells. One-half a culture of Yac-1 cells was labeled with 10^{-5}M PKH-1, taking care to remove all unbound dye. The labeled and non-labeled cell populations were mixed at a 1:1 ratio and incubated at 37° in a 7.5% CO_2 atmosphere. Each preparation was sampled for flow cytometric analysis periodically. As seen in Figure 1, the dynamic range between positively stained cells and unstained cells is so large that it is easy to calculate the average fluorescence intensity of both populations in the same culture. Furthermore, the percent labeled cells can be accurately and rapidly determined for each population in the culture.

A.

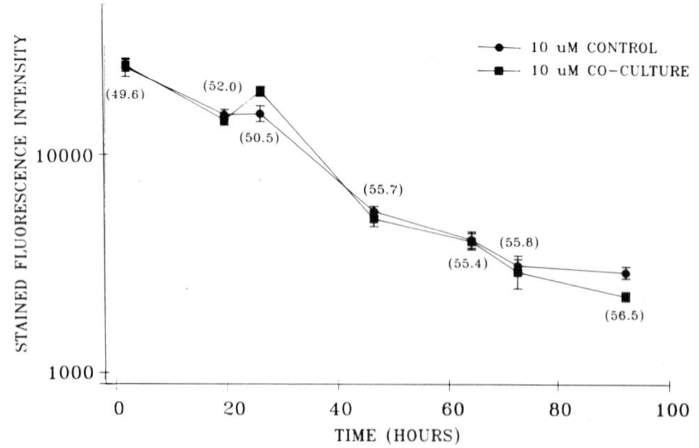

B.

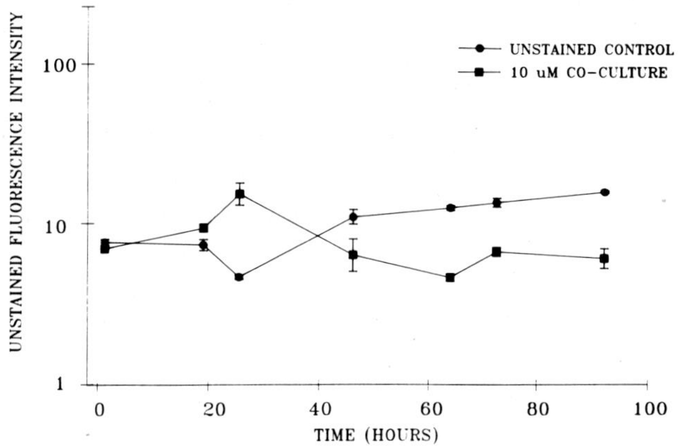

FIGURE 3: Fluorescence intensity of labeled (panel A) and non-labeled cells (panel B) as a function of time. Three cultures of Yac-1 cells were set at the same cell concentration on day 0. One culture consisted of 100% PKH-1 (10^{-5} M) stained Yac-1 cells (control), a second control culture contained non-stained Yac-1 cells and a third culture contained 50% stained and 50% non-stained cells. Each day an aliquot was removed from each sample and the fluorescence intensity was measured by Flow Cytometry as described in Figure 1. The percent labeled cells in the 50:50 culture is listed in parentheses for each time point. Standard deviations of the fluorescence intensity were calculated and plotted at each time point having been determined on quadruplicate samples.

In the upper panel of Figure 3, the fluorescence kinetics of a 100% stained control culture (10^{-5}M PKH-1) is compared with that of the stained population in the 50:50 mixture of stained and unstained cells described above. The fluorescence kinetics are identical for these two different culture conditions. In the lower panel, the fluorescence kinetics of an unstained control culture is compared with that of the unstained population in the 50:50 mixture of stained (10^{-5}M PKH-1) and unstained cells. One would expect that loss of dye or direct intercellular transfer of dye would cause the unstained fraction in the co-culture to become more fluorescent with time, leading to a mis-match in the kinetic responses in both panels of Figure 3. It can be concluded from these studies that there is little loss of the dye from the stained cells and there is no transfer of the dye from the labeled cells to the non-labeled cells in the same culture. These studies have been carried out with both suspension and adherent cell lines with identical results. In Figure 3 the percent of the population which remains labeled at each time point is listed in parenthesis. Note that the percent positively stained is essentially the same at all time points (within measurement error), indicating that both the labeled and non-labeled cells are growing at the same rate.

With this technique it is possible to label a population of tumor cells at the beginning of an experiment and, subsequently, follow their growth patterns. Non-cycling cells will maintain the fluorescence intensity observed at the beginning of the experiment as they are not dividing and therefore no change in the fluorescence per cell is noted (Slezak, S.E. and Horan, P.K., 1989). Cycling cells will continue to divide, which in turn will reduce their fluorescence until the intensity is not distinguishable from background levels. It should be noted that the dye, PKH-1, is excited at 488nm and emits with a maximum intensity at 520nm. This means that it is compatible with immunofluorescence staining when using phycoerythrin labeled antibodies (Loken, M.R., Lanier, L.L., 1984). When using heterogeneous cell populations such as bone marrow, uniform staining is observed for each cell type, however, differing levels of fluorescence are observed amongst the various cell types, presumably due the differences in plasma membranes. In addition, it is possible to use a different staining protocol and specifically label peritoneal macrophages IN SITU (Melnicoff, M.J., Morahan, P.S., Jensen, B.D., Breslin, E.W. and Horan, P.K., 1988).

The ability to use this procedure to follow the proliferative activity of cells in culture depends on three factors: 1) the level of autofluorescence exhibited by the cells of interest, 2) the maximum tolerated staining level of the cell of interest and 3) the number of divisions required for observation of a specific biological effect. To date this technology has been applied to many different cell systems successfully.

REFERENCES

Crissman, HA, Steinkamp, JA (1987) Experimental Cell Res. 173:256-261

Darzynkiewicz, Z (1984) In Growth, Cancer and the Cell Cycle (P. Skehan and S.J. Friedman, eds.). Humana Press, Inc., Clifton, NJ, pp. 249-277

Darzynkiewicz, Z, Traganos, F and Kimmel, M (1986) Assay of Cell Cycle Kinetics by Multivariate Flow Cytometry Using the Principle of Stathmokinesis, In Techniques in Cell Cycle Analysis (JE Gray and Z Darzynkiewicz, eds.), Humana Press, Inc.

Dolbeare, F, Beisker, W, Pallavicini, MG, Banderlaan, M, and Gray, JW (1985) Cytometry 6:521-530

Lanier, LL and Warner, NL (1981) J. Immunol. Methods 47:25-30

Loken, MR, Lanier, LL (1984) Cytometry 5:151-158 (1984)

Melnicoff, MJ, Morahan, PS, Jensen, BD, Breslin, E W, and Horan, PK (1988) J. Leuk. Biol. 43:387-397

Muirhead, KA, Horan, PK and Poste, G (1985) Biotechnology 3:337-356

Slezak, SE and Horan, PK (1989) Nature 340(6229):167-168

IDENTIFICATION AND ISOLATION OF BONE MARROW STEM CELLS

J.W.M. Visser and M.G.C. Hogeweg-Platenburg
Department of Molecular Pathology
TNO-Institute of Applied Radiobiology and Immunology
P.O.Box 5815
2280 HV Rijswijk
The Netherlands

INTRODUCTION

All blood cells and a variety of other cells in most tissues and organs of the body are derived from pluripotent haemopoietic stem cells which reside in the liver during fetal life and in the bone marrow later on. These stem cells are very rare; they constitute less than 1% of haemopoietic organs. They can nevertheless be studied by application of cell separation and culture techniques, and, if they are purified, by biochemical analysis.

The results of series of cell separation experiments indicated, that fractions of sorted cells gave different numbers of stem cells in a variety of different assays which were all thought to be specific for detecting stem cells (Baines and Visser, 1983; Jones et al., 1990; Jordan et al., 1990; Mulder et al., 1985; Mulder and Visser, 1987; Müller-Sieburg et al., 1986; Neben et al., 1991; Okada et al., 1991; Pallavicini et al., 1985; Ploemacher and Brons, 1988; Ploemacher and Brons, 1989: Spangrude et al., 1988; Spangrude and Johnson, 1990; Spangrude et al., 1991; Szilvassy et al., 1989; Terstappen et al., 1991; Verfaille et al., 1990; Visser et al., 1984; Visser and de Vries, 1988; Visser et al., 1991). This suggested that the haemopoietic stem cells compartment is heterogeneous. The question has become now which assay system detects the pluripotent haemopoietic stem cell that is capable of self renewal and of long-term repopulation of haemopoiesis. Long-term chimaerism is generally measured by determination of the haemopoietic reconstitution after transplantation into lethally irradiated recipients. The offspring of transplanted stem cells is detected by making use of endogenous markers,

NATO ASI Series, Vol. H 67
Flow Cytometry
Edited by A. Jacquemin-Sablon

such as subtle strain differences between donor and recipients that can be measured with immunological or biochemical techniques, or Y-chromosome-probes *in situ* hybridization and blotting methods for sex-mismatched transplantations, or genetic markers inserted into the sorted pluripotent stem cells. The latter has contributed significantly to our recent knowledge about the identity of the stem cells and about their response to growth factors.

Culture of sorted subpopulations of bone marrow cells showed high plating efficiencies of up to 50 % for cell types that were considered to be stem cells. However, more detailed analysis and comparison with *in vivo* repopulation experiments strongly suggests that these high plating efficiencies are only obtained with activated and committed daughter cells of the pluripotent haemopoietic stem cells. The presently known growth factors do not readily induce the stem cells to proliferate and differentiate. The only culture system that sustains stem cells *in vitro* in a manner that seems to be correlated to the long-term maintenance *in vivo* is the Dexter culture of stem cells inoculated on stromal layers. The assay measures colonies of haemopoietic cells in the stroma ("cobble-stone areas") or production of colony-forming cells (GM-CFU) and their presence in the culture supernatant. Successful long-term production of GM-CFU has been described for such cultures by several laboratories. In addition, transfection of stem cells with genetic markers is also performed in the best way in such co-culture systems. These findings indicate that adherence plays a key role in the regulation of the proliferation and differentiation of pluripotent haemopoietic stem cells.

The long-term production of GM-CFU on stroma is not a clonal assay, whereas the cobble-stone areas are also produced by committed progenitor cells. Therefore, these assays may detect the stem cells, but they do not facilitate their quantitation. The purity of stem cell sorts and even the absolute number of stem cells in total bone marrow therefore can not be determined accurately.

New methods have become available to assess the expression of genes in single cells and in rare, sorted cells such as the stem cells. Using PCR techniques cDNA libraries are now obtained from highly purified subpopulations of cells from the early haemopoietic compartments. These will serve to discover new cytokine receptors and adhesion molecules that are specific for the regulation of haemopoiesis at the stem cell level.

METHODS FOR STEM CELL ENRICHMENT

Stem cells are rare, and, therefore, the isolation of sufficient numbers of these cells often requires the availability of large numbers of cells as starting material. Processing of large cell numbers is not efficiently performed by flow cytometers. Therefore, density gradient centrifugation (Visser et al., 1984), elutriation (Jones et al., 1990) and immunotechnological methods such as panning and immuno-magnetic bead separation (Bertoncello et al., 1991; Spangrude et al., 1988) are often employed to bring down the size of the sample before cell sorting by flow cytometry. The methods have to be tested and optimized in order to prevent the loss of stem cells.

Two methods are to be distinguished when using the flow cytometer to further enrich for stem cells: negative selection and positive selection. Negative selection makes use of antibodies and labels that are known to bind mature haemopoietic cells and not stem cells. In the case of positive selection, the stem cells are labeled and no or only few other cells are labeled too. Figures 1 and 2 present the combinations of labels that are employed in the procedures to purify stem cells by Weissman and collaborators (Müller-Sieburg et al., 1986; Spangrude et al., 1988; Spangrude et al., 1991; Tsukamoto et al., 1991). The antibody Sca-1 was introduced by these investigators for positive selection of murine stem cells. It was later described to be directed against a member of the Ly-6 antigen family (Van de Rijn et al., 1989) that does not occur on stem cells exclusively. For positive selection of human stem cells the CD34 antigen (Civin et al., 1984) is used by most investigators.Terstappen et al. (1991) carefully studied many antibodies in double-labeling experiments with CD34 and found a combination that is relatively simple: CD34 for positive and simultaneously CD38 for negative selection. This finding facilitates further unraveling of the stem cell compartment with additional antibodies and stains.

Figure 3 illustrates our present procedure for sorting murine bone marrow stem cells (Visser et al., 1991; De Vries et al., 1991; Miggliaccio et al., 1991). This procedure has been developed in the course of a series of experiments to analyse the stem cell by flow cytometry that included a large variety of stains, lectins and antibodies, as well as light scatter intensities (for a review see: Visser and Van Bekkum, 1990). Wheat germ agglutinin (WGA) serves as positive marker. It has the advantage that it can be removed from the cells

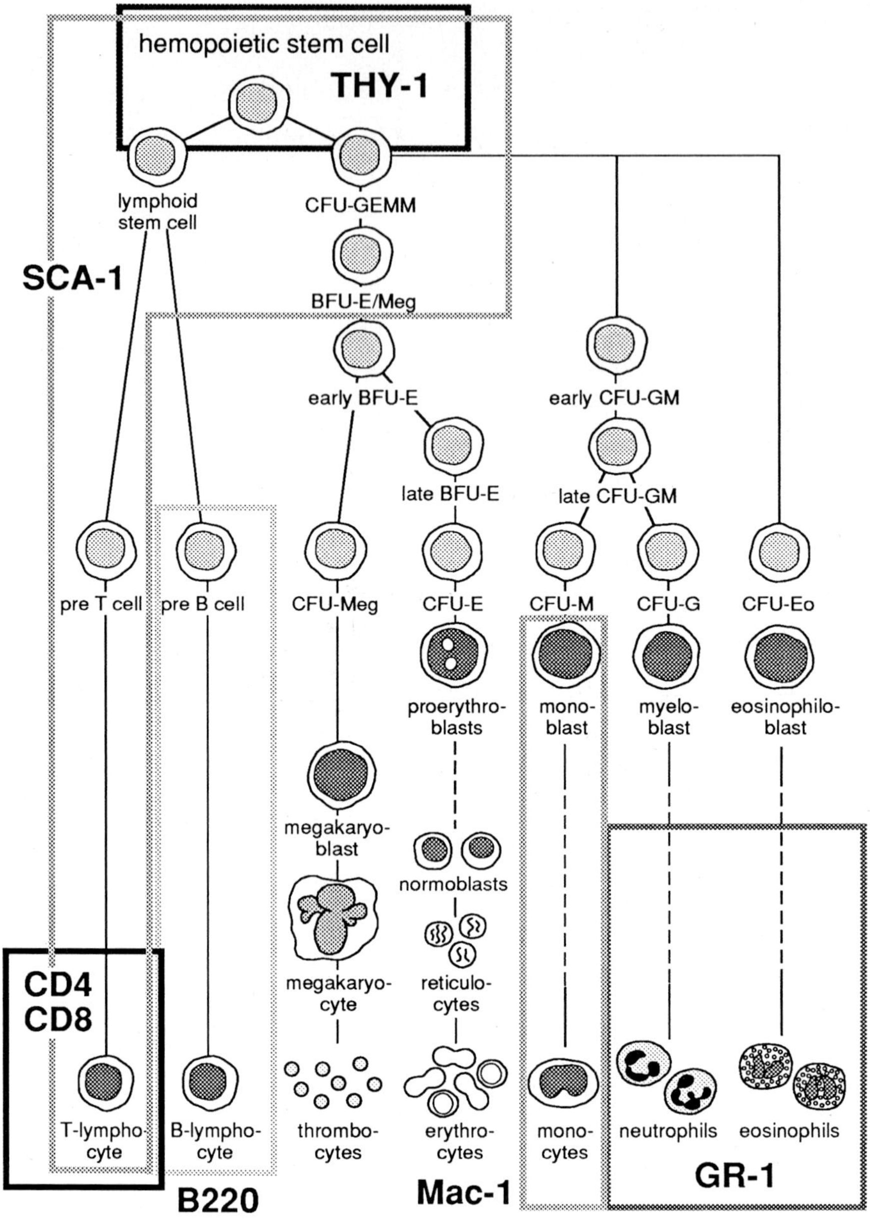

Figure 1.
Schematic representation of the procedure to isolate stem cells from mouse bone marrow as described by Spangrude et al. (1988). Negative selection was performed using immunomagnetic beads as well as flow cytometry.

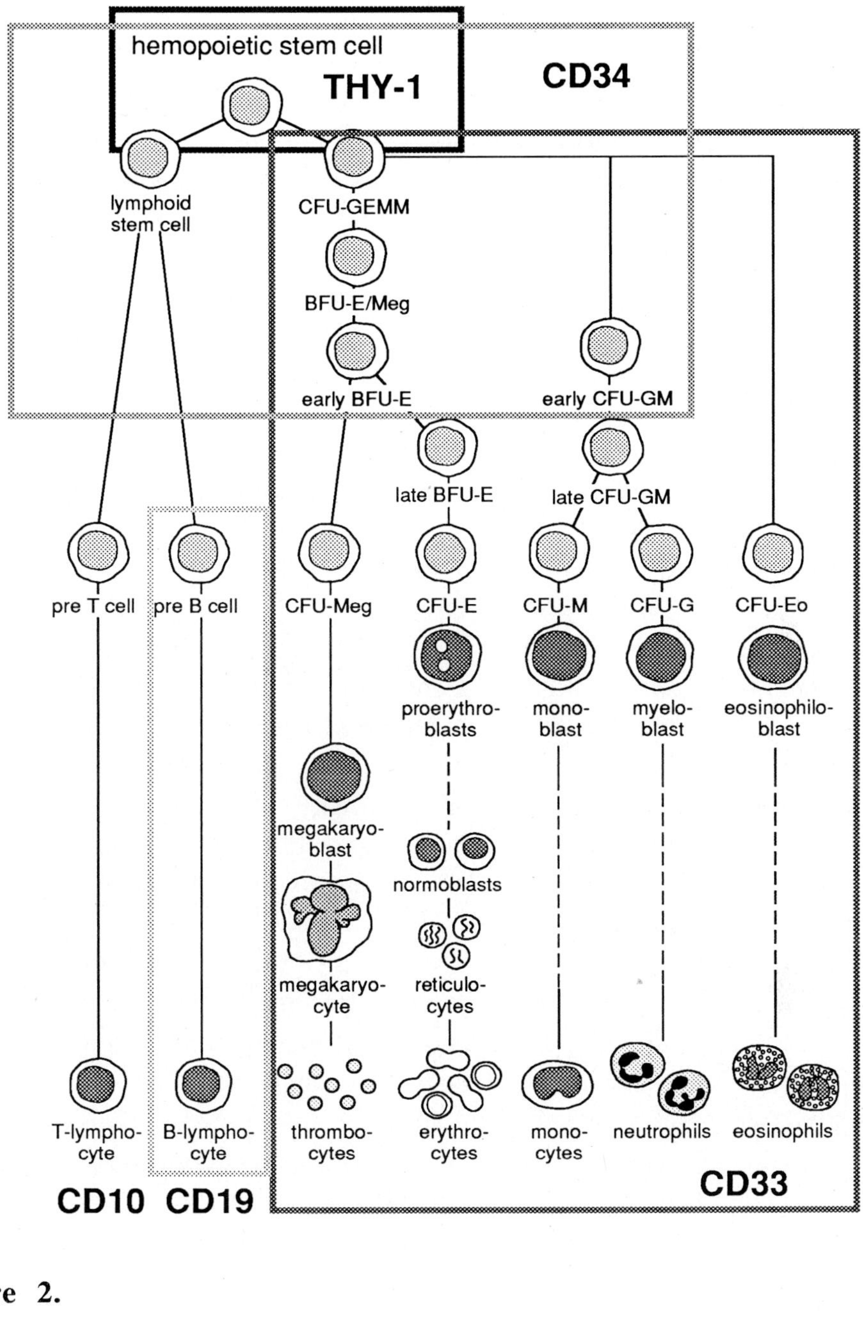

Figure 2.

Schematic representation of the procedure to isolate stem cells from human bone marrow as patented by Tsukamoto et al. (1991). A number of other antibodies are recommended in the description to be used in combination with the above: Class II HLA markers, anti CD3,-8, -15, -20, and finally also Rhodamine 123.

after sorting using the competing sugar. Negative selection is performed using the antibody 15-1.1 which does not bind to pluripotent stem cells in long-term repopulation assays *in vivo*. Rhodamine 123 is used as a supravital stain in the last step. This dye can be used to discriminate between cell types in the earliest haemopoietic compartment. It was found to make a distinction between primitive haemopoietic cells (Bertoncello et al., 1985; Mulder and Visser, 1987; Ploemacher et al., 1988; Visser and De Vries, 1988) with active and with resting mitochondria (Visser et al., 1990). The heterogeneity of Rh123 labeling of bone marrow cells is due to expression of the MDR-1 gene in some cell types, including the stem cells (Chaudhary and Roninson, 1991). Recently, antibodies have become available against the extracellular part of the MDR-1 gene product, the efflux pump P-gp. This facilitates the supravital sorting of early cells and further culture and analysis of the first steps in the regulation of haemopoiesis.
Figure 4 gives an overview of the negative and positive selection labels used by Lemischka and coworkers to enrich for stem cells from mouse fetal liver (Jordan et al., 1990; Mathews et al., 1991). The cocktail of negative selection markers used by these investigators reduces the long-term repopulation by the sorted cells. Similarly, probably one or more of the negative markers described by Gabbianelli et al. (1990) binds to the pluripotent stem cells so that these investigators using a cocktail with many antibodies obtain a suspension highly enriched for BFU-E. These results demonstrate that it is important to test each of the antibodies in a negative selection cocktail for its binding of the pluripotent stem cells.

A new antibody, Stro-1, was recently described by Simmons and Torok-Storb (1991). It binds to the stromal stem cells in the human bone marrow. It is important because these cells are also $CD34^+$, and therefore their presence may strongly affect experiments analysing the regulation of the differentiation of sorted $CD34^+$ cells. Since the antibody also binds to CD34-negative erythroid precursors and not to haemopoietic stem cells, as far as can be concluded from the present stem cell assays, it is a very useful new antibody for negative selection of pluripotent stem cells.

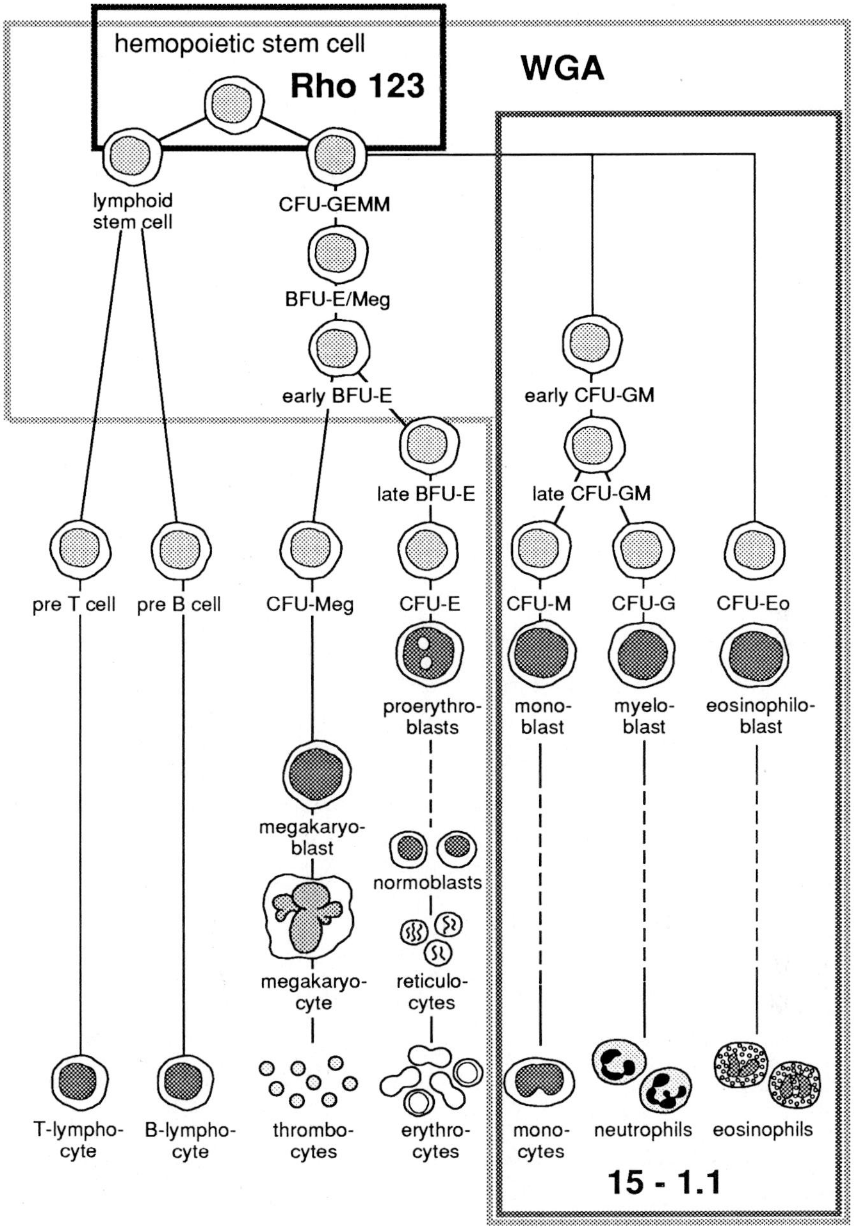

Figure 3.
Schematic representation of the procedure to isolate stem cells from mouse bone marrow as described by Visser et al. (Visser and De Vries, 1990; Visser et al., 1991). A density gradient separation preceeds this protocol. The low density cells (10% of total mouse bone marrow) are processed by the flow cytometer.

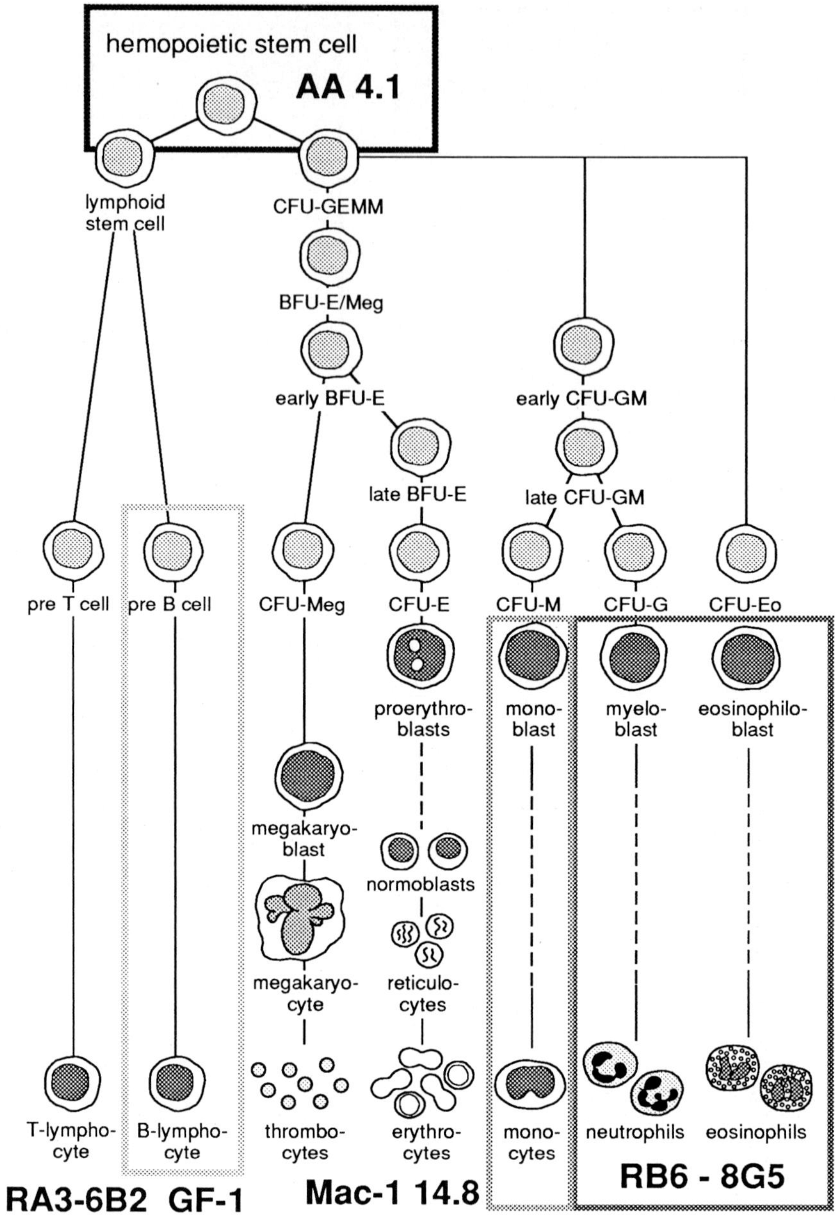

Figure 4.
Schematic representation of the procedure to isolate stem cells from mouse fetal liver as described by Jordan et al. (1990).

CULTURE OF SORTED STEM CELLS

The pluripotent stem cell could not be cultured efficiently in serum free media or serum containing media supplemented with single growth factors or combinations of these (Visser et al., 1990). A plating efficiency of 15% was obtained maximally with IL-1+IL-3+CSF-1. The Rh123 bright cells showed a plating efficiency of up to 50% with IL-1+IL-3+CSF-1 whereas also single growth factors already gave good efficiencies (20%). The low plating efficiency of the pluripotent stem cells (the Rh123 dull fraction) indicates that the right growth factor was not present amongst the ones we tested, or that the stem cells require other signals for optimal growth. Such signals may be derived from adherence molecules or bound growth factors. Therefore, the sorted cells were also cultured on stromal cell layers. This yielded a higher plating efficiency. It could be estimated that 30% of the RH123 dull and 50% of the Rh123 bright sorted cells gave so called "cobblestone areas" in stromal cultures. The sorted PHSC fraction showed long term production of primitive haemopoietic cells on these layers indicating that the stem cells replicated. The cobblestone area like colonies of haemopoietic cells derived from Rh123 bright sorted cells disappeared within three weeks (Visser et al., 1991). It may be speculated that these activated cells are highly responsive to haemopoietic growth factors and that they therefore rapidly differentiate both *in vivo* (providing short term repopulation) and *in vitro*, whereas the quiescent Rh123 dull cells express less growth factor receptors and therefore have more time to find the sites where a combination of adherence and bound growth regulating molecules favors self renewal.

DETECTION OF HAEMOPOIETIC CHIMAERISM

Long term haemopoietic repopulation can be determined quantitatively in female recipients of male bone marrow cells using fluorescent in situ hybridization with a mouse Y-chromosome specific DNA probe (Visser et al., 1990). Similar procedures were developed earlier to detect chimaerism after bone marrow transplantation in leukaemia patients (Van Dekken et al., 1989). A Rhodamine 123 (Rh123) bright fraction could be sorted which produced

many spleen colonies (CFU-S) in lethally irradiated recipients (up to 10 per 100 transplanted cells) but no long term repopulation of haemopoiesis as measured with the Y-probe. Another fraction, Rh123 dull, yielded considerably less spleen colonies, whereas it gave rise to nearly 100% male blood cells in female recipients for one year and longer (Visser et al., 1991). Spleen, lymph nodes and bone marrow showed a similar pattern of chimaerism one year after transplantation of these two fractions. This indicates that the chimaerism occurs in all differentiation lineages: erythroid, myeloid and lymphoid. Consequently, the CFU-S assay cannot be used reliably to detect pluripotent haemopoietic stem cells and our knowledge about stem cells as it is based on nearly thirty years of CFU-S determinations has to be carefully reevaluated. As a consequence even the frequency or incidence of the pluripotent stem cell is not known and the purity of the sorted preparations can only be roughly estimated.
Interestingly, another test system which is in use to detect stem cells, *viz.* the radioprotection or 30-day survival assay, is also shown to be misleading in this experiment. Mice which received the Rh123 bright sorted primitive cells lived for more than one year without any sign of chimaerism. Apparently, these sorted primitive cells produced sufficient offspring (together with spleen colonies) for the short time period that the recipients were at risk due to radiation induced anemia, whereas endogenous stem cells recovered and subsequently reconstituted haemopoiesis. About three times more of the Rh123 bright sorted cells were needed for 30-day survival than of the Rh123 dull sorted cells. A clinical implication of this finding is that apparently cells can be isolated from the bone marrow which provide radioprotection without giving long term repopulation and graft-versus-host disease. It has recently become evident that the pancytopenic period in patients after high-dose chemotherapy and radiation can be significantly shortened by transfusion of autologous peripheral blood obtained earlier and after growth factor treatment (Siena et al., 1989; Siena et al., 1991). This rapid reconstitution of the patient's mature haemopoietic compartments could be obtained with and without simultaneous bone marrow transplantation and it such significantly reduced the time to treat patients with solid tumors. The precise composition of the early haemopoietic compartment in primed blood has not been established. However, it is most likely that the rapid restoration is not caused by PHSC and that other cells than PHSC are responsible for radioprotection and the benificial fast reconstitution.

cDNA LIBRARIES OF SORTED STEM CELLS.

Since there is now no clonal assay for stem cells we decided to change our strategy with respect to the purification and characterization of the pluripotent haemopoietic stem cell.

At present *in vivo* reconstitution of haemopoiesis in experimental animals is the only valid detection assay for these cells. Therefore, it is our strategy to base the analysis of the phenotype of stem cells on description of the regulation of this cell at the molecular level. We study the phenotype in mice so that the presence of the pluripotent stem cell can be verified by transplantation. Then we test whether our findings can be extrapolated to monkey and we determine the occurrence of cells with the specific phenotype in human bone marrow and cord blood suspensions.
We develop and apply procedures to isolate fractions of cells from mouse bone marrow using biotinylated growth factors, monoclonal antibodies against antigens on mature haemopoietic and immunological cells, and adhesion molecules. The phenotype of the cells is carefully checked after the isolation to prevent and quantitate any degree of contamination. Part of these cells are further characterized by culture and by injection in lethally irradiated mice to control their capacity for long-term repopulation of haemopoiesis. The other cells of this well-characterized suspension are then analyzed by PCR amplification and cloning of their cDNA as described by Belyavsky et al. (1989) to identify the expressed genes and molecules that are unique for pluripotent stem cells. Using this method a number of cdc-2 related genes were recently discovered in this way (Ershler et al., 1992).

REFERENCES

Baines P, Visser JWM (1983) Analysis and separation of mouse bone marrow stem cells by H33342 fluorescence-activated cell sorting. Exp Hematol 11: 701-708

Belyavsky AV, Vinogradova T, Rajewsky K (1989) PCR-based cDNA library construction: General cDNA libraries at the level of a few cells. Nucl Acid Res 17: 2919-2932

Bertoncello I, Hodgson GS, Bradley TR (1985) Multiparameter analysis of transplantable hemopoietic stem cells: The separaton and enrichment of stem cells homing to marrow and spleen on the basis of rhodamine 123 fluorescence. Exp Hematol 13: 999-1006

Bertoncello I, Bradley TR, Watt SM (1991) An improved negative immunomagnetic selection strategy for the purification of primitive hemopoietic cells from normal bone marrow. Exp Hematol 19: 95-100

Chaudhary PM, Roninson IB (1991) Expression and activity of P-glycoprotein, a multidrug efflux pump, in human hematopoietic stem cells. Cell 66: 85-94

Civin CI, Strauss LC, Brovall C, Fackler MJ, Schwartz JF, Shaper JH (1984) Antigenic analysis of hematopoiesis. III A hematopoietic progenitor cell surface antigen defined by a monoclonal antibody raised against KG-1a cells. J Immunol 133: 157-165

De Vries P, Brasel KA, Eisenman JR, Alpert AR, Williams DE (1991) The effect of recombinant mast cell growth factor on purified murine hematopoietic stem cells. J Exp Med 173: 1205-1212

Ershler MA, Nagorskaya TV, Visser JWM, Belyavsky AV (1992) Identification of novel cdc2-related protein kinase genes expressed in murine stem cells. Dokl Akad Nauk 324: 893-897

Gabbianelli M, Sargiacomo M, Pelosi E, Testa U, Isacchi G, Peschle C (1990) "Pure" human hematopoietic progenitors: permissive action of basi fibroblast growth factor. Science 249: 1561-1564

Jones RJ, Wagner JE, Celano P, Zicha MS, Sharkis SJ (1990) Separation of pluripotent haemopoietic stem cells from spleen colony-forming cells. Nature 347: 188-189

Jordan CT, McKearn JP, Lemischka IR (1990) Cellular and developmental properties of fetal hemopoietic stem cells. Cell 61: 953-963

Matthews W, Jordan CT, Wiegand GW, Pardoll D, Lemischka IR (1991) A receptor tyrosine kinase specific to hematopoietic stem and progenitor cell-enriched populations. Cell 65: 1143-1152

Migliaccio G, Migliaccio AR, Valinsky J, Langley K, Zsebo K, Visser JWM Adamson JW (1991) Stem cell factor (SCF) induces proliferation and differentiation of highly enriched murine hematopoietic cells. Proc Natl Acad Sci US.A 88: 7420-7424

Mulder AH, Visser JWM (1987) Separation and functional analysis of bone marrow cells separated by rhodamine-123 fluorescence. Exp Hemat 15: 99-104

Mulder AH, Van den Engh, GJ, Visser JWM (1985) Thymus regeneration by bone marrow cell suspensions differing in the potential to form early and late spleen colonies. Exp Hematol 13: 768-775

Müller-Sieburg CE, Whitlock CA, Weissman IL (1986) Isolaton of two early B lymphocyte progenitors from mouse marrow: A committed pre-preB

cell and a clonogenic Thy-1^{lo} hematopoietic stem cell. Cell 44: 653-659
Müller-Sieburg CE, Townsend K, Weissman IL, Rennick D (1988) Proliferation and differentiation of highly enriched mouse hematopoietic stem cells and progenitor cells in response to defined growth factors. J Exp Med 167:1825-1840
Neben S, Redfearn WJ, Parra M, Brecher G, Pallavicini MG (1991) Short- and long-term repopulation of lethally irradiated mice by bone marrow stem cells enriched on the basis of light scatter and hoechst 33342 fluorescence. Exp Hematol 19: 958-967
Okada S, Nakauchi H, Nagayoshi K, Nishikawa S, Nishikawa S-I, Miura Y, Suda T(1991) Enrichment and characterization of murine hematopoietic stem cells that express c-kit molecule. Blood 78:1706-1712
Pallavicini MG, Summers LJ, Dean PN, Gray JW (1985) Enrichment of hemopoietic clonogenic cells by multivariate analysis and sorting. Exp Hematol 13:1173-1181
Ploemacher RE, Brons NHC (1988) Cells with marrow and spleen repopulating ability and froming spleen colonies on day 16, 12 and 8 are sequentially ordered on the basis of increasing rhodamine 123 retention. J Cell Physiol 136: 531-536
Ploemacher RE, Brons RHC (1989) Separation of CFU-S from primitive cells responsible for reconstitution of the bone marrow haemopoietic stem cell compartment following irradiation: Evidence for a pre-CFU-S cell. Exp Hematol 17: 263-266
Siena S, Bregni M, Brando M, Ravagnani F, Bonadonna G, Gianni AM (1989) Circulation of CD34+ hematopoietic cells in the peripheral blood of high-dose cyclophosphamide-treated patients: Enhancement by intravenous recombinant human granulocyte-macrophage colony-stimulating factor. Blood 74:1905-1914
Siena S, Bregni M, Brando B, Belli N, Ravagnani F, Gandola L, Stern C, Lansdorp PM, Bonadonna G, Gianni AM (1991) Flow cytometry for clinical estimation of circulating hematopoietic progenitors for autologous transplantation in cancer patients. Blood 77: 400-409
Simmons PJ, Torok-Storb B (1991) Identification of stromal cell precursors in human bone marrow by a novel monoclonal antibody, STRO-1. Blood 78: 55-62
Spangrude GJ, Johnson GR (1990) Resting and activated subsets of mouse multipotent hematopoietic stem cells. Proc Natl Acad Sci USA 87: 7433-7437
Spangrude GJ, Heimfeld S, Weissman IL (1988) Purification and characteristics of mouse hematopoietic stem cells. Science 241: 58-62
Spangrude GJ, Smith L, Uchida N, Ikuta K, Hreimfeld S, Friedman J, Weissman IL (1991) Mouse hematopoietic stem cells. Blood 78: 1395-1402
Szilvassy SJ, Lansdorp PM, Humphries RK, Eaves AC, Eaves CJ (1989) Isolation in a single step of a highly enriched murine hematopoietic stem

cell population with competitive long-term repopulating ability. Blood 74: 930-939

Terstappen LWMM, Huang S, Safford M, Lansdorp PM, Loken MR (1991) Sequential generations of hematopoietic colonies derived from single nonlineage-committed $CD34^+CD38^+$ progenitor cells. Blood 77: 1218-1227

Tsukamoto A, Baum M, Alhara Y, Weissman I (1991) Human hemopoietic stem cell. US patent 5,061,620

Van Dekken H, Hagenbeek A, Bauman JGJ (1989) Detection of host cells following sex-mismatched bone marrow transplantation by fluorescent in situ hybridization with a Y-chromosome specific probe. Leukemia 3: 724-728

Van de Rijn M, Heimfeld S, Spangrude GJ, Weissman IL (1989) Mouse hematopoietic stem-cell antigen Sca-1 is a member of the Ly-6 antigen family. Proc Natl Acad Sci USA 86: 4634-4638

Verfaillie C, Blakolmer K, McGlave P (1990) Purified primitive human hematopoietic progenitor cells with long-term in vitro repopulating capacity adhere selectively to irradiated bone marrow stroma. J Exp Med 172: 509-520

Visser JWM, De Vries P (1988) Isolation of spleen-colony forming cells (CFU-s) using wheat germ agglutinin and rhodamine 123 labeling. Blood Cells 14: 369-384

Visser JWM, Van Bekkum DW (1990) Purification of pluripotent hemopoietic stem cells: past and present. Exp Hemat 18: 248-256

Visser JWM, De Vries P (1990) Identification and purification of murine hematopoietic stem cells by flow cytometry. In: Methods in Cell Biology, Vol. 33, Flow Cytometry. (Darzynkiewicz, Z. and Crissman, H.A., eds.) pp 451-468 Academic Press Inc San Diego

Visser JWM, Bauman JGJ, Mulder AH, Eliason JF, de Leeuw AM (1984) Isolation of murine pluripotent hemopoieitic stem cells. J Exp Med 59: 1576-1590

Visser JWM, Hogeweg-Platenburg MGC, De Vries P, Bayer J, Ploemacher RE(1990) Culture of purified pluripotent haemopoietic stem cells. In: The Biology of Hematopoiesis (Dainiak N, Cronkite EP, McCaffrey R, Shadduck R.K, eds) pp. 1-8 Wiley-Liss Inc New York

Visser JWM, De Vries P, Hogeweg-Platenburg MGC, Bayer J, Schoeters G, Van Den Heuvel R, Mulder AH (1991) Culture of hemopoietic stem cells purified from murine bone marrow. Seminars in Hematology 28: 117-125

T LYMPHOCYTES RECOGNITION MOLECULES IN HOMING : A FLOW CYTOMETRY STUDY OF LECTIN-GLYCOCONJUGATES INTERACTIONS.

Claudine Kieda, Nadine Bizouarne, Véronique Denis and Michèle Mitterrand

Département de Biochimie des Glycoconjugués et Lectines Endogènes
Centre de Biophysique Moléculaire du CNRS
1, rue Haute, 45071, Orléans CEDEX 2

Introduction

Intercellular recognitions are fundamental processes which allow tissue organization and organ development as well as the exchange of informations among cells. This happens through the interaction of specific molecules which are located on the surface of one cell and can recognize other specific molecules present on the surface of another cell or in the extracellular matrix.

High interest is devoted to the identification of the molecular mechanisms which permit those specific interactions to occur. This is especiallly important in the case of the immune system which can indeed function only if many such cell-cell recognitions may be achieved properly.

Most of the work so farpublished concerns the detection and the isolation of adhesion molecules using monoclonal antibodies which happen to react with the appropriate cell surface antigen.

Studying the cell surface glycoconjugates as well as their biological role, we have hypothesized that in the above processes, glycoconjugates and their specific receptors are involved to assign specific recognitions (Kieda and Monsigny, 1986). The bases for such an hypothesis are the presence of lectins in lymphocytes (Kieda *et al.*, 1978) and their distribution on subpopulations (Kieda *et al.*, 1979) ; furthermore, lectins are highly modulable molecules in their surface expression during a stimulation by activating molecules or upon cell-cell cooperation (Grillon *et al*, 1990, 1991).

This paper will present a short summary of some of the peculiar properties of cells in the immune system and of the molecules which take part in lymphocytes and endothelial cell recognitions. Special emphasis will be made on what is known about the part played by the lectin/glycoconjugate interaction in the immune system. Attention will be devoted to both the circulating lymphoïd cells and the receiving endothelial cells and to the contribution brought by flow cytometry methods to the cell identification and to molecules responsible for the adhesion properties.

NATO ASI Series, Vol. H 67
Flow Cytometry
Edited by A. Jacquemin-Sablon

Lymphocyte Homing

One of the peculiarities of the immune system is the fact that lymphocytes travel throughout the whole body. They circulate in the blood and lymph as non adherent, freely moving cells. This provides the opportunity to all immunological responders to be able to meet foreign antigens. But, because of the great variety of antigens, only a limited number of lymphocytes can specifically respond and give reactive clones. To increase the efficacity of the response, competent cells have to meet at points where antigens are drained : namely the secondary lymphoïd organs such as spleen, peripheral lymph nodes and gut associated lymphoïd organs (see Butcher, 1986, for a review).

Lymphocytes actively enter and are selected to enter into secondary lymphoïd organs. Indeed, among all the blood borne cells, only lymphocytes enter given lymph nodes ; furthermore a selection also occurs among lymphocytes subpopulations at the level of secondary lymphoïd organs. Such an organization is necessary to make all immunological responders available and to drain antigens which appear anywhere in the organism. Indeed the various types of secondary lymphoïd organs drain different portals of the organism. Lymphocytes, on the other hand, enter a lymphoïd organ *via* the blood, they either undergo an immune response or leave the organ by efferent lymphatics. They are further collected in the lymphatics and return to the main ducts which carry them to return to the blood. This is a very active process which provides lymphoïd organs with high number of cells. The turn over is 30 minutes and 5×10^{12} cells are turning in the blood of a human being.

The high specificity of the mechanism by which lymphocyte enter the secondary lymphoïd organs is a key point to explain the homing pattern. Gowans and Knight (1964) and Marchesi and Gowans (1964) have described the site of entry and the special type of endothelium lining the post capillary venules : high endothelial veinules, where the key recognition takes place to allow proper lymphocytes to bind, become adherent onto the endothelial cells and extravasate between them towards the lymphoïd tissue. The first step during which lymphocytes bind to endothelium is characterized by very deep changes in the lymphocyte properties because, from a freely circulating cell, a lymphocyte has to change to become an adherent migrating cell. This adhesion can be studied *in vitro* (Stamper and Woodruff, 1977). The pattern and the selectivity of the migration could be further elucidated by Weissman and Butcher using mainly the *in vitro* frozen section assay and raising monoclonal antibodies which blocks homing. This led to the first identification of molecules : homing receptor, like the MEL14 antigen which governs the recognition between lymphocytes and endothelial cells of peripheral lymphoïd organs (see the review by Butcher, 1986). This led also to the idea of corresponding molecules on the receiving endothelium : the addressins (Berg *et al.*, 1989).

Molecules involved in Homing

Adhesion molecules are largely described and classified into three main families (see Springer, 1990, for a review) : the immunoglobulin superfamily, the integrin superfamily and the new selectins family. Selectins, with their three members described up to now, have the special feature of being lectin-like molecules. They are present either on endothelial cells or lymphocytes and their is modulated. They have mainly been studied with monoclonal antibodies but also with our approach which follows the activity of lectins directly using **neoglycoproteins** as ligands (Monsigny *et al.*, 1983).

The selectin family of molecules have sequence homologies between each other and their sequences present homologies with C-lectins (Drickamer, 1988), with epidermal growth factor and with complement regulatory proteins repeats (Streeter *et al.* 1988, see Springer 1990).

The lectin activity of these molecules has been reviewed by Brandley (1991).

To find out what their role may be in homing, we took advantage of our knowledge about lectin-like molecules in lymphocytes. As we had previously shown that lectins are present in thymus and spleen cells extracts (Kieda *et al.*, 1978) and that they are distributed on cell subpopulations by the use of fluorescent neoglycoproteins (Kieda *et al.*, 1979). The use of neoglycoproteins was not only for cytological purposes, they were also tools for biological assays (Monsigny *et al.*, 1983). Neoglycoproteins were prepared as described previously (Monsigny *et al.*, 1983) and derivatized to provide fluoresceinylated molecules, allowing to detect the presence of membrane associated sugar binding molecules : membrane lectins.

During the homing process, the possible role of membrane lectins was tested in both the *in vitro* and the *in vivo* assays. It appears that, by the frozen sections adhesion assay (Stamper and Woodruff, 1977), inhibitions were obtained when neoglycoproteins bearing galactoside residues were used. The *in vivo* assay leads to complex homing patterns which could be strongly modified by coinjecting neoglycoproteins with the cells.

In order to find out what is the role of membrane lectins of lymphocytes in their homing pattern, the separation of lymphocytes on the basis of the sugar receptors they have on their surface was undertaken (Grillon *et al.* 1991) and the isolated subpopulations were fluoresceinylated and injected. As an example, peripheral lymph nodes cells were separated into Fuc_{lectin}^{+} T cells (25 $\pm$ 3 %) and Lac_{lectin}^{+} T cells (60 $\pm$ 5 %). During a short term *in vivo* homing assay, T lymphocytes having receptors for galactosides (Lac_{lectin}^{+} T cells) are mainly recovered in peripheral lymph nodes, while cells with receptors for fucosides (Fuc_{lectin}^{+} T cells) are found preferentially in Peyer's patches.

The presence of lectins on the surface of lymphoïd cells may be a sufficient characteristic to assign specific homing. Inhibitions of the homing into a given organ can be obtained by coinjection of the cells with an appropriate neoglycoprotein.

Membrane lectins are not only on lymphoïd cells ; they are also present on **endothelial cells** and play an active role in stopping lymphocytes and thus participate to the homing process. Endothelial cells bind neoglycoproteins ; when immobilized neoglycoproteins onto fluorescent beads where injected to mice, 16 h later, they selectively localize into lymphoïd organs, indicating a role for endothelial cell surface lectins.

Such data show the necessity to obtain specialized endothelial cells from high endothelial veinules in culture, then to establish an immortalized cell line.

Using these tools and lymphoïd cells it is possible to study the adhesion properties of cells in an *in vitro* assay and to identify by flow cytometry the adherent cells as well as to study the kinetics and the extent of the interaction using PKH 26-GL as a dye (Horan and Slezak, 1989), which is a stable liposoluble molecule, the exchangeability of which can be a reporter of the interaction.

Adhesion assay : Identification of cells by their membrane lectins.

Intercellular interactions can be followed *in vitro* by measurement of adherent cells provided they have been radioactively labelled or substituted with fluorescent dyes.

Flow cytometry allows the identification of the cell types and the quantification of the interaction.

In the following experiments, lymphoïd cells are murine lymphoma cells : EL4 cell line which is "homing deficient" ; these cells do not adhere on endothelial cells and, the derived cell line : EL4IL2, which has receptors for IL2 and has efficient adhesion capacities towards endothelial cells, while EL4 do not adhere onto endothelial cells. Endothelial cells are of the high endothelial cell type from peripheral lymph node capillaries, they have been immortalized by a plasmid from SV40; they are designated as SVHEC.

Here we give an example of an adhesion assay of lymphoïd cells to endothelial cells from murine peripheral lymph nodes and **flow cytometry techniques** are used to identify both adhering and recipient cells (figure 1).

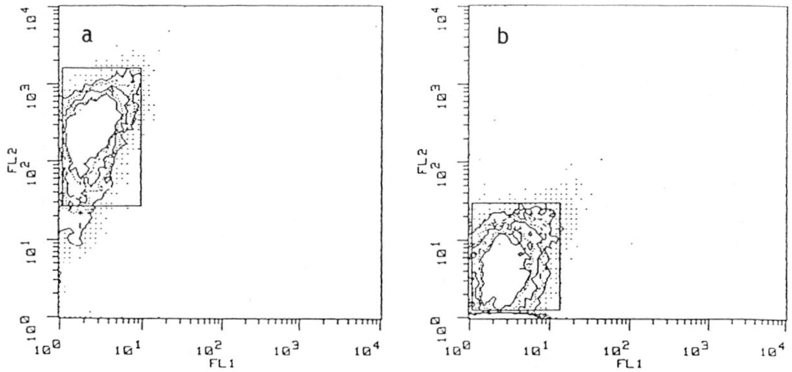

Figure 1 : Identification of the cells during an adhesion assay.
a) Lymphoïd cells are labelled with PKH 26-GL, before the adhesion step.
b) SVHEC were grown as monolayers and detached mechanically.
Fl1 : fluorescence emitted at 520 nm, Fl2 : fluorescence emitted at 575 nm.

Endothelial cells were identified on the basis of their specific labelling with rabbit anti-endothelial cell antibodies and anti-rabbit antibodies coupled to Phycoerythrin and lymphoid cells with rabbit anti EL4IL2 cell antibodies or anti CD's antibodies.

Fluorescent NEOGLYCOPROTEINS are used to study membrane lectins :

FITC-labelled neoglycoproteins (Monsigny *et al.*,1984) are utilized to label the **endothelial** cells membrane lectins (see figure 2b).

When isolated lymphoïd cells were labelled by PKH 26-GL, they are identified by their fluorescence at 575 nm as Fl2 (figure 1a, 2a and 3).

Lymphoïd cells do adhere to endothelial cells by a mechanism which involves sugar / protein **membrane lectins** interactions as an early step of the adhesion. The molecules which are involved can be identified by using antibodies and, neoglycoproteins to characterize the endogeneous lectins. Both cell types can be distinguished by their forward and side scattered lights as well as by their surface antigens and their membrane lectins ; for this purpose, double labellings are often necessary (figure 2a).

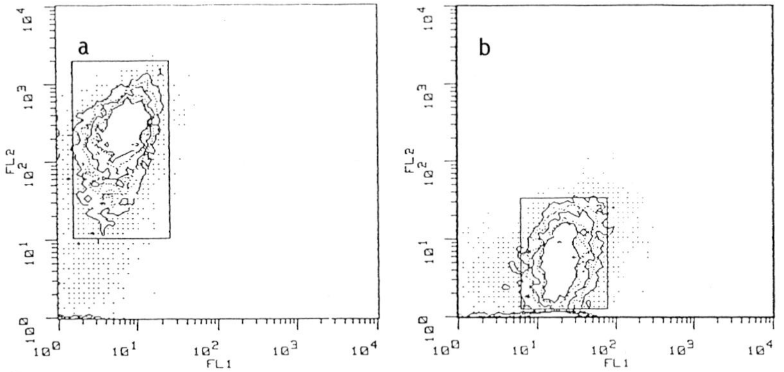

Figure 2 : Adhering cells (EL4IL2) and recipient cells (SVHEC) distinguished by their membrane lectins.
a) El4IL2 are doubly labelled by PKH 26-GL (Fl2) and FTC-Fuc-BSA (Fl1).
b) SVHEC are labelled by FTC-Fuc-BSA.
Fl1 : fluorescence emitted at 520 nm from fluorescein residues
Fl2 : fluorescence emitted at 575 nm from PKH 26-GL dye.

Adhesion experiment :

EL4 and EL4IL2 are labelled by the PKH 26-GL (Horan and Slezak, 1989) as recommended. In the adhesion experiments, endothelial and lymphoïd cells have to be detached by mechanical treatment by a rubber policeman to get all cells.

The cell suspension can then be labelled, either as described above with antibodies revealed by phycoerythrin-labelled second antibodies ; or, endogenous lectins are detected by the use of FTC-neoglycoproteins such as fluoresceinylated fucosylated BSA revealing a fucose-specific lectin on SVHEC (figure 2b, figure 3).

Flow cytometry analyses performed upon excitation at 488 nm, the fluorescence emitted by FITC is collected at 520 nm, fluorescence emitted by phycoerythrin is collected above 600 nm, and by the PKH 26-GL dye at 575 nm.

For time course studies of the adhesion, experiments are performed for a long enough time (figure 3b, 3d), the exchange of the PKH 26-GL dye can be used to follow intercellular interactions. This can be assessed by flow cytometry : by measuring the fluorescence intensity and the side scattered light), after 4 h (figure 3 a, b) and after 16 h incubation (figure 3 c, d). The PKH 26-GL dye is indeed finally exchanged after 16 h, while after 4 h it is still possible to distinguish and separate both cell types on the basis of labelling lymphoïd cells by PKH 26-GL (Fl2, figure 3), and specific labelling of SVHEC by fluorescent fucosylated BSA. During the long term assay (figure 3 c and d) the two cell types have interacted in such a way that they cannot any longer be distinguished or separated on the basis of Fl2 (PKH 26-GL) and Fl1 (FTC-Fuc-BSA) labellings (figure 3 c, d).

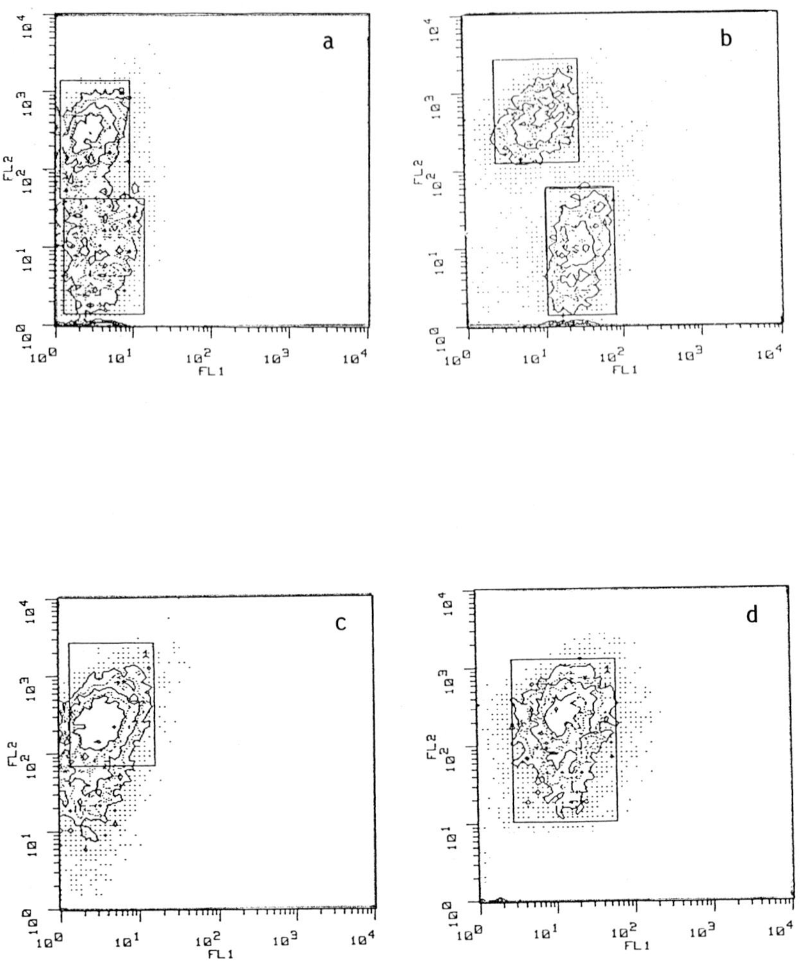

Figure 3 : Time course adhesion assay.
EL4IL2 were labelled wiyh PKH 26-GL before the adhesion, the two cell types are distinguished after adhesion by labelling with fluorescent neoglycoproteins.
(a, b) 4 h adhesion, (c, d) 16 h adhesion : analyses of PKH 26-GL fluorescence - Fl2 (a, c) and of FTC-Fuc-BSA - Fl1 (b, d).
Fl1 : fluorescence emitted at 520 nm from fluorescein residues
Fl2 : fluorescence emitted at 575 nm from PKH 26-GL dye.

Remarks

Comparisons of labellings obtained in both conditions i.e. when both cells are labelled separately and when cells are labelled after the adhesion step, allows the evaluation of the modulation effects due to cell-cell recognitions and cooperation in the expression of antigens and/or membrane lectins which are very sensitive to environmental conditions.

CONCLUSION

The identification of molecules involved in the process of cell/cell recognition is a crucial requierement that can be fulfilled by the help of flow cytometry.

The use of specific tools such as monoclonal antibodies and neoglycoproteins provides a direct means to identify cells and molecules responsible for adhesion. Membrane lectins appear to play a role during early steps of the adhesion process. The design of an *in vitro* cellular model for the adhesion allows an approach to the study of modulation of cell surface molecules expression during the various steps which lead to recognition, adhesion and extravasation. In long term experiments, the dye exchange appears to provide a way to approach cell/cell interaction.

References

BERG, E.L., L.A. GOLDSTEIN, and M.A. JUTILA (1989) Homing receptors and vascular addressins : cell adhesion molecules that direct lymphocyte traffic. Immunol. Rev. 108: 1-18.

BRANDLEY, B.K., S.J. SWIEDLER, and P.W. ROBBINS (1990) Carbohydrate ligands of the LEC cell adhesion molecules. Cell 63: 861-863.

BUTCHER, E.C. (1986) The regulation of lymphocyte traffic. Curr. Topics Microbiol. Immunol. 128: 85-122.

DRICKAMER, K. (1988) Two distinct classes of carbohydrate-recognition domains in animal lectins. J. Biol. Chem. 263: 9557-9560.

GOWANS, J.L., and E.J. KNIGHT (1964) The route for the recirculation of lymphocytes in the rat. Proc. R. Soc. Lond. B. Biol. Sci. 159: 257-282.

GRILLON, C., M. MONSIGNY, and C. KIEDA (1990) Cell surface lectins of human granulocytes : their expression is modulated by mononuclear cells and GM-CSF. Glycobiol. 1: 33-38.

GRILLON, C., M. MONSIGNY, and C. KIEDA (1991) Changes in human T lymphocyte membrane lectin expression upon mitogenic stimulation. Carbohyd. Res. 213: 283-292.

HORAN, P.K., and S.E. SLEZAK (1989) Stable cell membrane labelling. Nature 340: 167-168.

KIEDA, C., D.J. BOWLES, A. RAVID, and N. SHARON (1978) Lectins in lymphocyte membrane. FEBS Letters 94: 391-396.

KIEDA, C., and M. MONSIGNY (1986) Involvement of membrane sugar receptors and membrane glycoconjugates in the adhesion of 3LL cell subpopulations to cultured pulmonary cells. Invasion and Metastasis 6: 347-366.

KIEDA, C., A.C. ROCHE, F. DELMOTTE, and M. MONSIGNY (1979) Lymphocyte membrane lectins. Direct visualisation by the use of fluoresceinyl-glycosylated cytochemical markers. FEBS Letters 99: 329-332.

MARCHESI, V.T., and J.L. GOWANS (1964) The migration of lymphocytes through the endothelium venules in lymph nodes. Proc. R. Soc. Lond. Biol. 159: 282-289.

MONSIGNY, M., C. KIEDA, and A.C. ROCHE (1983) Membrane glycoproteins, glycolipids and membrane lectins as recognition signals in normal and malignant cells. Biol. Cell 47: 95-110.

MONSIGNY, M., A.C. ROCHE, and P. MIDOUX (1984) Uptake of neoglycoproteins via membrane lectin(s) of L1210 cells evidenced by quantitative flow cytofluorometry and drug targeting. Biol. Cell 51: 187-196.

SPRINGER, T.A. (1990) Adhesion receptors of immune system. Nature 346: 425-434.

STAMPER, H.B.J.r., and J.J. WOODRUFF (1997) An in vitro model of lymphocyte homing. 1 Characterization of the interactionbetween thoracic duct lymphocytes and specialized high endothelial venules. J. Immunol. 119: 772-779.

STREETER, P.R., E.L. BERG, B.N. ROUSE, R.F. BARGATZE, and E.C. BUT CHER (1988) A tissue specific endothelial cells molecule involved in lymphocyte homing. Nature 331, 41-46.

THREE- AND FOUR-COLOR IMMUNOFLUORESCENCE ANALYSIS BY FLOW CYTOMETRY

P. CARAYON
Immunology Department
SANOFI RECHERCHE
371, rue du Professeur Joseph-Blayac
34184 MONTPELLIER CEDEX 04 (France)

INTRODUCTION

Multicolor immunofluorescence analysis by flow cytometry has been used extensively to study leukocyte subsets in hematology and immunology. In this chapter, no attempt has been made to review all the work that has been done on the subject, but rather to describe recent technological developments and to discuss specific problems linked to multicolor immunofluorescence analysis.

Over the last 10 years, new fluorochromes having spectral properties well adapted to the excitation sources used in flow cytometry have been purified or synthetized. Increasing interest in flow cytometry for both fundamental and clinical applications have conducted several companies to develop these fluorochromes directly linked to monoclonal antibodies (Mabs). In parallel, flow cytometer manufacturers have improved the performances of machines: by increasing the number of parameters simultaneously measured on a cell and the sensitivity of fluorescence detection; by introducing electronic systems to correct spectral overlaps of the dyes; by developing softwares making multiparametric analysis easier. These improvements mean it is now possible to carry out three-color immunofluorescence analyses using flow cytometers equipped with one argon-ion laser or four-color immunofluorescence analyses using flow cytometers equipped with two ion lasers as excitation sources. All these analyses

NATO ASI Series, Vol. H 67
Flow Cytometry
Edited by A. Jacquemin-Sablon

are relatively easy to perform because of the commercial availability of several fluorochromes linked to different Mabs.

DYES FOR IMMUNOFLUORESCENCE

A fluorochrome must have the following properties to be used for immunofluorescence studies by flow cytometry. It should:

- absorb light at wavelengths available with existing laser-excitation sources;
- have a large Stokes shift, that is, the wavelength difference between the emission maximum and the absorption maximum;
- have a good quantum yield (Q), that is, the energy ratio between emitted and absorbed light;
- be easily linked to Mabs and should maintain its spectroscopic properties after conjugation. Biochemical and spectral properties of dyes used for cell surface immunofluorescence are shown in table I and figure 1.

In 1977, the first dual-fluorescence flow cytometric analysis was described using fluorescein-5-isothiocyanate (FITC) and tetramethylrhodamine-5-isothiocyanate (TRITC) (LOKEN et al, 1977). Before their introduction in flow cytometry, both fluorochromes had widely been employed for dual-immunofluorescence studies using fluorescent microscopes equipped with mercury arc-lamps. Flow cytometers essentially use ion lasers as excitation sources. Argon-ion laser displays high power at the wavelengths of 488 nm and 514 nm. The wavelength of 488 nm is well suited to FITC excitation (Fig. 2) whereas the wavelength of 514 nm is not optimum for TRITC excitation. Simultaneous excitation of both fluorochromes at 514 nm is a bad compromise because it results in a poor excitation of TRITC and a low separation between the excitation and the emission of FITC. Moreover, spectral emissions of the two dyes overlap, creating an additional difficulty when they are used simultaneously for flow cytometric analysis.

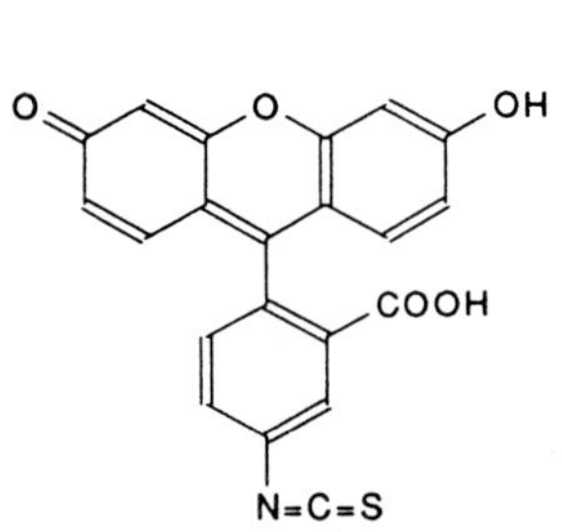

Fluorescein-5-isothiocyanate
(FITC)

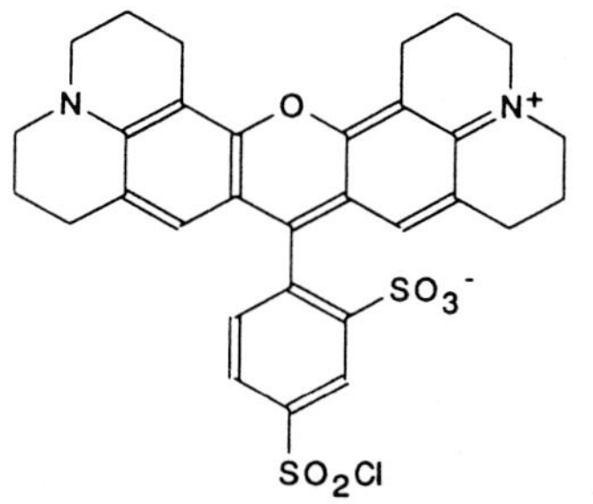

Texas Red
(sulfonyl chloride derivative)

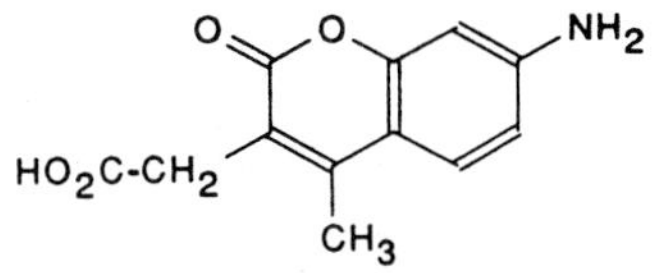

7-amino-4-methylcoumarin-3-acetic acid
(AMCA)

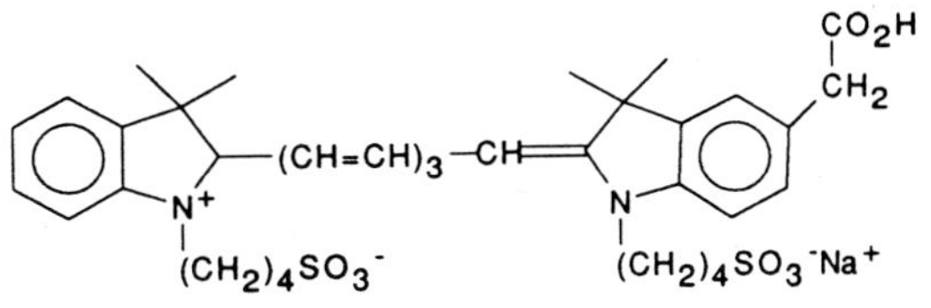

Carboxymethylindotricarbocyanine
(CY5.18)

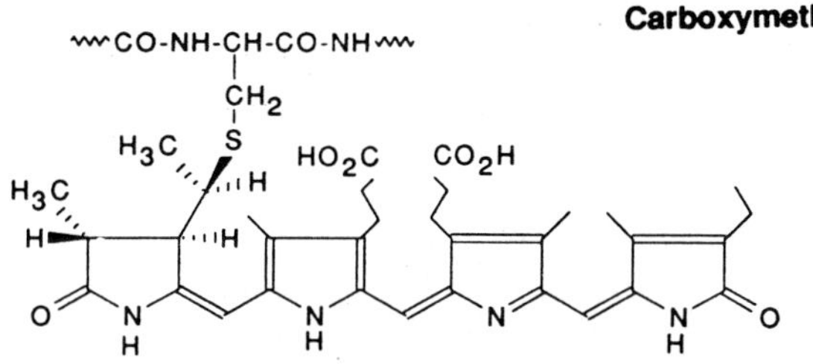

Peptide-linked PHYCOCYANOBILIN
Allophycocyanin
R-Phycocyanin

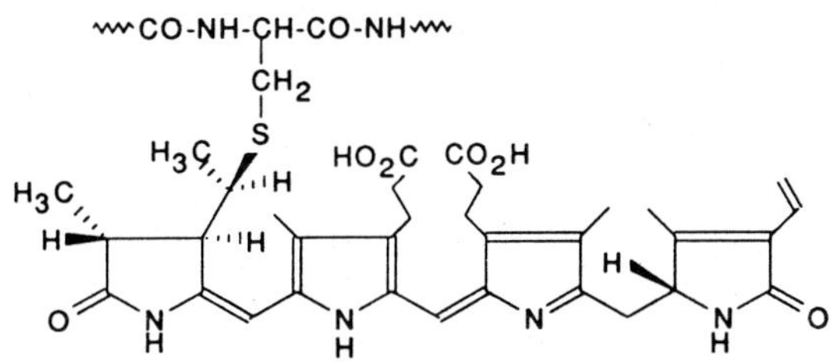

Peptide-linked PHYCOERYTHROBILIN
B-Phycoerythrin
R-Phycoerythrin

Fig. 1: Structure of fluorochromes used for cell surface immunofluorescence

Table I: Spectroscopic properties of fluorochromes used for cell surface immunofluorescence

Fluorochromes	Molecular weights	Absorption λ max (nm)	Absorption $\epsilon (M^{-1} cm^{-1})$	Emission λ max (nm)	Emission Q*	Comments
AMCA	175	351	$1.7x10^{4}$	430	0.8	pH sensitive
Cascade Blue	596	376;399	$2.8x10^{4}$	423	0.5	
BODIPY	~1000	505	$8.0x10^{4}$	512	0.9	
FITC	389	494	$7.2x10^{4}$	520	0.5	pH sensitive
X-RITC	584	578	$8.0x10^{4}$	604	>0.3	
Texas Red	625	596	$8.5x10^{4}$	615	>0.3	
B-Phycoerythrin	240000	545;565	$2.4x10^{6}$	575	>0.9	
R-Phycoerythrin	240000	488;545; 565	$2.0x10^{6}$	578	0.7	
C-Phycocyanin	110000	620	$1.7x10^{6}$	650	0.5	
Allophycocyanin	100000	650	$7.0x10^{5}$	660	0.7	
DuoCHROME		488		615		Energy transfer (Phycoerythrin/ Texas Red)
TRI-COLOR/ CyCHROME		488		667		Energy transfer (Phycoerythrin/ Cy5)
PerCP	32000	438;478		672		
CY5	857	658	$2.2x10^{5}$	685	0.2	

A majority of data in this table are from data sheets of Molecular Probes, Inc. Junction City,OR. DuoCHROME and PerCP data are from data sheets of Becton Dickinson. TRI-COLOR data are from data sheets from CALTAG. CyCHROME data are from data sheets from PHARMINGEN.

* The fluorescence quantum yield (Q) can be decreased when fluorochromes are linked to proteins.

Fig. 2 : Absorption and excitation spectra of fluorochromes employed in flow immunofluorescence analysis

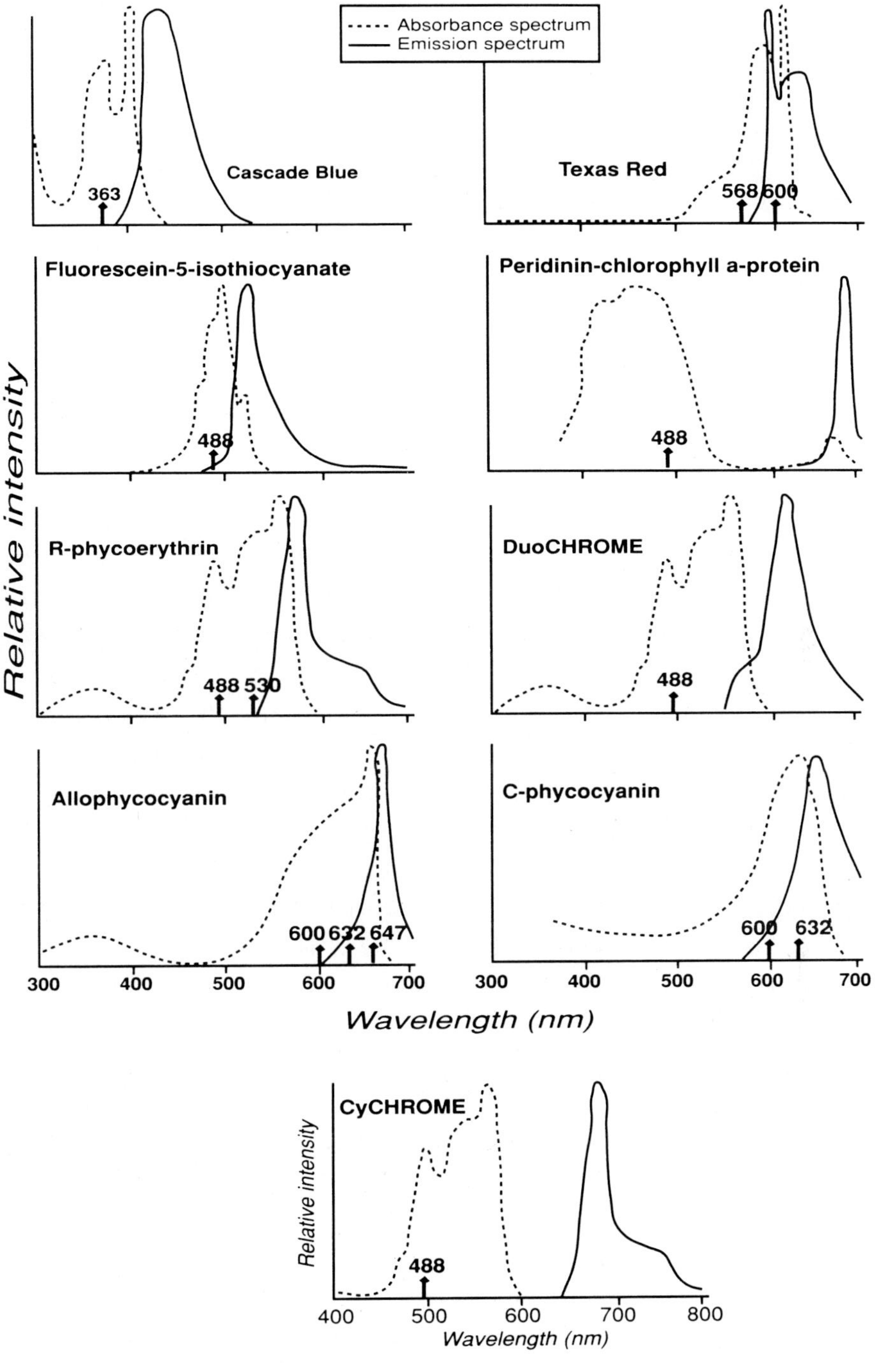

A good alternative for dual-immunofluorescence analysis has been the use of rhodamine-X-isothiocyanate (X-RITC) in conjunction with FITC. FITC was excited by an argon laser at the wavelength of 488 nm and X-RITC by a krypton laser at 568 nm (SEGAL et al, 1981). Although the use of two laser beams as excitation sources is complex, the problem of spectral emission overlap is avoided since the emissions of the two fluorochromes are spatially and electronically separated. Texas Red, a derivative of sulforhodamine 101, has spectroscopic properties close to the X-RITC ones (Fig. 2 and Table I). It is more hydrophilic than X-RITC making its attachment to Mabs more successful (TITUS et al, 1982).

Multicolor analysis was greatly improved following the isolation of a family of natural highly fluorescent chromoproteins called phycobiliproteins. These dyes were purified from cyanobacteria and red algae (GLAZER and STRYER, 1984). One of these fluorochromes, phycoerythrin (PE), can be used in combination with FITC. In fact, FITC and PE absorb argon laser light at the wavelength of 488 nm (Fig. 2), but their emissions are different enough to be separated by optical filters and electronic compensation (OI et al, 1982). Another member of this family, the allophycocyanin (APC), allows three- or four-color immunofluorescence using argon laser at the wavelength of 488 nm for FITC and PE excitations and dye laser at 590 nm for Texas Red and/or APC excitations (HARDY et al, 1983; LANIER and LOKEN, 1984; LOKEN and LANIER, 1984; HARDY et al, 1984; PARKS et al, 1984) (Fig. 2 and table II).

Recently, a new fluorochrome called DuoCHROME has been introduced (STRYER and GLAZER, 1986). It is a complex of two fluorochromes PE and Texas Red which are linked together. Based on the principle of fluorescence energy transfer, argon laser light at the wavelength of 488 nm is absorbed by PE whose emission, in the range of 550-595 nm, is absorbed in turn by Texas Red (Fig. 2). Texas Red emits red light in the range of 597-640 nm, at higher wavelength than FITC or PE

Table II: Multicolor dye combinations for immunofluorescence flow cytometry

Dyes	Excitation light source	Emission (nm)	Multicolor analysis
FITC/PE	Argon-ion laser	488	Two-color immunofluorescence analysis
FITC/PE/PI	Argon-ion laser	488	Two-color immunofluorescence analysis + dead cells
FITC/PE/PerCP DuoCHROME CyCHROME TRI-COLOR	Argon-ion laser	488	Three-color immunofluorescence analysis
FITC/PE Texas Red/APC	Argon-ion laser Dye laser	488 590	Four-color immunofluorescence analysis
FITC/PE/PerCP APC	Argon-ion laser Krypton-laser or He/Ne-laser	488 647 632	Four-color immunofluorescence analysis

emission (525 and 575 nm respectively). The major problem with this reagent is its high spectral overlap with PE (Fig. 3A). Alternatives to this fluorochrome are CyCHROME and TRI-COLOR (Fig. 2) which exploit the same principle of fluorescence energy transfer using a cyanine fluorochrome instead of Texas Red. The cyanine used in these complex is CY5.18 which is a carboxymethylindocyanine (SOUTHWICK et al, 1990) (Fig. 1). The cyanine fluorochrome emits red light at a wavelength higher than 650 nm and, therefore, avoids the problem of spectral overlap with PE (Fig. 3B). This PE-CY5.18 complex is available directly linked to Mabs allowing three-color analysis using a direct immunofluorescence procedure (Table II).

The latest development of fluorochrome useful for multicolor immunofluorescence flow cytometry is the peridinin-chlorophyll a-protein (PerCP) conjugated to Mabs (RECHTENWALD, 1989). PerCP, a chromoprotein isolated from marine dinoflagellates (PREZELIN, 1976), has been recently crystallized (STECK, 1990). PerCP makes it possible to carry out three-color flow cytometric analysis with a single laser using a direct immunofluorescence procedure (AFAR et al, 1991) (Table II, Fig. 2 and 3C). It can also be used in conjunction with FITC, PE and APC fluorochromes excited with two laser beams to simultaneously label four cell surface antigens (Table II).

Other fluorochromes such as 7-amino-4-methylcoumarin-3-acetic acid (AMCA) and cascade blue (Fig. 2) can be used for three-color immunofluorescence analysis but require UV excitation source (AUBRY et al, 1990). Different carboxymethylindocyanines other than CY5.18, and aluminium phthalocyanine (Fig. 2) may be good candidates for multicolor immunofluorescence analysis soon (SOUTHWICK et al, 1990; KELLY et al, 1991).

Fig. 3 : Comparison of the emission spectra of DuoCHROME, CyCHROME and PerCP when excited by 488 nm light in conjunction with FITC and R-PE

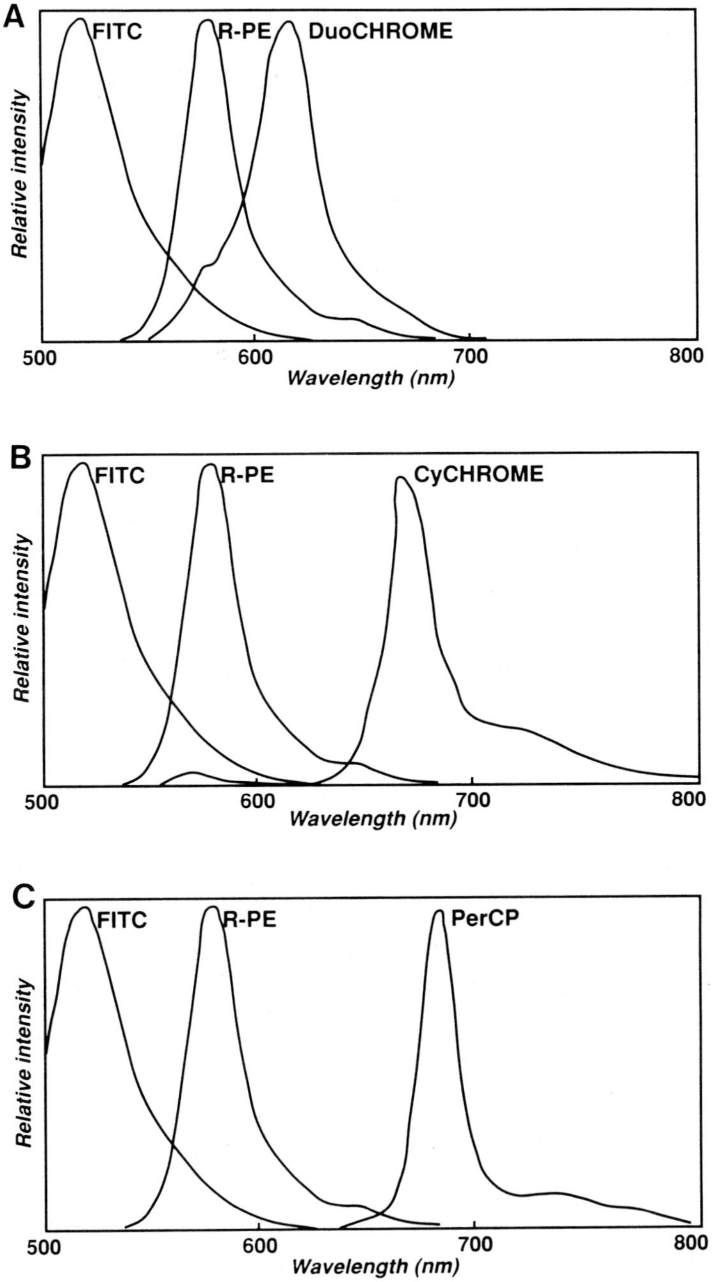

PROBLEMS ENCOUNTERED WITH MULTICOLOR IMMUNOFLUORESCENCE ANALYSIS

Several problems are associated with multicolor immunofluorescence analysis. They can be clustered in two groups which are:

- the problems encountered with immunofluorescence analysis even when only one-color is used (dead cells, Fc receptors, autofluorescence),

- the problems linked to multicolor immunofluorescence analysis which are due either to spectral overlap when one laser excites two fluorochromes or to resonance energy transfer when dye molecules are in close proximity at cell surface.

In most analyses, dead cells are a potential source of false positive cells and can be excluded from the analysis using light scatter parameters or propidium iodide (PI). In general, dead cells display lower forward light scatter and higher perpendicular light scatter than live cells. PI stains the DNA of cells whose plasma membrane is damaged. It can be used in conjunction with FITC since it absorbs light at 488 nm. It cannot be used for multicolor immunofluorescence analysis with DuoCHROME because its emission spectrum completely overlaps the one of DuoCHROME. The association of PI with CyCHROME/TRI-COLOR, PerCP or PE requires high correction to compensate for its spectral overlap with these fluorochromes. However, PI is particularly useful when analysing rare cells.

It should be pointed out that Mabs can bind cells through Fc receptors. Three receptors for the Fc portion of IgG have been clustered under the names CD16, CD_W32 and CD64. CD16 and CD_W32 are low affinity receptors for IgG, but bind aggregated IgGs with high affinities. CD64 is a high affinity receptor for IgG and is only expressed on monocytes and polymorphonuclear cells activated with interferon-gamma (MADJIC O, 1989). Lyophilized and old reagents may contain Mab aggregates and should be avoided when working with flow cytometry. Binding Mabs by their Fc portion on high affinity receptors can be excluded when using F(ab) or $F(ab')_2$ fragments of Mabs. Unfortunately, Mabs are not commercially available under these forms without Fc portion. Non specific controls must be used in order to identify binding on Fc receptors. These controls must be Mabs which are non relevant for the studied cells. They must have the same Ig isotype as the test reagent, they must contain the same fluorochrome and be used at approximately the same Ig concentration. These controls are also useful for the characterization of a cell subset with high autofluorescence.

Light emitted by the different fluorochromes used in multicolor immunofluorescence analysis are separated by an optical system made of dichroic mirrors, bandpass and/or long pass filters. These filters placed in front of photomultiplier tubes only transmit a selected narrow band of wavelengths corresponding to each fluorochrome. If the emission of one dye overlaps with that of another, this signal will not be removed by the filters. In the example shown in figure 4A, the quantity of FITC fluorescence reaching the R-PE filter of 575 ± 15 nm (black area) is proportional to the quantity of fluorescence reaching the FITC filter of 530 ± 15 nm. For each cell, it is possible to compensate for this spectral overlap by electronically subtracting that portion of contaminating emission as already described (LOKEN and LANIER, 1984; PARKS and HERZENBERG, 1984). When different choices of fluorochrome combination are possible, the best is to use dyes whose spectral overlaps are minimum. As illustrated in figure 3, CyCHROME or PerCP are easier to use in conjunction with FITC and R-PE than DuoCHROME, as DuoCHROME emission greatly overlaps with R-PE emission and needs high correction. On the contrary, CyCHROME and PerCP emissions slightly overlap with R-PE emission and need very low compensation corrections. The adjustment of

Fig. 4 : Spectral overlap correction

A : Emission spectra overlap between FITC and R-PE (black area corresponds to the approximative emission of one fluorescence reaching the optical filter of the other).

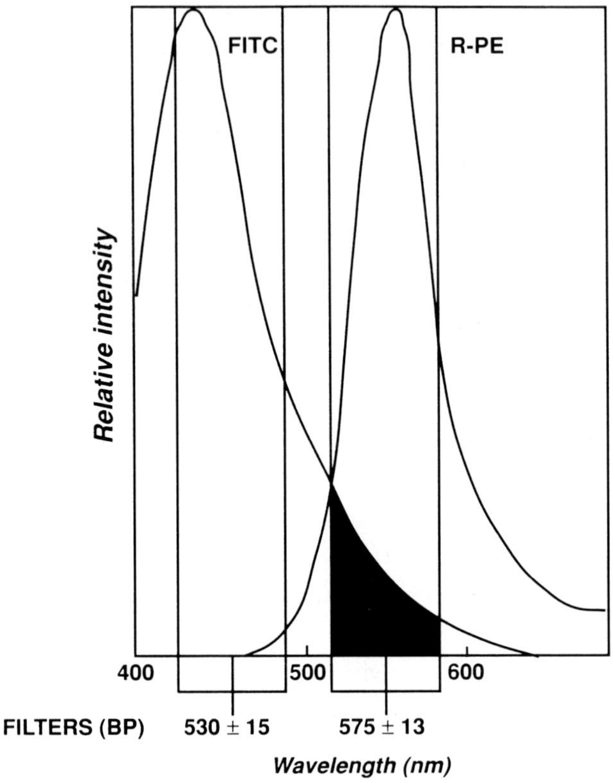

B : Dual-fluorescence compensation

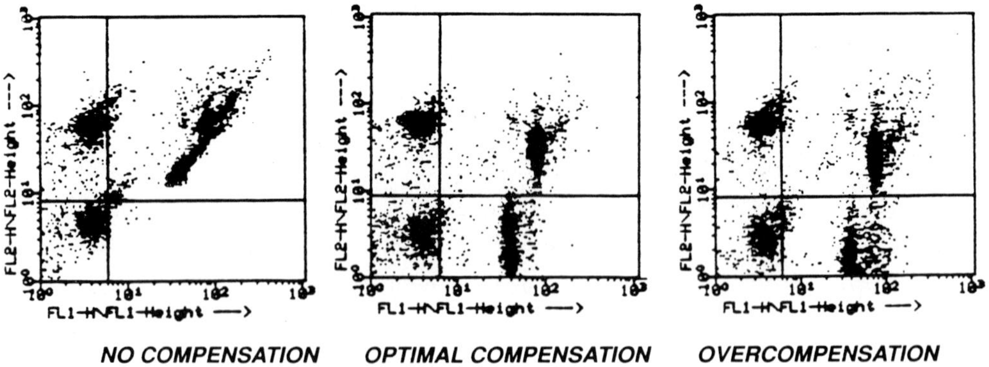

color compensation must be accurately carried out as shown in figure 4B. High compensation corrections may lead to render negative those cell subsets which display a low level of fluorescence.

The second problem linked to multicolor immunofluorescence analysis is the quenching of the fluorescence emission of one fluorochrome by another. This phenomenon, called resonance energy transfer occurs when:

- the emission spectrum of one dye corresponds to the absorption spectrum of another. This is the case for example of FITC whose emission can be absorbed by R-PE, DuoCHROME and CyCHROME/TRI-COLOR (Fig. 2),

- the dye molecules are separated by less than about 10 nm at the cell surface.

The incubation of cells with reagents at 4° C generally avoids the cross-linking of receptors at the cell surface. In this condition, resonance energy transfer is unusual in practice. However, fluorescence quenching of FITC emission by R-PE has already been reported on human tonsillar B cells (CHAPPLE et al, 1988). In this study, the fluorescence of cells labeled by FITC-coupled CD37 Mabs was quenched by R-PE-coupled anti- HLA-Dr Mabs. Moreover, when tonsillar B cells were dual-labeled by FITC-coupled CD37 Mabs and R-PE-coupled CD19 Mabs, a majority of cells were dual-colored. When the fluoro-chromes used for detection were reversed (FITC-coupled CD19 Mabs and R-PE coupled CD37 Mabs), only 50% of cells were dual-colored; the remaining others had lost the FITC fluorescence due to CD19 Mabs. This last example shows that it is better to detect weakly expressed receptors with the acceptor fluorochrome (R-PE) and to detect strongly expressed receptors with the donor fluorochrome (FITC). In order to avoid the pitfall of energy transfer in multiple color immunofluorescence, control tests can be included which consist of cells labeled only with the donor fluorochrome.

On the other hand, resonance energy transfert can be used to study the proximity of receptors at the cell surface (DAMJANOVICH et al, 1983). The tandem fluorochromes such as DuoCHROME, CyCHROME and TRI-COLOR operate on this principle.

THREE-COLOR IMMUNOFLUORESCENCE ANALYSIS: simultaneous identification of nine leukocyte subsets of human peripheral blood

As described in table II, several fluorochromes make it possible to carry out three-color immunofluorescence analysis using a direct immunofluorescence procedure. In general, the identification of leukocyte subsets is based on the specificity of Mabs linked to each of the different fluorochromes. One way to optimize this identification is to use the relative fluorescence intensity obtained for each subset in combination with the specificity of Mabs which is usually used alone. This approach has already been used for the simultaneous identification of five leukocyte subsets by dual-color immunofluorescence analysis (HORAN et al, 1986). We have found this technique to be particularly useful for the analysis of samples containing low numbers of cells and we have improved it using three-color immunofluorescence analysis (FITC, PE and TRI-COLOR). It can be applied to whole blood cells after erythrocyte lysis and permits the simultaneous identification of nine subsets of human peripheral blood leukocytes (PBL), namely polymorphonuclear (PMN) cells, monocytes, T4 cells, bright T8 cells, dim T8 cells, T null ($CD3^+$ $CD4^-$ $CD8^-$) cells, B cells, $CD8^+$ NK cells and $CD8^-$ NK cells (CARAYON et al, 1991). The combination of Mabs used for this labeling is shown in table III.

Before staining, buffy coat was removed after centrifugation of 2 ml human peripheral blood. Erythrocytes were lysed using 2 ml ammonium chloride lysing solution (8.99 g/l NH_4Cl, 1 g/l $KHCO_3$, 37 mg/l disodium EDTA, in distilled water, pH 7.4). After 5 minutes of incubation at room temperature and two

Table III: Monoclonal antibodies

Monoclonal antibodies	Volumes (μl)	Fluoro-chromes	Targets
Anti-Leu-M3 (CD14)	10	FITC	Monocytes/PMN
Anti-Leu-16 (CD20)	20		B-cells
Anti-Leu- 4 (CD3)	1*		T-cells
Anti-Leu-19 (CD56)	20	PE	NK-cells
Anti-Leu-M3 (CD14)	10		Monocytes/PMN
IOT8 (CD8)	20		T8-cells/NK-cells
Anti-Leu- 4 (CD3)	15*	Unconjugated Ab	T-cells
S3.5 (CD4)	5	TRI-COLOR (PE/Cy5.18)	T4-cells

The anti-Leu-M3-FITC, anti-Leu-16-FITC, anti-Leu-4-FITC, anti-Leu-19-PE, anti-Leu-M3-PE, anti-Leu-4 Mabs were purchased from Becton Dickinson, Le Pont-de-Claix, France.
The IOT8-PE Mab was purchased from Immunotech, Marseille, France.
The S3.5-TRI-COLOR Mab was purchased from Caltag, San Françisco, USA.

* Volumes indicated correspond to the amount of each commercial solution used to stain the cell pellet from 2 ml human blood. These values are only indicative and may vary from one lot to another.

washes by centrifugation in cold PBS, cell pellet was resuspended in the mixture of Mabs. The incubation was carried out for 30 minutes at 4° C. After one wash in cold PBS, cells were analyzed by flow cytometry. FITC, PE and TRI-COLOR fluorochromes were excited by an argon-ion laser at the wavelength of 488 nm. Fluorescence emissions were respectively collected using 530, 575 and 660 nm band pass filters plus dichroic mirrors. Electronic compensation was used to avoid the contamination of fluorescence of one fluorochrome in the channel of another.

Figure 5 shows the results obtained on PBL (A), spleen (B) and tonsillar (C) cells (spleen and tonsils were obtained from surgical procedures performed for therapeutic purposes. Tissues were gently dissociated and filtered through gauze to form a single suspension. The cells were labeled as described for peripheral blood leukocytes). PMN cells were determined using only the two light scatter parameters. They were removed from the analysis shown in figure 5 on the basis of their light scatter properties. The unconjugated CD3 Mab (Table III) is used to dilute the FITC-conjugated Mabs of the same specificity. This dilution results in decreasing the binding of FITC- coupled CD3 Mab although binding saturation of receptors is maintained. Consequently, T cells are well separated from the other subsets on the FITC (Fl-1) axis (Fig. 5A and B). However, the large number of B cells observed in tonsils did not allow a good separation between T and B cells on the FITC (Fl-1) axis (Fig. 5C). In this case, the dilution of FITC-coupled CD3 Mab by unconjugated Mab needed to be increased in order to achieve a good resolution between T and B cells. In addition, the specific activity of each reagent may vary from batch to batch. Consequently it may be necessary to optimize the separation between T cells and the other subsets by adjusting the dilution factor of FITC-conjugated CD3 Mab by its unconjugated counterpart.

The determinations of cell subsets using this method is well correlated with those obtained with the classical dual-

Fig. 5: Identification of leukocyte subsets in human blood, spleen and tonsils (Fl-1 = FITC; Fl-2 = R-PE; Fl-3 = TRI-COLOR)

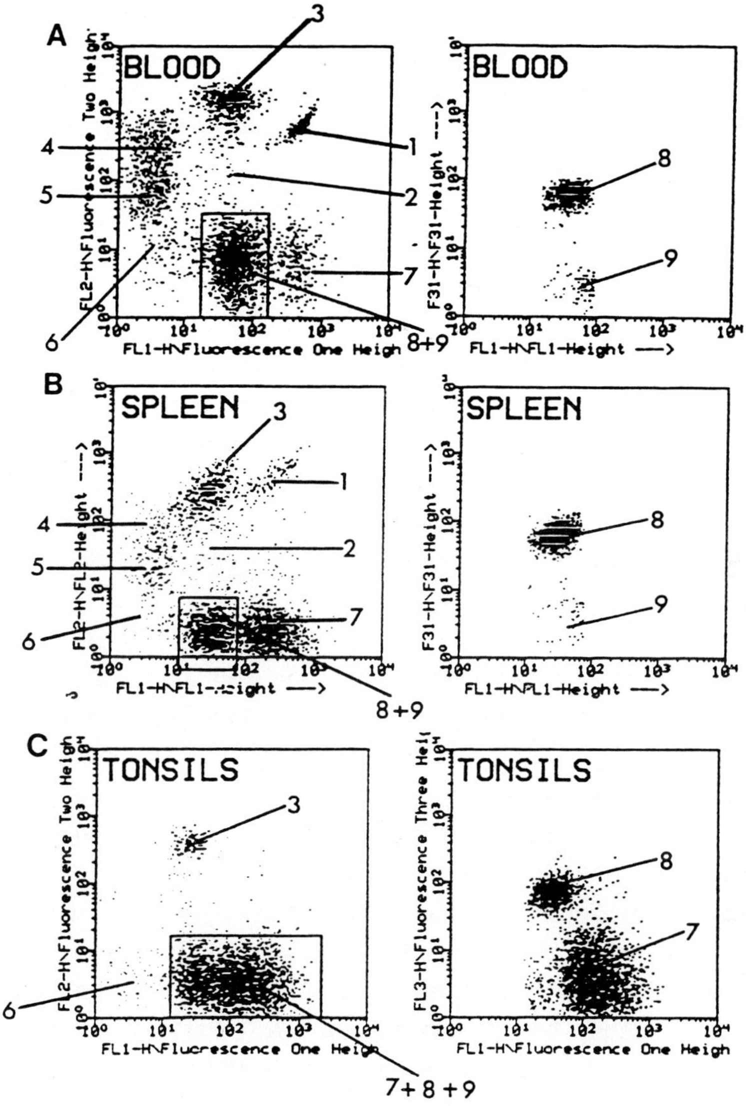

Leukocyte subsets

1. Monocytes
2. Dim T8 cells
3. Bright T8 cells
4. $CD8^+$ NK cells
5. $CD8^-$ NK cells
6. Non identified cells
7. B cells
8. T4 cells
9. T null ($CD3^+$, 4^-, 8^-) cells

immunofluorescence analysis. The variation of fluorescence intensities among the different donors is not so wide as to cause failure of this method. To date, more than one hundred different samples have been analyzed without any problem.

FOUR-COLOR IMMUNOFLUORESCENCE ANALYSIS USING COMMERCIALLY AVAILABLE REAGENTS

Until now, four-color immunofluorescence analysis has been carried out using FITC, R-PE, Texas Red and APC (Table II). In general, this type of analysis requires three fluorochromes directly linked to Mabs and one linked to streptavidin which binds biotin-coupled Mabs. Unfortunately two reasons make this configuration difficult to use:

1. Texas Red and APC fluorochromes are excited by a dye-laser at a wavelength of 590 nm. This wavelength is weakly absorbed by APC (Fig. 2).

2. Mabs directly linked to Texas Red or APC are not commercially available. Attachment of these fluorochromes to Mabs is not easy to perform and requires technical skill in biochemistry.

An alternative to Texas Red has appeared with the availability of peridinin-chlorophyll a-protein (PerCP) conjugated to Mabs. FITC, R-PE and PerCP directly linked to Mabs are excited by an argon-ion laser at a wavelength of 488 nm. APC linked to streptavidin is excited by a krypton-ion laser or a He/Ne-laser at wavelengths of 647 nm or 632 nm, respectively (Table II). The spectral overlap between the emissions of APC and PerCP is quite strong (Fig. 6A). However, these fluorochromes are excited by two different laser beams resulting in a spatial and electronic separation of both emissions.

In order to perform this four-color immunofluorescence analysis, peripheral blood mononuclear cells were labeled by

Fig. 6 : Four color immunofluorescence analysis

A : Emission spectra of the dyes

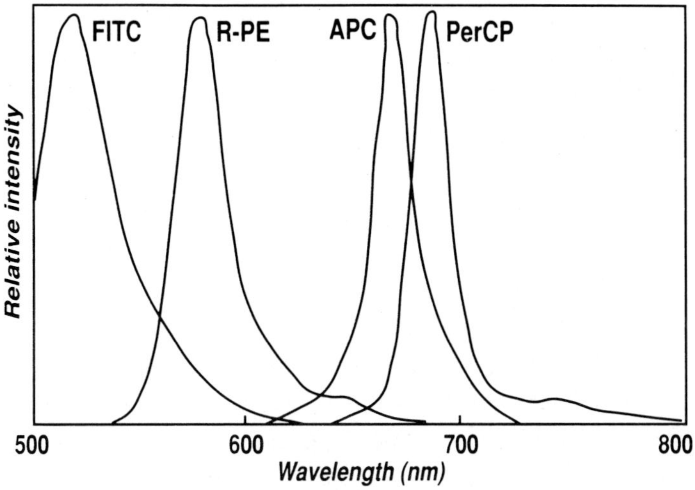

B : Optical system

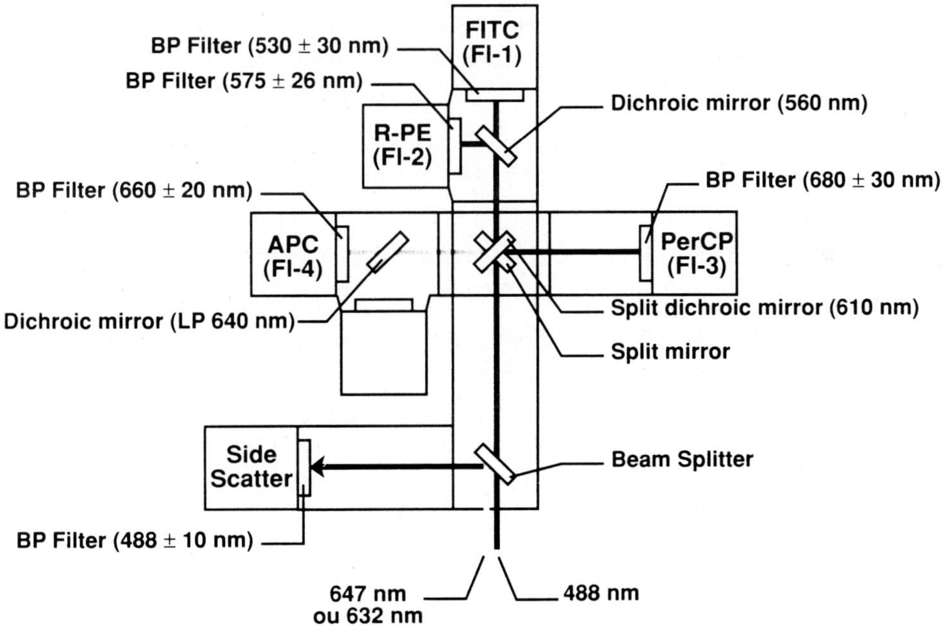

the following reagents: FITC-coupled Leu-3 (CD4) Mab (20 μl); R-PE-coupled Leu-12 (CD19) Mab (20 μl); PerCP-coupled Leu-4 (CD3) Mab (20 μl) and biotin-coupled Leu-2 (CD8) Mab (20 μl) + APC-streptavidin (20 μl). Before staining, mononuclear cells were isolated from whole blood by standard Ficoll-Hypaque density gradient centrifugation. Following two washes in cold PBS, the cell pellet was resuspended with the mixture of Mabs. The incubation was carried out for 30 minutes at 4° C and followed by two washes in cold PBS. The cell pellet was resuspended by APC-streptavidin and incubated for another 15 minutes at 4° C. Cells were washed once and analyzed by flow cytometry using spectral configuration as shown in figure 6B. The analysis allowed the determination of six leukocyte subsets in one step (Fig. 7), namely: B cells ($CD19^+$), T4 cells ($CD3^+$ $CD4^+$), bright T8 cells ($CD3^+$ $CD8^+$), dim T8 cells ($CD3^+$ $CD8^{+/-}$), T null cells ($CD3^+$ $CD4^-$ $CD8^-$) and $CD8^+$ NK cells. The intensity of fluorescence obtained with PerCP on $CD3^+$ T cells was not as high as the ones usually observed with FITC or R-PE fluorochromes. However, the resolution was sufficient to distinguish this leukocyte subset from the others.

CONCLUSIONS

During past years, multicolor immunofluorescence analysis has become increasingly easy to perform. The development of new fluorochromes linked to Mabs and the generation of more efficient flow cytometers allow immunofluorescence analysis using three- or four-color without any major problem. New applications of flow cytometry may be intended in order to determine not only the different leukocyte subsets in a sample, but also to assess their functions. This is well illustrated by the simultaneous labeling of cell surface antigens and of bromodeoxyuridine incorporated into DNA of cells in S phase (CARAYON et al, 1992) or by the intracellular labeling of cytokines (SANDER et al, 1991).

Fig. 7: Four-color immunofluorescence analysis of peripheral blood mononuclear cells.

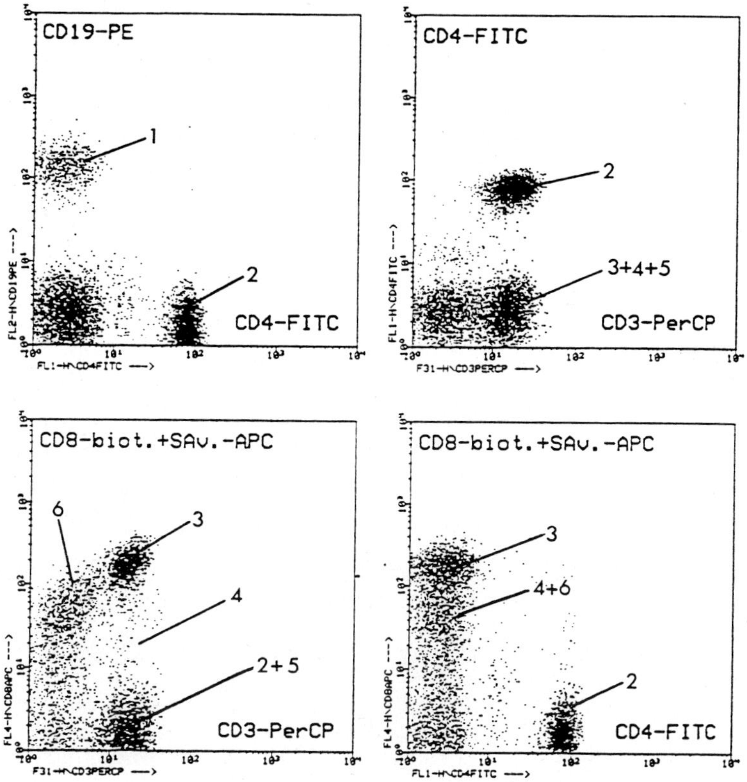

1. B cells
2. T4 cells
3. Bright T8 cells
4. Dim T8 cells
5. T null cells ($CD3^+$ $CD4^-$ $CD8^-$)
6. $CD8^+$ NK cells

REFERENCES

Afar B, Merrill J, Clark EA (1991). Detection of lymphocyte subsets using three-color/single laser flow cytometry and the fluorescent dye peridinin chlorophyll-a protein. J. Clin. Immunol. 11: 254-261.

Aubry JP, Durand I, de Paoli P, Banchereau J (1990). 7-Amino-4-methylcoumarin-3-acetic acid-conjugated streptavidin permits simultaneous flow cytometry analysis of either three cell surface antigens or one cell surface antigen as a function of RNA and DNA content. J. Immunol. Methods 128: 39-49.

Carayon P, Bord A, Raymond M (1991). Simultaneous identification of eight leucocyte subsets of human peripheral blood using three-colour immunofluorescence flow cytometric analysis. J. Immunol. Methods 138: 257-264.

Carayon P, Bord A. (1992). Identification of DNA-replicating lymphocyte subsets using a new method to label the bromo-deoxyuridine incorporated into the DNA. J. Immunol. Methods. 147: 225-230.

Chapple MR, Johnson GD, Davidson RS (1988). Fluorescence quenching of fluorescein by R-phycoerythrin. A pitfall in dual fluorescence analysis. J. Immunol. Methods 111: 209-217.

Damjanovich S, Tron L, Szollosi J, Zivodtzki R, Vaz WLC, Regateiro F, Arndt-Jovin DJ, Jovin TM (1983). Distribution and mobility of murine histocompatibility $H2K^k$ antigen in the cytoplasmic membrane. Proc. Natl. Acad. Sci. USA 80: 5985-5989.

Glazer AN, Streyer L (1984). Phycofluor probes. TIBS 9: 423-427.

Hardy RR, Hayakawa K, Parks DR, Herzenberg LA (1983). Demonstration of B-cell maturation in X-linked immuno-deficient mice by simultaneous three-colour immuno-fluorescence. Nature 306: 270-272.

Hardy RR, Hayakawa K, Parks DR, Herzenberg LA, Herzenberg LA (1984). Murine B cell differentiation lineages. J. Exp. Med. 159: 1169-1188.

Horan PK, Slezak SE, Poste G (1986). Improved flow cytometric analysis of leukocyte subsets: simultaneous identification of five cell subsets using two-color immunofluorescence. Proc. Natl. Acad. Sci. U.S.A. 83: 8361-8365.

Kelly TA, Hunter CA, Schindele DC, Pepich BV (1991). Aluminium phthalocyanine-streptavidin: new, sensitive fluorescent tracer for immunoassay. Clin. Chem. 37: 1283-1286.

Lanier LL, Loken MR (1984). Human lymphocyte subpopulations identified by using three-color immunofluorescence and flow cytometry analysis: correlation of Leu2, Leu3, Leu7, Leu8 and Leu11 cell surface antigen expression. J. Immunol. 132: 151-156.

Loken MR, Parks DR, Herzenberg LA (1977). Two-color immuno-fluorescence using a fluorescence activated cell sorter. J. Histochem., Cytochem. 25: 899-907.

Loken MR, Lanier LL (1984). Three-color immunofluorescence

analysis of Leu antigens on human peripheral blood using two lasers on a fluorescence activated cell sorter. Cytometry 5: 151-158.
Majdic O (1989). Cluster report: CD64. In: Knapp W (ed) Leucocyte typing IV; white cell differentiation antigens. Oxford University Press, Oxford-New York-Tokyo, p. 835.
Oi VT, Glazer AN, Streyer L (1982). Fluorescent phycobiliprotein conjugates for analysis of cells and molecules. J. Cell Biol. 93: 981-986.
Parks DR, Herzenberg LA (1984). Fluorescence-activated cell sorting: theory, experimental optimization, and applications in lymphoid cell biology. In: Di Sabato G, Langone JJ, Van Vunakis H (eds). Methods in enzymology, vol. 108. Academic Press, Inc. Orlando p. 197.
Parks DR, Hardy RR, Herzenberg LA (1984). Three-color immunofluorescence analysis of mouse B lymphocyte subpopulations. Cytometry 5: 159-168.
Prezelin BB, Haxo FT (1976). Purification and characterization of peridin-chlorophyll a-proteins from the marine dinoflagellates Glenodinium dp. and Gonyaulax polyedra. Planta 128: 133-141.
Rechtenwald D (1989). Peridinin-chlorophyll complex as fluorescent label. US patent 4, 876, 190.
Sander B, Andersson J, Andersson U (1991). Assessment of cytokines by immunofluorescence and the paraformaldehyde-saponin procedure. Immunol. Reviews 119: 65-93.
Segal DM, Sharrow SO, Jones JF, Siraganian RP (1981). Fc(IgG) receptors on rat basophilic leukemia cells. J. Immunol. 126: 138-145.
Southwick PL, Ernst LA, Tauriello EW, Parker SR, Mujumdar RB, Mujumdar SR, Clever HE, Waggoner AS (1990). Cyanine dye labeling reagents-Carboxymethylindocyanine succinimidyl esters. Cytometry 11: 418-430.
Steck K, Wacker T, Welte W, Sharples FP, Hiller RG (1990). Crystallization and preliminary X-ray analysis of a peridinin-chlorophyll a-protein from Amphidinium carterae. FEBS Letters 268: 48-50.
Stryer L, Glazer AN (1986). Phycobilin protein fluorescent conjugates for use in fluorescent analysis of molecules and cells. US patent 4, 542, 104.
Titus JA, Haugland R, Sharrow SO, Segal DM (1982). Texas Red, a hydrophilic, red-emitting fluorophore for use with fluorescein in dual parameter flow microfluorometric and fluorescence microscopic studies. J. Immunol. Methods 50: 193-204.

Energy Transfer

Wolfgang G. Eisert
Department of Biophysics
University of Hannover and
Department of Clinical Pharmacological Research
Dr. Karl Thomae GmbH
W-7950 Biberach/Riß
Germany

Introduction

In most instances classification of biological cells is based on a morphological description. This was established by conventional microscopy using different staining procedures and in many instances still is the prime source for the discrimination of different cell types. However, staining reveals or highlights the spatial organization of mostly intracellular chemistry and biochemistry. Antibodies raised against different protein structures and labeled with various chromophores increased the level of discrimination of different cell classes even further. In most cases, cytometry uses the very same principles to discriminate different cell types. Cells are stained and/or labeled (after fixation) and the intensity of emitted fluorescence light is measured. Based on the assumption, that fluorescence emission intensity is proportional to the number of molecules bound and binding is stoichiometric to molecules/structures of interest, a quantitative analysis of cell constituents serves as discrimination of cell classes or subclasses.

These approaches therefore will reveal structural rather than functional information. In an increasing number of investigations functional rather than structural properties of cells are now being investigated. Furthermore, in most cases this requires time lapsed studies in living cells. A fixation of cells therefore is not acceptable, which results in limitations of applicable dyes, markers and procedures for cell labeling.

NATO ASI Series, Vol. H 67
Flow Cytometry
Edited by A. Jacquemin-Sablon

Dynamic changes of cell properties therefore will often rely on optical measurements other than quantifying fluorescence emission intensities. Multiangle and multifrequency light scattering (Schrader, Salzman) as well as fluorescence emission anisotropy and fluorescence energy transfer (Eisert, Arndt-Jovin) are methods of choice allowing continuous monitoring of a number of cell function. This can be done in an ensemble of cells in suspension, as well as in cells attached to surfaces and is of particular interest for studies in cell clusters, where cell properties are influenced by neighboring cells such as in a meshwork of neurons (Kater et al.). Energy transfer measure-ments are often used for measuring distances between molecules or within one molecule showing different tertiary structures. In our own hands fluorescence energy transfer also proved to be useful to follow transport phenomena within a single cell (Birk).

Theory

It is well known in physics that energy is transfered between colliding molecules. This holds also true in the case of excited molecules, where electronic excitation energy is transfered during collision. In a similar way over long distances energy transfer can be observed at high concentration of excitable molecules (of the same kind) such as in a crystal structure or very highly concentrated solutions.

As early as 1924 and 1927 J. Perrin for the first time described a direct electronic energy transfer between molecules in solution as alternative to emission and reabsorption of energy. J. and later F. Perrin developed a classical theory for the transfer of excitation energy to which later a quantum mechanical approach was added. They followed the theoretical description of energy transfer between molecules in gas-phase as proposed by Kallmann. These theories were tested by measuring the fluorescence of fluorescein in highly viscous solvents such as glycerin. Polarized excitation of fluorescent molecules results in the emission of polarized fluorescence. This is true for diluted solutions of fluorescein. However, if the concentration of fluorescein is increased, an increased depolarization of fluorescence emission can be observed.

This must be explained by a radiationless transfer of excitation energy to neighboring molecules, which then emit light with a different direction of polarization according to the present orientation of their emission vector. This observation was and still is gaining increasing attention in modern biology. It enables to measure molecular distances within one large molecule or between molecules labeled with matching pairs of fluorescent markers.

It can be excluded that the half life time of fluorescence emission is decreased with increased concentration, which alternatively could explain a decrease in polarization.

Th. Foerster in 1944 and later in 1947 extended Perrin`s approach to the general problem of radiationless energy transfer between different molecules with different optical properties. He derived in his theory the equation which relates distances between donor and acceptor to the effectiveness of energy transfer. With an excitation probability ρ (t) and the mean half life of the excited state τ with and τ_s without quenching, the quantum efficiency η of fluorescence emission is:

$$\eta = 1/\tau_s \int \rho(t)\, dt = \tau/\tau_s$$

For the transition of energy from molecule k to another one l of the same kind during excitation the excitation probability of molecule l depends on the number F of transitions $k \rightarrow l$ which will get smaller as distance between k and l gets larger. With $F_{k\rightarrow l} = F_{l\rightarrow k}$. the total emission probability for both molecules becomes

$$\eta_{1,2} = 1/\tau_s \int \rho_{1,2}(t)\, dt \quad \text{with}$$

$$\eta_1 = \eta\ (1 + \tau F) / (1 + 2\tau F) \qquad \text{and} \qquad \eta_2 = \eta\ \tau F / (1 + 2\tau F)$$

For $\tau F << 1$, (weak interaction during excitation) only the originally excited molecule radiates,
For $\tau F >> 1$, (strong interaction during excitation) both molecules radiate with equal probability.

Under the assumption that movements are small within the period of observation the theoretical approach can neglect movements of the nucleus and can deal with the interaction of the electron system of the atoms only. Following Dirac's formula a matrix description of Coulomb's interaction energies.can be used including the necessary terms for the relative orientation of the molecular dipoles and the dielectric constant of the medium, which is given by the refractive index n. Furthermore in viscous solvents and if energy transfer is slow with respect to the Brownian motion the transition probability has to average all orientations and all energy values W (expressed as fluorescence and excitation spectra) of both molecules. This leads to:

$$F_{kl} \approx 3\ (h/2\pi)\ c^2\ J\ (W)\ /\ 4\pi^3\ n^2\ N'^2\ W_0^2\ R_{kl}^6$$

with: $J\ (W) = (\ln 10)^2 \int_{W=0}^{\infty} \varepsilon\ (W)\ \varepsilon\ (2W_o - W)\ dW$

if absorption and emission spectra are symmetric.

If spectra show overlap, than $W_0 \approx W$ and

$F_{kl} = 1/\tau_0\ (R_0/R_{kl})^6$ with

$$R_0 = (3\ (h/2\pi)\ t_0\ c^2\ J(W)\ /\ 4\pi^3\ n^2\ N'^2\ W_0^2)^{1/6}$$

with $N' = 6.02 * 10^{20}$ molecules per millimole.

R_0 being the critical distance of molecules below which energy will be transferred during the excitation time interval. In addition fluorescence energy transfer will also be observed over similar distances between different molecules, if the fluorescence emission spectrum overlaps the fluorescence excitation spectrum.

For most of the commonly used inorganic or organic molecules, energy transfer can be observed at molecular distances ranging from approximately 10 A to 70 A. In these cases the absorption spectrum of the donor molecule overlaps the fluorescence excitation spectrum of the acceptor molecule. By exciting the donor molecule the emission by the donor itself, as well as the fluorescence emission by the acceptor is measured. Furthermore the degree of fluorescence depolarization of the emission by the donor as well as by the acceptor can therefore be used to probe inter- and intramolecular distances. Ideally suited for these measurements are dual beam and dual frequency flow cytometers. In a sequential experimental protocol this configuration than allows determination of donor as well as acceptor fluorescence intensities with and without energy transfer from which the efficiency of the transfer can be calculated. By using Förster's equation as derived above, the distance from donor to the acceptor molecule can be easily calculated.

Application in biology, cytometry and biochemistry

In biological and biochemical applications measurements of distances between different types of structures either within a single macromolecule or between macromolecules are of interest. This information can be obtained by energy transfer measurements using matching pairs of fluorescent molecules, i.e. donor molecules with a fluorescence emission spectrum overlapping the excitation spectrum of acceptor molecules. One molecular structure will be labeled with the donor the other structure of interest with the acceptor molecule. In many cases intrinsic chromophores within a macromolecule such as a single tryprophane residue in the galactose binding protein (Zukin) or in human serum albumin as well as the fluorescent Y base at the anticodon of yeast phenylalanine transfer RNA can serve a donor (Stryer). On the other hand heme groups in heme proteins as well as the retinal group in rhodopsin are suitable energy acceptors. For many applications fluorescent analogs may be considered. This has been demonstrated with adenosine derivatives serving as energy donor and replacing AMP, ADP, ATP, NAD+, FAD, CoA and other adenine nucleotides or with fluorescent analogs of fatty acids. In a number of cases fluorescent lanthanides (Tb^{3+} and Eu^{3+}) can replace Ca^{2+}. Zn^{2+} may be replaced by Co^{2+} serving as energy acceptor. In energy transfer studies of immunoglobulin G dansyl and fluorescein haptens were used (Werner et al.).

Alternatively suitable fluorescent groups may be inserted by covalent binding such as the labeling of a single sulfhydryl group by fluorescent iodoacetamide derivatives (Hudson et al.).

Stryer describes energy transfer measurements in rhodopsin, galactose receptor protein, metalloproteins, the cytochrome c oxidase complex as well as transfer RNA. Tron et al. describe steady state energy transfer measurements of cell surface proteins such as lectin receptors, as well as of the histocompatibility complex.

The structure of chromosomes are an other area where fluorescence energy transfer measurements were reported to give useful information about the spatial arrangement and orientation of specific chromosomal regions (Sahar et al.). Either quinacrine or 33258 Hoechst were used as energy donor with 7-amino-actinomycin D as energy acceptor. Alternatively the pairs chromomycin A_3 with methyl green (Sahar) and fluorescamine with ethidium bromide were used to probe chromatin structure.
It is worth mentioning that distance measurements between actin and myosin by using fluorescence energy transfer measurements have been the subject of a number of publications allowing dynamic measurements. The majority of static data were later confirmed by crystallography. A review of this area was given by dos Remedios et al.

The beauty of energy transfer measurements however seems to lay with dynamic distance measurements in living cells. This has been demonstrated in flow cytometry studies of the lateral movement of lectin receptors after stimulation of cells forming clusters (Chan et al.).

In another application using flow cytometry as well as fluorescence polarization image analysis of single cells the turn over of the outer cell membrane and the uptake of parts of the cell membrane in intacellular stores was studied (Birk). The outer cell membrane was labeled by TMA - Diphenyl-hexatriene (TMA - DPH) and the intracellular lipid stores by nile red. By monitoring the intensity as well as the degree of fluorescence polarization of TMA - DPH and nile - red the turn-over of the cellular membrane during and after various treatments of the cells can be studied directly. If excited with polarized light the fluorescence emission of TMA - DPH is also polarized whereby if excited with polarized light nile red will emit depolarized fluorescent light. If therefore fluorescence emission is observed at the emission wavelength of nile red fluorescence band which is depolarized, transfer of energy from TMA - DPH must be concluded.

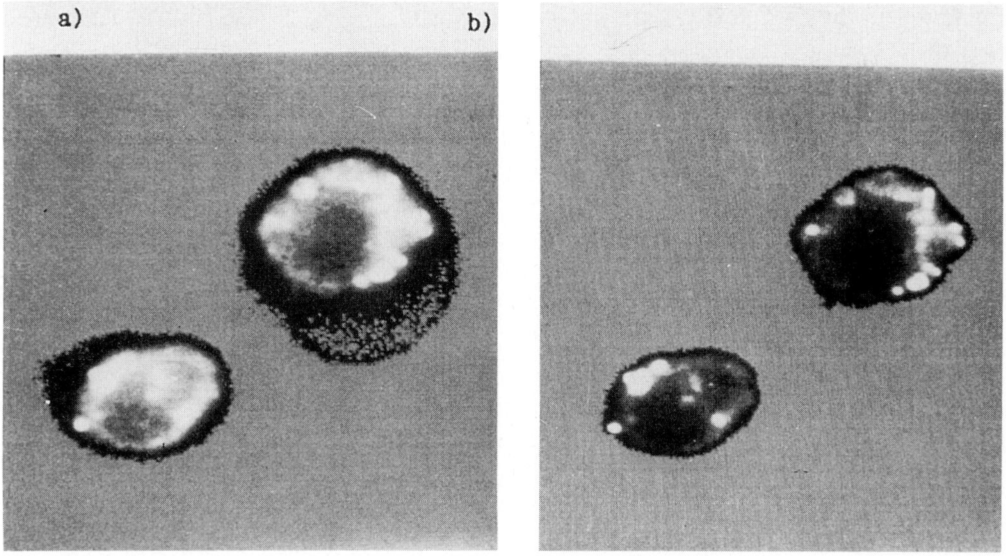

Fig. 1: Cells stained with TMA - DPH and nile red and analyzed in a fluorescence polarization image analysis system using epi - illumination and analyzing spatial distribution of polarized fluorescence emission. The direction of polarization of fluorescence is parallel (left picture) or perpendicular (right picture) to the polarization of the excitation. TMA-DPH fluorescence is highly polarized and mainly found on the cell surface. As parts of the cellular membrane get in close proximity with intracellular lipid stores energy is transfered and the fluorescence emission is partially or entirely depolarized.

An example is given in fig.1. Cells were labeled simultaneously with TMA - DPH and nile red. Fluorescence polarization image analysis is done on cells with either parallel (left picture) or orthogonal polarization analyzer (right picture). The spectral bandwidth of the detectors has been rather wide. The analysis of fluorescence intensities of parallel polarization primarily shows the distribution of TMA - DPH, whereas orthogonally polarized fluorescence intensities are primarily emitted from nile red molecules. While TMA - DPH fluorescence is mainly homogeneously distributed in the outer cell membrane, depolarized fluorescence intensity is found in small intracellular clusters, representing cellular lipid stores (right picture).

These studies revealed that a cell population will respond to short term chemical and physical changes in their environment quite inhomogeneously and that subpopulations show a distinctly different response (Birk, Beisker and Eisert). Fluorescence polarization image analysis now also allow monitoring changes in the spatial arrangement and distance measurements in single living cells or in small cell clusters. This will support flow cytometry studies in many instances.

Discussion

Although fluorescence anisotropy, as well as energy transfer has been known in macromolecular biochemistry and cytometry for more than two decades its use is reported only sparsely. One may argue that instrumental and experimental difficulties may have limited this technique to only those laboratories involved in technological and instrumentation research. In the past, however, modern flow cytometry equipment has become rather reliable and allows stable measurements of complex optical parameters.

Earlier papers even trying to derive malignancy markers from cytoplasmic structures as measured by fluorescence depolarization and in some instances subsequent energy transfer measurements have not found a more general follow - up indicating that potentially some results are depending on specific laboratory set-ups and preparations.

Presently information on spatial arrangements of macromolecular structures have become more readily available by crystallography using highly purified in most cases recombinant material, which can be made available in large quantities. On the other hand structural information can be gathered sometimes more easily using modern approaches to protein engineering. Energy transfer measurements with its high demand on instrumentation and labeling technology seems to have lost in general interest. However energy transfer measurements is one of the few diagnostic technologies allowing a continuous observation and measurement of the spatial arrangements and rearrangements within a living cell. By extending the approach to energy transfer measurements using fluorescence polarization image analysis these spatial rearrangements can be followed in a single cell or a small cluster of cells over time.

One of the biggest advantages of this approach seems to be the availability of information of spatial and structural arrangements in between cells, i.e. the cell junctions. This opens the field to the observation of transport, exchange and membrane interaction studies under physiological conditions.

The biggest limitation however seems be be the availability of suitable marker. As the number of groups using this technique is still small, the development of new fluorescent markers or analogs is slow. One can hope, that with a stronger demand for better and or more specific label academic institutions as well as specialized industry will provide better tools allowing to make this approach more widespread usable.

References

Arndt-Jovin DJ, Jovin TM (1976) Cell separation using fluorescence emission anisotropy. In: Membranes and Neoplasia. New Approaches and Strategies, Vol. 9, Progress in Clinical and Biological Research. Marchesi VT (ed). Alan Liss, New York pp. 123-136

Auld DS, Latt SA, Vallee BL (1972) An approach to inhibition kinetics. Measurement of enzyme-substrate complexes by energy transfer. Biochemistry 11:4994-4999

Beisker W, Eisert WG (1984) Probing macromolecular structures by flow cytometric fluorescence polarization measurements. In: Eisert WG and Mendelsohn ML (eds.) Biological Dosimetry, Springer - Verlag, Berlin Heidelberg pp.235-242

Beisker W and Eisert W G (1985) Double beam autocompensation for fluorescence polarization measurements in flow cytometry. Biophysical Journal 47: 607 - 612

Birk G (1990) Hochauflösende Fluoreszenzpolarisationsmessungen membrangebundener Farbstoffe an Einzelzellen, Dissertation Universität Hannover

Cantor CR, Pechuks P (1971) Determination of distance distribution functions by singlet-singlet energy transfer. Proc. Natl. Acad. Sc. U.S.A. 68:2099-2101

Cathou RE, Bunting JR Resonance energy transfer studies on the conformation of antibodies, In: Chen RF, Edelhoch H (eds) (1976) Biochemical Fluorescence Concepts, Vol.2, Marcel Dekker, New York, pp. 845-878

Cercek L, Cercek B, Ockey CH (1973) Structureness of the cytoplasmic matrix and Michaelis-Menten constants for the hydrolysis of FDA during the cell cycle in Chinese hamster ovary cells. Biophysik 10: 187-194

Cercek L, Cercek B, Franklin CIV (1974) Biophysical differentiation between lymphocytes from healthy donors, patients with malignant diseases and other disorders. Br. J. Cancer 29: 345-352

Chan SS, Arndt-Jovin DJ, Jovin TM (1979) Proximity of lectin receptors on the cell surface measured by fluorescence energy transfer in a flow system. J. Histochem. Cytochem. 27: 56-64

Chance B., Lee C-P, Blasie JK (eds) (1971) Probes of Structure and Function of Macromolecules and Membranes. Academic Press, Vol. 1: New York

Dale RE, Bauer RK (1971) Concentration depolarization of the fluorescence of dystuffs in viscous solution. Acta Phys. Pol. A40:853-882

Dale RE, Eisinger J (1976) Intramolecular energy transfer and molecular conformation. Proc. Natl. Acad. Sc. U.S.A. 73 : 271-273

Damjanovich S, Somogyi B, Balasz M, Kertai P, Redai I (1980) Fluorescence Double Labeling and Energy Transfer in Studying Intracellular Interactions. Karger, Basel. Antibiotics Chemother. 28 : 142-146

Darnall DW, Abbot F, Gomez JE, Birnbaum ER (1976) Fluorescence energy transfer measurements between the calcium binding site and the specificity pocket of bovine trypsin using lanthanide probes. Biochemistry 15 : 5017-5023

Eisert W G, Beisker W (1980) Epi - illumination optical design for fluorescence polarization measurements in flow systems. Biophys. J. 31 : 97-112

Fernandez SM Berlin RD (1976) Cell surface distribution of lectin receptors determined by resonance energy transfer. Nature 264 : 411-415

Foerster T (1946) Naturwiss. 88 : 166

Foerster T (1948) Zwischenmolekulare Energiewanderung und Fluoreszenz. Annalen der Physik. 2 (6) : 55-75

Foerster T (1951) Fluoreszenz Organischer Verbindungen. Vandenhoeck & Ruprecht, Goettingen

Grinvald A, Haas E, Steinberg IZ (1972) Evaluation of the distribution of distances between energy donors and acceptors by fluorescence decay. Proc. Natl. Acad. Sci. U.S.A. 69:2273-2277

Hudson EN, Weber G (1973) Biochemistry 12: 4154-4161

Inbar M, Shinitzky M, Sachs L (1973) Rotational relaxation time of concanavalin A bound to the surface membrane of normal and malignant transformed cells. J. Mol. Biol. 81 : 245-253

Inbar M, Shinitzky M (1975) Decrease in microviscosity of lymphocyte surface membrane associated with stimulation induced by concanavalin A. Eur. J. Immunol. 5 : 166-170

Jablonski A (1970) Anisotropy of fluorescence of molecules excited by excitation transfer. Acta Phys. Pol. A38:453-458

Kallmann H, London F (1928) Z. physik. Chem. (B) 2:207

Kater S B, Cohan C S, Jacobs G A, Miller J P (1986) Image intensification of stained, functioning, and growing neurons. In: De Weer P, Salzberg B M (eds) Optical methods in cell physiology, Wiley-Interscience, New York, p. 31

Kawski A, Kaminsky J (1974) Energieübertragung zwischen gleich- und ungleichartigen Molekülen in Lösung. Theorie Z. Naturforsch. A 29 : 452-456

Keller PM, Person S, Snipes W (1977) A fluorescence enhancement assay of cell fusion. Biophys. J. 17:30a

Latt SA, Cheung HT, Blout FR (1965) Energy Transfer. A system with relatively fixed donor - acceptor separation. J. Am. Chem. Soc. 87:995-1003

Latt SA, Auld DS, Vallee BL (1970) Surveyor substrates: Energy transfer gauges of active center topography during catalysis. Proc. Natl. Acad. Sc. U.S.A. 67:1383-1389

Levinson SA, Dandliker WH, Brawn RJ, Vanderlaan WP (1976) Fluorescence polarization measurement of the hormone-binding site interaction. Endocrinology 99:1129-1143

Luk GK (1971) Study of the nature of the metal-binding sites and estimate of the distance between the metal-binding sites in transferrin using trivalent lanthanide ions as fluorescent probes. Biochemistry 10:2838-2843

Matsumoto S, Hammes GG (1975) Fluorescence energy transfer between ligand binding sites on aspartate transcarbamylase. Biochemistry 14:214-224

Perrin F (1926) Polarization de la lumiere de fluorescence Vie moyenne de molecules dans l`etat excite. J. Phys. 12:390-401

Perrin F (1932) Ann. Chim. Phisique 17:283

Perrin J (1924) 2me conseil de Chimie Solvay Bruxelles. Gauther-Villars Paris 1925:322

Perrin J (1927) C. R. Acad. Sci. Paris 184:1097

Peters R (1971) Study of membrane thickness by energy transfer. Biochem. Biophys. Acta 233:465-468

Radda GK, Vanderkooj J (1972) Can fluorescent probes tell us anything about membranes? Biochem. Biophys. Acta. 265 : 509-549

Radda GK (1975) Fluorescent probes in membrane studies. Phil. Trans. R. Soc. London B 270:539-549

Remedios CG, Miki M, Barden JA (1987) Fluorescence resonance energy transfer measurements of distances in actin and myosin. A critical evalution. J. of Muscle Research and Cell Motility 8:97-117

Sahar E, Latt SA (1980) Energy Transfer and Binding Competitionbetween Dyes used to enhance Staining Differentiation in Metaphase Chromosomes. Chromosoma, Berlin 79:1-28

Schiller PW The measurement of intramolecular distances by energy transfer in Chen RF, Edelhoch H (eds) (1975) Biochemical Fluorescence Concepts, Vol. 2 Marcel Dekker, New York pp. 285-304

Schrader H W and Eisert W G (1985) High resolution particle sizing using the combination of time-of-flight and light-scattering measurements.. Appl. Optics 25 : 4396 - 4401

Shinitzky M, Dianoux A-C, Gitler C, Weber G (1971) Microviscosity and order in the hydrocarbon region of micelles and membranes determined with fluorescent probes. Synthetic micelles. Biochemistry 10: 2106-2113

Stryer L (1960) Energy Transfer in proteins and polypeptides. Radiat. Res. Suppl. 2:432-451

Stryer L, Haugland RP (1967) Energy Transfer. A spectroscopic ruler. Proc. Natl. Acad. Sci. U.S.A. 58 : 719-726

Stryer L (1978) Fluorescence energy transfer as a spectroscopic ruler. Ann. Rev. Biochem. 47:819-846

Stryer L, Thomas DD, Meares CF (1982) Diffusion-Enhanced Fluorescence Energy Transfer. Ann. Rev. Biophys. Bioeng. 11 : 203-222

Stunitzky M (1973) Rate of non radiative energy transfer between identical molecules. Chem. Phys. Lett. 18 : 247-249

Tasaki I, Warashina A, Pant H (1976) Studies of light emission, absorption and energy transfer in nerve membranes labelled with fluorescent probes. Biophys. Chem. 4 : 1-13

Thaer AA, Sernetz M (eds) (1973) Fluorescence Techniques in Cell Biology. Springer Verlag, Berlin

Vaz WLC, Kaufmann K, Nieksch A (1977) Use of energy transfer to assay the association of proteins with lipid membranes. Anal. Biochem. 83 : 85-393

Weber G (1952) Polarization of the fluorescence of macro-molecules. 2. Fluorescent conjugates of ovalbumin and bovine serum albumin. Biochem. J. 51:155-167

Weber G (1973) Polarized fluorescence. In: Thaer AA, Sernetz M (eds) Fluorescence Techniques in Cell Biology. Springer Verlag, Berlin pp. 5-13

Werner TC, Bunting JR, Cathou RE (1973) Proc. Natl. Acad. Sci. USA 69: 795-799

Wright K, Takahashi M (1977) Fluorescence energy transfer between heterologous active sites of affinity-labeled aspartokinase of Escherichia coli. Biochemistry 16 : 1548-1554

Zukin RS, Hartig PR, Koshland DE (1977) Use of a distant reporter group as evidence for a conformational change in a sensory receptor. Proc. Natl. Acad. Sci. U.S.A. 74 :1932-1936

FLUID MECHANICAL PROPERTIES OF FLOW CYTOMETERS AND ASSESSMENT CELL-CELL ADHESION FORCES

F. AMBLARD
Department of Immunology
Institut Pasteur
25 rue du Docteur Roux
75015 Paris
France.

INTRODUCTION

The work described here originates from an effort aimed at analyzing the molecular basis of cell-cell adhesion between T lymphocytes and antigen-presenting cells. Classical techniques used by cell biologists and immunologists to "measure" cell-cell adhesion are based on counting conjugates[(1)] and assume that the extent of conjugate formation reflects the adhesion "efficiency"; flow cytometric versions of this conjugate-counting approach have already been used [Segal 1984; Shaw 1986]. While studying the adhesion between human peripheral resting T cells and B cells, we reproducibly failed, in all flow cytometers tested thus far, to detect conjugates that could be seen otherwise under the microscope; this led us to surmise that flow cytometers could disrupt fragile conjugates. In the first section, we describe the mechanical stress to which cells are subjected when being analyzed in standard flow cytometers.

Hydrodynamic focusing, the essential principle of flow cytometry, is realized by forcing a fluid stream into a strongly constricting flow chamber. Every cell, when injected along the center of this converging stream experiences a set of tremendous acceleration and high velocity gradients. The resulting elongational stress tends to flatten cells and/or to disrupt cell-cell conjugates. The magnitude of this stress is proportional to the flow rate and depends on the geometry of the flow chamber. In order to minimize it, a new flow chamber was designed with a "softly" constricting geometry together with accompanying modifications of

---

[1] Conjugates (or cell doublets) refer here to the adhesive association of two cells; conjugates made between two cell populations can be detected by staining each of the two populations with distinct dyes.

NATO ASI Series, Vol. H 67
Flow Cytometry
Edited by A. Jacquemin-Sablon

the signal processing circuitry and the fluidic system. The modified instrument when operated at low stream velocity, was compatible with the detection of fragile conjugates [Amblard, 1992a]. In addition, the possibility of operating the above modified instrument over an extended range of stream velocity, permitted a very wide range of calibrated stresses to be generated in the flow chamber [Amblard 1992b,c]. This opened the possibility, for a given conjugate suspension, of assessing the critical stress above which half of the conjugates are disrupted, and thus to measure the mean adhesion strength between conjugated cells. Practical and theoretical aspects of this methodology are presented in the second section.

1. WHAT HAPPENS IN A FLOW CYTOMETER ?

Particles to be analyzed in flow cytometers (cells, nuclei, chromosomes, synthetic beads, etc.) are handled in fluid suspensions(1), and the mechanical stress they experience mainly originates from the forces exerted by the surrounding fluid. This stress can be split into two components: (*1*) the isotropic forces caused by the hydrostatic pressure, and (*2*) the viscous forces resulting from the local motion of the fluid relative to the particles. The aim of the present analysis is to quantify the latter component of the stress(2). The actual motion of the fluid around a given particle is extremely complex, but the bulk force it imparts upon flowing particles can be related to the ambient local relative velocity field at the position where the particle is located(3). This local relative velocity field can be directly obtained from the space-derivatives of the velocity field of the overall stream.

The analysis of viscous stresses experienced by particles in each part of their trajectory through a flow cytometer can be performed as follows:

1- description of the overall velocity field(4),

---

1 Carrier fluids are usually aqueous fluids with or without physiological concentrations of salts. Their viscosities are close to that of water.

2 Practically, the effect of isotropic hydrostatic pressure forces on the particles will be neglected, because most particles (cells, nuclei,..) can be considered as incompressible bodies.

3 The ambient local relative velocity field in the vicinity of a particle located at a given position is defined in the particle-fixed reference frame as the relative velocity the fluid would have in the vicinity of that position in the absence of the particle.

4 Because the particles are much smaller than the size of the stream, and since they usually occupy a negligible fraction of the stream volume (10^{-3} or much less), the overall velocity of the suspension can be considered to be identical to that occuring in the absence of particles. Moreover, the mean distance between neighbour particles is much larger than their size, and

2- description of the ambient local relative velocity field, which will be now simply referred to as relative velocity field, and

3- description of the mechanical effects of the relative velocity field upon flowing particles: rotation, extension/flattening, doublet disruption, etc.

For a given particle, order-of-magnitude comparisons between stresses experienced in different parts of the trajectory can be made by comparing the rates of strain which characterize the different relative velocity fields[1]. These rates (velocity gradients, shear rates, or extension rates) are all expressed in reciprocal seconds. A basic assumption of the following treatment is that flows are everywhere laminar and steady-state. Instabilities will not be discussed here (see Pinkel (1985) for a general discussion).

Let us first consider what happens to particles when the sample suspension is driven through the sample tubing toward the flow chamber. In typical analysis conditions[2], Reynolds numbers[3] in this part of the trajectory amount to several hundred; therefore the flow is laminar, and a parabolic velocity profile is rapidly installed downstream of the inlet region (figure 1). In this simple overall velocity field, which is the classical Poiseuille flow, each particle located at a distance r from the tubing axis undergoes a linear shear with a rate S which is proportional to r. $S=\frac{4rV_0}{R^2}$ where V_0 is the mean fluid velocity through the tubing, and R the tubing radius. This transversal shear rate generates a torque that causes tumbling of non-spherical particles; in the case of two attached particles, a force is also generated along the line of centers which is proportional to S and tends to separate them [Arp 1977]. Noteworthy, the mean value of S over the tubing cross-section is of the same order of magnitude as the velocity gradient at the tubing inlet[4]. In standard conditions, the rates of strain at the inlet of and through the sample tubing and are roughly 3 orders of magnitude smaller than those experienced downstream when the flow is constricted through the flow chamber (see below). Therefore, the major stress occurs in the flow chamber.

---

each particle (or pair of associated particles) can be considered to be free of hydrodynamic interactions with its neighbours.

1 Schematically, for a given particle, the stresses are proportional to the rate of strain and to the viscosity of the carrier fluid, which falls off in the comparison.

2 In standard analysis conditions, typical orders of magnitude of the main analysis parameters are: suspension concentrations: 10^6 particles/ml; diameter of the sample tubing: 150 to 300 μm; analysis rates: 10^3 particles/sec.

3 The Reynolds number is a dimensionless parameter that represents the ratio of the inertia forces to the viscous forces. It is used to characterize different regimes of flow: laminar or turbulent. It is computed as $R_e=Ud/\nu$, where U and d are a characteristic velocity and a characteristic linear dimension of the stream, and ν the dynamic viscosity [Batchelor 1967]. Here $R_e=2RV_0/\nu$.

4 An order of magnitude of the velocity gradient due to the stream acceleration at the tubing inlet is given by $V_0/2R$, and the mean value of S over the tubing cross-section is $8V_0/3R$.

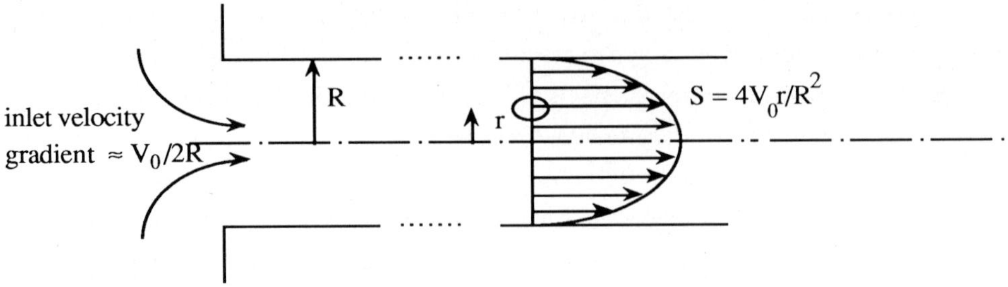

Figure 1: Flow through the sample tubing.
The flow of the sample suspension through the sample tubing is laminar. The velocity gradient at the tubing inlet and the radial shear rate S due to the parabolic velocity profile developped downstream are of the same order of magnitude. The corresponding stresses imparted onto sample particles are 3 order of magnitude smaller than those experienced in the focusing flow chamber.

When leaving the sample tubing, the particle suspension is precisely injected along the axis (z) of the flow chamber, or nozzle, as a fine stream in the middle of a larger sheath of flowing fluid. Figure 2 schematically shows two typical realizations of this constriction. A extended presentation of various flow chamber geometries is given by Pinkel and Stovel [Pinkel 1985].

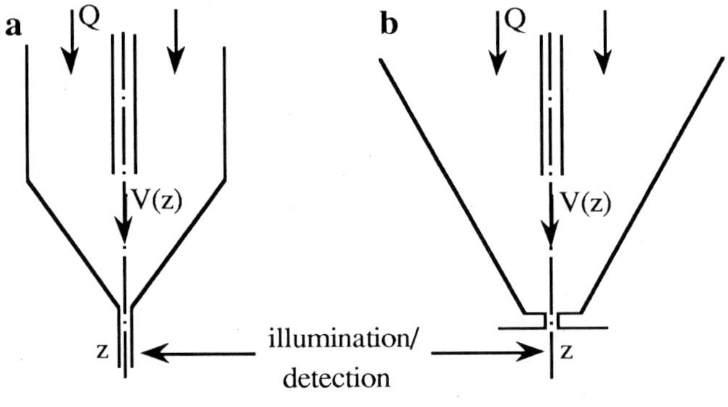

Figure 2: Flow chamber geometries.
Axial cross-section of two typical flow chamber geometries. The sheath flow, together with the central sample core, is hydrodynamically focused toward the sensing devices. The total flow rate Q of the stream and the sample is simply measured at the chamber exit, and easily gives the mean exit velocity (V_e). Illumination/detection can occur either within a channel with square cross-section (a), or as a circular jet in the air (b). In the last case, chamber boundaries are usually axisymmetric (FACS™ series).

The converging geometry of the flow chamber reduces the cross-section to a circular or rectangular cross-section with typical linear dimensions between 50 and 300 μm. The overall reduction ratio of the crossectional area might range from 100 to 1500. Along the axis z of the chamber, the resulting reduction of the sample stream to a very narrow and well centered cross-section, with typical core diameters(1) below 10 μm, is called hydrodynamic focusing. This principle, first realized by Crosland-Taylor [Crosland-Taylor 1953], has the essential advantage that all particles pass through a region of uniform sensitivity of the detectors [Kachel 1990]. Consequently, all particles will be subjected to the same set of hydrodynamic stresses, that will be analyzed now.

The overall pattern of the velocity field strongly depends on the geometry of the chamber; its quantitative description is in most cases very complex, especially when the slope of the chamber boundaries display discrete variations. Nevertheless, since we are essentially interested in computing the maximum of the stress along the z axis, rough approximations may lead to useful order of magnitude estimates.
In the case of geometries similar to that shown in figure 2a (i.e. with continuous variation of the cross-section size), the velocity V(z) at position z along the axis can be obtained in a first approximation by computing the mean axial velocity $V_m(z)$ through the normal cross-section at position z, and assuming that the ratio $V(z)/V_m(z)$ does not depend on z (2). Because the fluids used in flow cytometry are incompressible, the mean axial velocity $V_m(z)$ through the normal cross-section at position z, is proportional to its reciprocal area $A^{-1}(z)$: $V_m(z)=Q.A^{-1}(z)$ where Q is the flow rate. The ratio $V(z)/V_m(z)$ equals 2 for Poiseuille flows through straight circular ducts, and can be computed for various cross-section shapes(3). This approximation might be safely used as far as the quasi-Poiseuille approximation is valid(4). Nevertheless, fluid inertia affects V(z) by a z-shift, but does not modify its overall shape and amplitude (data not shown). Therefore, the quasi-Poiseuille approximation is useful over a wide validity domain for most geometries of type a (figure 2a), as far as one is only interested in the maximum of V(z) and of its gradient.

---

1 The core refers to the portion of the stream containing the sample.

2 V/V_m reflects the shape of the transversal distribution of the velocity, and can be practically assumed to be constant provided the area or the shape of the cross-section does not change too abruptly with z. This is the case for chambers with geometries similar to that shown on figure 2a but not 2b.

3 Exact solutions for the velocity of viscous flows through straight ducts with various shapes have been established and can be used to compute V/V_m [White 1991].

4 The quasi-Poiseuille approximation consists in neglecting the inertia of the fluid. It is valid when the crossectional area A(z) varies slowly with z. Strictly speaking, this is almost never the case with usual commercial flow chambers.

In the case of abruptly constricting geometries such as the "hole" shown on figure 2b, the velocity field along the axis may be computed using the exact solution developped for the viscous flow through a circular aperture in a plane wall [Happel 1983].

From the velocity function V(z) computed in any of the two above cases, a quantitative description of the relative velocity field can be made as follows in the particle-fixed reference frame that moves with the fluid. Let M be the point at position z on the axis, $M+\overrightarrow{\delta M}$ a point in the vicinity of M, and δx, δy and δz the (small) cartesian coordinates of $\overrightarrow{\delta M}$. The difference in velocity between M and $M+\overrightarrow{\delta M}$ can be linearized as:

$$\vec{V}(M+\overrightarrow{\delta M}) - \vec{V}(M) = G(z)\begin{bmatrix} -p_1 & 0 & 0 \\ 0 & -p_2 & 0 \\ 0 & 0 & 1 \end{bmatrix} \cdot \begin{bmatrix} \delta x \\ \delta y \\ \delta z \end{bmatrix}$$

where G(z)=dV(z)/dz is the gradient of the axial velocity, and p_1 and p_2 are dimensionless parameters that depend on the variations of the shape of the cross-section. p_1 and p_2 are such that $1-p_1-p_2=0$. In the case of circular cross-sections, $p_1=p_2=1/2$. Noteworthy, the diagonal form of the above matrix is lost if M is not on the central axis. In that case, non diagonal terms appear which correspond to transversal shearing. This terms increase in a linear fashion with the reduced radius r/R. They can be neglected provided the sample core is kept well centered and with a small size. Usually, stable sample cores are achieved with reduced radii as small as 1/10 to 1/30. At all events, the relative velocity field around M is a pure extension with rate G(z), that streches the fluid in the direction of the flow and compresses it in the transversal plane -figure 3a. From the quasi-Poiseuille approximation, an order of magnitude of the velocity gradient is given by: $G(z) = \frac{-2Q}{A^2}\frac{dA}{dz}$. In standard instruments, the maximum values reached by G(z) along the trajectory can be as high as $10^5\ s^{-1}$ [Pinkel 1985].

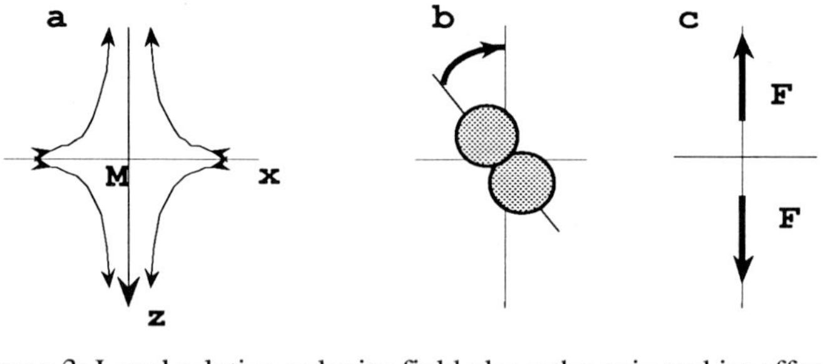

Figure 3: Local relative velocity field along the axis and its effects.
Figure 3a shows an axial cross-section of the stream lines of the ambient fluid velocity relative to the particle-fixed reference frame moving with an axial point M. The central axis z is oriented as shown on figure 2, and x represents a transversal axis. Figure 3b and 3c represent the orientation and the extension effects of the relative velocity field.

A particle sitting in the center of the above extension (figure 3a) will experience two effects:

- if it is a non-spherical particle, it will be progressively oriented with its longest axis parallel to the direction of the flow (figure 3b).
- in all cases, the particle will experience an extensional stress that tends to flatten it; in the case of two attached particles, this stress can detach them (figure 3c).

The orientation of particles is critical in several techniques, like chromosome analysis, in which the optical information detected by the instrument has to be resolved along the longest axis of the particle. In the case of particles with prolate shapes and of doublets of equal spheres attached to each other, we have modelled in a previous work the kinetics of the orientation effect [Amblard 1992c]. Interestingly, it was found that the orientation efficiency does not depend on the flow rate but only on the shape of the particles and of the chamber boundary. In other words, if particles with a known geometry are injected with a random orientation at position z_0, their orientation at position z_1 with respect to the z axis will only depend on the ratio $A(z_1)/A(z_0)$ between the two crossectional areas[(1)], and not on the flow rate. Axial orientation can be considered to be completed for most particles as soon as the dimensionless number $\mathrm{Log}\{A(z_1/A(z_0))\}$exceeds 3 to 4 [Amblard 1992c]. Practically, in standard instruments, this condition is met and most particles get oriented before exiting the constriction[(2)].

The extension effect is very useful because it separates particles that are weakly attached in the suspension, and allows them to be analyzed individually. A simple way for obtaining an order of magnitude estimation of the extension effect is to introduce the notion of bulk extension pressure defined by $P(z)=2\mu G(z)$, where μ is the fluid viscosity. Using the above estimates of G(z) in standard instruments, P(z) can be as high as 10^2 Pa. For any particle, the extension force can be very roughly estimated from P(z) using $F=A_pP(z)$ where A_p is the crossectional area the particle opposes to the extension. Beyond this order of magnitude estimation, the extension force was computed more precisely in the case of two attached spherical particles, thus leading to a linear relationship between the extension rate G(z) and the force that tends to detach them -see below and [Amblard 1992c].

---

[1] A prediction of our model is that increased flow rates do not improve the orientation efficiency.

[2] Direct microscopic observations have been reported for the orientation of particles beyond the constriction [Fulwyler 1977]. In the case of discoidal particles like chicken erythrocytes sometimes used for instrument calibration, the orientation leads to two equilibrium positions because these particles have two principal axis.

Particles subjected to high extension rates can be deformed if they are not rigid enough. To estimate the extent of particle deformation, one must take into account their mechanical characteristics. In the case of viscoelastic particles[1], the viscous time constant for large deformations has to be compared with the duration of the stress[2], and the elastic moduli with the bulk extension pressure P(z) defined above. Generally, if the duration of the extension is shorter than the viscous time constant, no viscous deformation can occur during the stress; consequently, the particle can be considered as an elastic body, and its relative deformation can be estimated from the ratio of P(z) to its elastic modulus. If particles have short viscous time constants, they will undergo both elastic and viscous deformations. For a discussion in the case of lymphocytes, see Amblard (1992b).

The above developments were made for particles ideally located on the axis of the chamber. Practically, the crossectional area of the sample core is determined by the ratio of the sample flow rate to the sheath flow rate, and can be made arbitrarily small by reducing the sample flow rate; with standard sample injection needles, stable sample cores can be usually achieved with diameters in the 1-10 μm range, i.e. with reduced diameters in the 1/10-1/30 range as stated above. Within this range, the amplitude of the transversal velocity gradient remains much smaller than the extension rate in the constricting region, and the tumbling of non-spherical particles is not induced in that region. Consequently, all particles can be considered to be subjected to the same orientation process and to the same set of elongating forces. This feature of flow cytometry, hydrodynamic homogeneity, represents an essential advantage for the assessment of adhesion forces as describe below.

2. ANALYSIS AT VARIABLE STREAM VELOCITY AND ASSESSMENT OF CELL-CELL ADHESION FORCES

In this section, instrumental modifications as well as experimental and theoretical results will be briefly described; a detailed presentation is given in Amblard (1992a,b,c).

As stated above, the major mechanical stress imparted onto particles originates from the extension G(z) experienced along the axis of the flow chamber. In order to control G(z) as

---

1 Most cells can be described according to a viscoelastic model, characterized by viscous time constants and elastic moduli [Schmid-Schönbein 1990].

2 The duration of the stress can be estimated by looking at the curve showing G(z) as a function of z [Amblard 1992b]. It is usually very brief, in the 10^{-3} to 10^{-4} sec range in standard instruments.

precisely as possible, we built a new quartz flow chamber adapted for the FACSTARPlus™ (1) [Amblard 1992a]. This chamber has a geometry similar to that of a converging funnel: it is made of a 30° revolution vertex connected to a straight channel with a 250 µm square cross-section(2). As opposed to most standard flow chambers used, the essential feature of the new chamber geometry is that the connection between the circular cross-section of the vertex and the square one of the channel is very progressive, so that the whole chamber can be considered to be hydrodynamically equivalent to a revolution funnel defined at each position z by a boundary radius b(z), b(z) being a "smooth" function of z (3). As a consequence, it was possible to compute the velocity field and the axial velocity gradient G(z) more precisely than just with the quasi-Poiseuille approximation. We found a linear relationship between the mean exit velocity of the flow (V_e)(4) and the maximum $G_m = \max_z G(z)$ of the extension rate: G_m ranges from 44 to 44,000 s^{-1} when V_e is between 0.01 and 10 $m.s^{-1}$ - i.e. 0.625 $mm^3.s^{-1} < Q < 625$ $mm^3.s^{-1}$. Accompagnying modifications of the fluidic system were necessary to permit analysis to be carried out at any position between 0.01 and 10 $m.s^{-1}$ [Amblard 1992a]. Indeed, commercial instruments are usually designed to be operated over a very narrow range of sheath flow rates, usually less than one order of magnitude, and the pressure system used to force the sheath fluid and the sample suspension through the constriction is calibrated accordingly. In order to control low flow rates with minimal adaptation of the fluidic system, a flow restrictor was inserted on the sheath line. It was therefore possible to obtain very stable flow conditions for the sheath fluid over long periods of time, i.e. two days. At very low velocity, correct hydrodynamic focusing and small sample cores can only be obtained if the sample flow rates are kept proportionally very low.

The whole electronic circuitry of commercial instruments is dedicated to the detection of brief optical signals generated when the particles cross the laser beam focus(5). Since the detection optics of the FACSTARPlus™ was not modified in our experiments, lower particle velocity

---

1 The new chamber has can be mounted on the standard FACS™ nozzle holder in the same way as standard nozzles.

2 Illumination and optical detection take place in the straight channel using the same optical setup as that of the standard FACSTARPlus™.

3 "hydrodynamically equivalent" means here that the fluid velocity in the vicinity of the axis for a given flow rate through this model funnel is identical to that in the actual chamber for the same flow rate. The radius b(z) is defined by $\pi b^2 = A(z)$, where A(z) is the crossectional area of the actual channel.

4 This parameter V_e can be easily and precisely measured as the ratio of the output flow rate to the exit crossectional area. On the standard FACS™ machine, V_e can be tuned between 5 and 10 $m.s^{-1}$. Below this range, the jet in the air becomes a series of droplets and the analysis is no longer possible.

5 The length of the beam focus is approximately 20 µm in the direction of the flow and it is crossed at 5-10 $m.s^{-1}$.

generated longer pulses, in the 100-1,500 μs range instead of the typical 3 μm pulses. The detection circuitry was therefore adapted to a wide range of temporal conditions; this was achieved by modifying the control of the detection time constants as already described [Amblard 1992a]. As a result, regular analysis capabilities of the FACSTARPlus™ were translated over the whole range of exit velocity (V_e). Incidentally, the increase in illumination and detection times at low velocity could be useful with respect to: (*1*) the improvement of signal to noise ratios, and therefore sensitivity levels, (*2*) the development of time- and frequency-resolved illumination and signal detection, and (*3*) the implementation of more refined analog or digital processing of optical signals.

Because our original interest was the detection of cell-cell conjugates[(1)], the modified instrument was first used for a classical conjugate-counting experiments at low and fixed velocity. As expected, conjugates[(2)] that were observed in fluorescence microscopy but were disrupted in all unmodified flow cytometers tested so far, could be detected in the modified flow cytometer provided the stream velocity was kept low [Amblard 1992a]. Such experiments at fixed velocity, like the classical conjugate counting technique, either with a microscope or a flow cytometer, indicate the "statistic efficiency" of conjugate formation, but do not yield any information on the strength of cell-cell adhesion.

The new instrumental concept introduced here is based on the possibility of operating the above modified instrument over an extended range of stream velocity, thus permitting a very wide range of calibrated stresses to be generated. This opens the possibility of assessing, for a given conjugate suspension, the critical value of the stress above which conjugates are disrupted, thus giving the mean adhesion strength between conjugate-forming cells. Such physical informations can usually be obtained by various methods, in which the disrupting force is caused either by micromanipulation, hydrodynamic shear, or centrifugal force [Bongrand 1988]. The unique advantages of using a flow cytometer are: (*1*) it is so far the most powerfull instrument for rapidly counting individual and associated cells [(3)], (*2*) the very principle of hydrodynamic focusing ensures that every individual cell will be subjected to the same set of hydrodynamic extensions, (*3*) the disrupting stress is much more brief than in any other technique[(4)], and (*4*) the intensity of the extensional stress can be much smaller than the

---

1 Conjugates between two cell types are pairs of cells of each type; they are formed by incubating the two suspensions for a precise conjugation time. Conjugates are detected in the flow cytometer by two distinctive parameters, either two dyes, or one dye and cell size, see figure 4.

2 Conjugates were made between human peripheral resting T lymphocytes and B lymphocytes.

3 Most other techniques (flow-induced detachment assays or micromanipulation) are very slow because they are based on the visual observation and counting of individual conjugate-disruption events.

4 The duration of the stress (typically 1 to 10 ms) is much shorter than duration of the stress in micromanipulation.

minimal stress developped in micromanipulation techniques[1]; (5) there is no practical lower limit for the size of particles, as long as the instrument can detect them.

Because the present approach is many orders of magnitude more rapid than any other previous technique, it was thus possible for the first time to assess cell-cell adhesion forces in different biological circumstances, and in particular to use monoclonal antibodies against cell surface adhesion molecules to dissect quantitatively the molecular basis of these forces [Amblard 1992b].

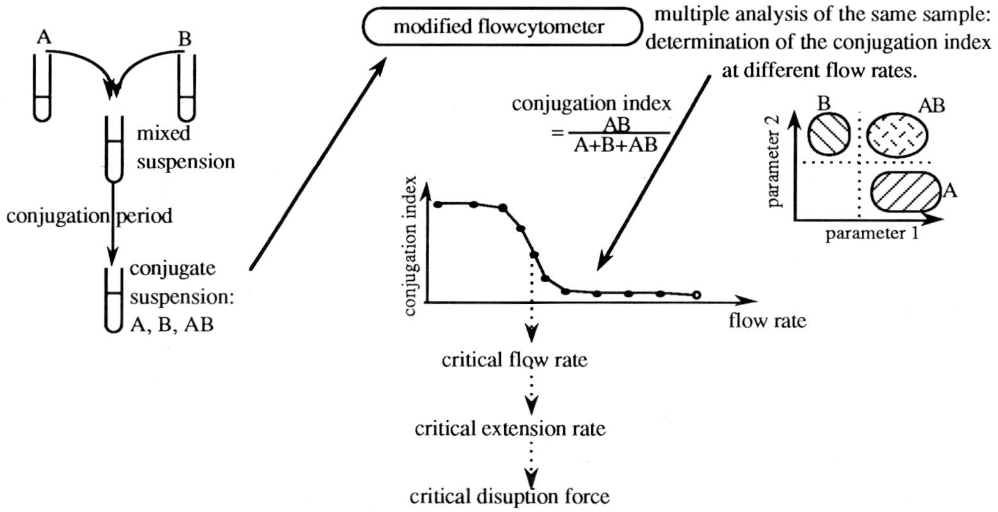

Figure 4: Schematic representation of the strategy used for the assessment of the hydrodynamic force required to rapidly disrupt cell-cell conjugates resulting from the adhesion between two cell types. Equal volumes of two suspensions containing equal density of two cell types (A and B) are mixed and allowed to form conjugate at rest for a given conjugation time. The conjugate-containing suspension is loaded onto the modified flow cytometer and successively analyzed at different flow rates. The conjugate index, i.e. the number of conjugates that resisted the hydrodynamic extension during the analysis, is then plotted as a function of the flow rate. The critical flow rate extracted from the "disruption" curve is then translated into a critical extension rate (or velocity gradient) G_m. If the geometric and mechanical properties of the cells are known, G_m might then be converted into a disruption force, the magnitude of which is a indication of the cell-cell adhesion strength.

Practically (figure 4), the conjugate suspension to be measured is loaded and successively analyzed at different values of the flow rate Q (which is proportional to the mean

---

[1] Cell micromanipulation involves traction forces generated by small aspiration pressures of at least 1mm H_20. This is roughly 1 order of magnitude higher than the lowest bulk extension pressure (2μG) that could be reached in our experimental conditions.

exit velocity V_e). The level of resistant conjugates is detected and plotted as a function of the flow rate. From the "dissociation curve" obtained, a critical flow rate is extracted. Using the above linear relationship between Q and the maximum velocity gradient G_m, a critical rate G_c is then obtained. Finally, G_c is translated into a disrupting force as already described [Amblard 1992b,c]. Briefly, the bulk extension force experienced by two contacting rigid and spherical cells with radii a_1 and a_2 is given by: $F = 4\pi\mu a_1 a_2 G_c L$ where μ is the fluid viscosity, and L a function of a_2/a_1 [(1)]. Noteworthy, the latter conversion of G_c into a force is only valid if cells can be considered as rigid bodies during the extension[(2)]. The condition on the sphericity of cells might be simplistic, but more realistic models can not yet be used because of the complexity of the problem. In our first experiments, the range of the disruption forces imparted upon T and B lymphocytes along the "disruption curve" was from 10^{-11} to 10^{-8} nN; this range comprises the lowest cell-cell adhesion forces measured so far [Bongrand 1988; Baltz 1990]. Future instrument designs should be adapted to perform measurements over a more extended range of forces.

CONCLUSION

The present work describes the mechanical treatment to which particles are subjected when analyzed in a flow cytometer. Based on this analysis, a commercial flow cytometer was modified to generate calibrated hydrodynamic extensions and to assess the critical conditions beyond which cell-cell conjugates are disrupted in the flow chamber. This novel method combines the unique power of flow cytometry for rapidly counting individual cells, with the fact that all cells are subjected exactly to the same calibrated extension while hydrodynamically focused through the flow chamber. As already demonstrated, the new strategy is very powerful for the precise and rapid quantification of cell-cell adhesion forces, and permits therefore, for the first time, these forces to be resolved at the molecular level of adhesion receptors. Multiple applications can be envisioned for the present technique for the study of the molecular and mechanical phenomena underlying cell-cell recognition and adhesion.

---

[1] Obviously $L(a_2/a_1)=L(a_1/a_2)$. L(1)=3.057, L(0.5)=2.630, L(0.25)=2.077 [Amblard 1992c].

[2] This condition depends both on the flow rate and on the viscoelastic characteristics of the cells. It was met in our experiments with lymphocytes at low and medium velocity but not at high velocity.

REFERENCES

Amblard F, Cantin, C, Durand J, Fischer A, Sékaly R, Auffray C (1992a) New chamber for flow cytometric analysis over an extended range of stream velocity and its application to cell adhesion measurements. Cytometry 13:15-22.

Amblard F, Auffray C, Sékaly R, Fischer A (1992b) Molecular analysis of antigen-independent adhesion forces between T and B lymphocytes. submitted.

Amblard F (1992c) Hydrodynamic extensions applied to the assessment of contact forces between suspended particles: a theoretical description. submitted.

Arp P A, Mason S G (1977) The kinetics of flowing dispersions: doublets of rigid spheres (theoretical). J of Colloïd and Interface Science 61:21-43.

Baltz J M, Cone R A (1990) The strength of non-covalent biological bonds and adhesion by multiple independent bonds. J Theor Biol 142:163-178.

Batchelor GK (1967) An introduction to fluid dynamics. Cambridge University Press, Cambridge.

Bongrand P (1988) Physical basis of cell-cell adhesion, CRC Press, Boca Raton.

Crosland-Taylor PJ (1953) A device for counting small particles suspennded in a fluid through a tube. Nature 171:37-38.

Fulwyler M J (1977) Hydrodynamic orientation of cells. J of Histochem Cytochem 25:781-783.

Happel J, Brenner H (1983) Low Reynolds number hydrodynamics, Kluwer Academic publishers, Dordrecht Boston London, p 96.

Kachel V, Fellner-Feldegg H, Menke E (1990) Hydrodynamic properties of flow cytometry instruments. In: Melamed MR, Lindmo T, Mendelsohn ML (eds) Flow cytometry and sorting. Wiley and Sons, New York, p 27.

Pinkel D, Stovel R (1985) Flow chambers and sample handling. In: Van Dilla M A, Dean P N, Laerum O D, Melamed M R (eds) Flow cytometry: instrumentation and data analysis. Academic press, p 77.

Schmid-Schönbein G (1990) Mechanical properties of leukocytes. Cell Biophysics 17:107-135.

Segal DM, Stephany DA (1984) The measurement of specific cell-cell interaction by dual parameter flow cytometry. Cytometry 5:169-181.

Shaw S, Ginther-Luce GE, Quinones R, Gress RE, Springer TA, Sanders ME (1986) Two antigen-independent adhesion pathways used by human cytotoxic T-cell clones. Nature 323:262-264.

White FM (1991) Viscous Flows. McGraw-Hill, New York, p 104.

Use of biotin-labeled growth factors for receptor studies

M.O. DeJong, H. Rozemuller, J.G.J. Bauman and J.W.M. Visser
Department of Molecular Pathology
Institute of Applied Radiobiology and Immunology TNO
P.O. Box 5815
2280 HV Rijswijk
The Netherlands

INTRODUCTION

Most receptor studies are performed using either radiolabeled growth factors (GF) (Foxwell et al. 1988, Nicola et al. 1988, Park et al. 1989), or antibodies against the receptor (Weil-Hillman et al. 1990). With radiolabeled GF, only functionally intact GF receptors are labeled. However, ^{125}I-labeled GF can not be used to sort cells for culture experiments. Cells, labeled with antibodies against the receptors, can be double-stained for other markers and sorted in a fluorescence activated cell sorter (FACS). Use of anti-receptor antibodies can cause problems if the labeled and sorted cells are used for in vivo assays (Bauman et al. 1985).

The advantages of radiolabeled GF and antibodies can be combined by using biotinylated GF (bGF). These bGF retained the capacity to induce proliferation of GF dependent cells. Several papers on bGF have been published (Foxwell et al. 1988b; Yamasaki et al. 1988; Newman et al. 1989; Peters and Norback 1990). In these studies, cells with high receptor densities (10^4 - 10^5 per cell) were used. The fluorescence signal of cells with low receptor numbers has to be amplified for analysis in a flow cytometer (Wognum et al. 1992). We incubated cells with bGF and fluorescently labeled (strept)avidin. After amplification of the fluorescence signal with alternate layers of biotinylated anti-(strept)avidin and fluorescently labeled (strept)avidin, it was possible to analyze cells with 100 cell surface receptors in a flow cytometer.

NATO ASI Series, Vol. H 67
Flow Cytometry
Edited by A. Jacquemin-Sablon

MATERIALS AND METHODS

Biotinylation of growth factors

There are some companies now that sell biotin-labeled GF (e.g.., British Bio-technology Products Ltd., Oxon, UK). We biotinylated interleukin 2, 3 and 6 (IL-2, IL-3, IL-6), granulocyte-macrophage colony stimulating factor (GM-CSF), and stem cell factor (SCF) ourselves. Biotin-*N*-hydroxy succinimide ester (Pierce, Rockford IL, USA) in dimethyl sulfoxide was added to solutions of GF in 0.2 M carbonate-bicarbonate buffer pH 8.5 to get a molar biotin/protein (B/P) ratio of 20 during the reaction (note that this is not the resulting conjugation ratio). After 3h of incubation on ice, the samples were dialyzed overnight against PBS. The bGF were stored at 4°C in the presence of 0.02 % (w/v) sodium azide.

Cell lines and biological activity assays

Cells were grown in RPMI 1640 supplemented with 10% (v/v) heat inactivated fetal calf serum, 0.1 mM β-mercapto ethanol, 2 mM L-glutamine, and 100 U of IL-3 or GM-CSF. Cultures were maintained at 37°C in a humidified atmosphere of 5% CO_2 in air. The biological activity of the bGF was assayed using murine bone marrow cells, or GF-dependent cell lines (CTLL-2 for IL-2, DA-1, 32D or FDC-P1 for IL-3, T1165 for IL-6, and FDC-P1 for GM-CSF). In 96 well microtiter plates, 10^4 cells per well were cultured in 200 µl RPMI medium containing different amounts of either bGF or unmodified GF. ^{3}H-thymidine (0.25 µCi per well) was added after 24 hours. After 16-18 hours of thymidine incorporation the cells were harvested, and the radioactivity was measured in a liquid scintillation counter.

Immunocytochemical staining and flow cytometry

Cells were labeled in 50 µl PBS/Serum/Azide (PSA): PBS containing 5 % (v/v) fetal or newborn calf serum and 0.02 % (w/v) sodium azide. First, 10^6 cells per sample were incubated with bGF (10-100 ng) for 60 minutes on ice or at room temperature. Next, the samples were labeled for 15 minutes on ice, depending on the desired fluorescence wavelength either with avidin/ FITC (1:300 v/v, Vector Laboratories, Burlingame CA, USA) or streptavidin/ Bodipy (1:100 v/v, Molecular Probes, Eugene OR, USA), or with streptavidin/ Phyco erythrin (1:5 v/v, Becton Dickinson, San Jose CA, USA; or 1:200-1:1000, Molecular Probes). After every incubation the samples were washed in PSA. The fluorescence signal was amplified by incubating the cells for 15 minutes on ice with alternate layers of biotinylated goat-anti-avidin or goat-anti-streptavidin (both 1:100 v/v, Vector) and fluorescently labeled (strept)avidin.

Samples were analyzed using a flow cytometer. Cells were illuminated with the 488-nm line of an argon ion laser. Green FITC or Bodipy fluorescence was measured through a 530-nm/ 30-nm bandpass filter (Becton Dickinson, Mountain View CA, USA). Red PE fluorescence was measured through a 577-nm/ 10-nm bandpass filter.

RESULTS AND DISCUSSION

Biological activity

The biological activities of bGF and native GF were compared in proliferation assays, using GF dependent cell lines. On DA-1 cells IL-3 was shown to have retained full growth promoting activity after biotinylation at B/P ratios of 20 or less (**figure 1**).

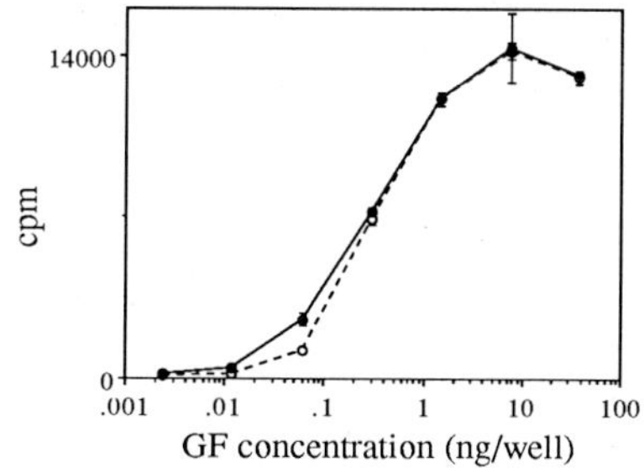

Figure 1.
Incorporation of ^{3}H-thymidine into DNA of DA-1 cells as a function of the concentration of native IL-3 (open circles) and biotinylated IL-3 (B/P 20, closed circles).

At B/P 50 the bGF was less active, while bGF at B/P 100 lost almost all biological activity (not shown). Similar results were obtained with other bGF on other GF dependent cell lines. For labeling experiments it is important that the bGF can bind to the receptors. These results show that bGF at B/P 20 can be used for flow cytometric studies.

Fluorescence amplification

Cells were incubated with bGF and fluorescently labeled (strept)avidin, and the effect of fluorescence amplification with alternate layers of biotinylated anti (strept)avidin and fluorescently labeled (strept)avidin was studied using a flow cytometer. With each amplification step the difference in fluorescence intensity between negative and positive cells was larger (**figure 2**), even if there was an increase in noise (background signal) along with the increase in signal.

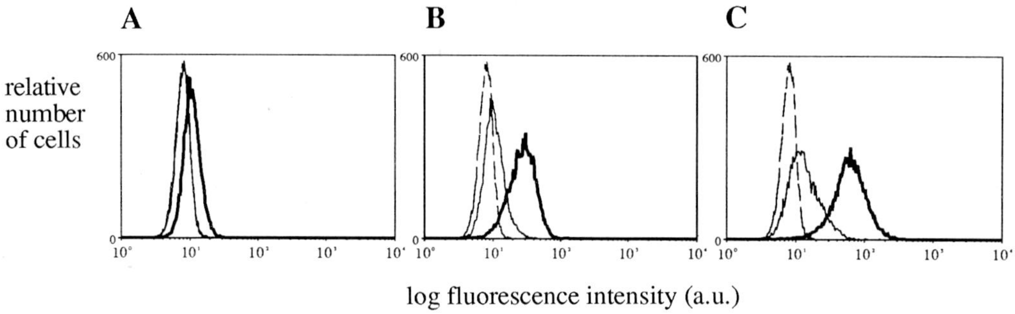

Figure 2.
Fluorescence histograms of DA-1 cells, incubated with (thick solid line) or without (thin solid line) biotinylated IL-3. Cells were labeled with avidin/FITC, and 1x (**A**), 2x (**B**), or 3x (**C**) amplified with biotinylated anti-avidin and avidin/FITC. Broken lines indicate unlabeled cells.

After several amplification steps it was possible to analyze cells from a subclone of the FDC-P1 cell line. These cells were shown by Scatchard analysis (Budel, personal communication) to have 100 GM-CSF receptors per cell. However, it was not possible to distinguish murine bone marrow (BM) cells, stained with biotinylated GM-CSF, from background stained cells. This could be due to the high background signal of murine BM cells, or it could be the result of a different processing of hemopoietic GF receptors on BM cells, as compared to cell lines.

The specificity of the labeling was studied by incubation with bGF in the presence of a 100-fold molar excess of the unbiotinylated GF. This resulted in a fluorescence signal that was approximately equal to the background signal of cells without bGF. A 100-fold molar excess of another GF did not change the fluorescence signal. Labeling of cells with bGF for 60 minutes on ice resulted in similar levels of fluorescence as labeling at room temperature.

Double-staining

Double-staining of cells with bGF and antibodies against the receptors or against other cell markers provides additional information about the characteristics of those cells. Labeling with bGF and anti-receptor antibodies is especially interesting for GF receptors that consist of different subunits, like the IL-2 receptor α and β (Foxwell et al. 1988a, Weil-Hillman et al. 1990). In our hands, incubation of human peripheral blood lymphocytes (PBL) with biotinylated IL-2 or with the antibody against the α subunit of the IL-2 receptor (Tac) resulted in different fluorescence histograms (**figure 3**).

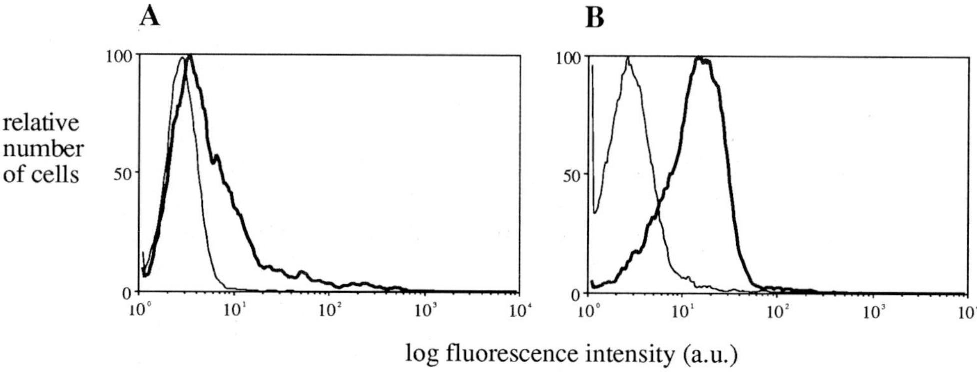

figure 3
Fluorescence histograms of human peripheral blood lymphocytes, incubated (**A**) with an antibody against the α subunit of the IL-2 receptor, or (**B**) with biotinylated IL-2. Thick lines represent stained cells, thin lines represent fluorescence controls, incubated without the first layer.

Occupied GF receptors are rapidly internalized into the cells (Nicola et al. 1988) It may be possible to target streptavidin-conjugated immunotoxins into cells, using bGF in stead of biotinylated antibodies (Meyer et al. 1991).

Other applications of bGF can be found in the study of receptors that can bind different GF. E.g.., competition between biotinylated IL-3 and unbiotinylated GM-CSF vice versa can be used in the study of the receptor or receptor subunit that can bind both IL-3 and GM-CSF (Park et al. 1989, Kitamura et al. 1991).

ACKNOWLEDGMENTS

We thank Behringwerke Aktiengesellschaft (Marburg, Germany) for supplying us with IL-3 and GM-CSF. We also thank Drs. R. Delwel and L. Budel (Daniël den Hoed Cancer Center, Rotterdam, the Netherlands) for the 32D and FDC-P1 cell lines, and the ^{125}I-GM-CSF Scatchard analysis data. This work was supported by TNO via grant 900-508-232 of the Netherlands Organization for Scientific Research NWO.

REFERENCES

Bauman JGJ, Mulder AH, van den Engh GJ (1985) Effect of surface antigen labeling on spleen colony formation: comparison of the indirect immunofluorescence and the biotin-avidin methods. Exp Hematol 13:760-767

Foxwell B, Taylor D, Ryffel B (1988a) Comparison of the structure of the murine interleukin 2 (IL 2) receptor on cytotoxic and helper T cell lines by chemical cross linking of ^{125}I-labeled IL 2. Eur J Immunol 18:1515-1519

Foxwell BMJ, Taylor D, Greiner B, Mihatsch MJ, Olivieri V, Ryffel B (1988b) Biotinylated recombinant interleukin-2. A tool for research on the interleukin-2 receptor. Journal of Immunological Methods 113:221-229

Kitamura T, Sato N, Arai K-i, Miyajima A (1991) Expression cloning of the human IL-3 receptor cDNA reveals a shared β subunit for the human IL-3 and GM-CSF receptors. Cell 66:1165-1174

Meyer BF, Stoner ML, Raphael CL, Davis RE, Herrmann RP (1991) Streptavidin-biotin immunotoxins: a new approach to purging bone marrow. Exp Hematol 19:710-713

Newman W, Beall LD, Bertolini DR, Cone JL (1989) Modulation of TGF-ß type 1 receptor: flow cytometric detection with biotinylated TGF-ß. Journal of Cellular Physiology 141:170-180

Nicola NA, Peterson L, Hilton DJ, Metcalf D (1988) Cellular processing of murine colony-stimulating factor (Multi-CSF, GM-CSF, G-CSF) receptors by normal hemopoietic cells and cell lines. Growth Factors 1:41-49

Park LS, Friend D, Price V, Anderson D, Singer J, Prickett KS, Urdal DL (1989) Heterogeneity in human interleukin-3 receptors. A subclass that binds human granulocyte/ macrophage stimulating factor. The Journal of Biological Chemistry 264:5420-5427

Peters DK, Norback DH (1990) Binding and internalization of biotinylated interleukin-2 in human lymphocytes. Blood 76:97-104

Weil-Hillman G, Voss SD, Fisch P, Schell K, Hank JA, Sosman JA, Sugamura K, Sondel PM (1990) Natural killer cells activated by interleukin 2 treatment *in vivo* respond to interleukin 2 primarily through the p75 receptor and maintain the p55 (TAC) negative phenotype. Cancer Research 50:2683-2691

Wognum AW, Krystal G, Eaves CJ, Eaves AC, Lansdorp PM (1992) Increased erythropoietin-receptor expression on CD34-positive bone marrow cells from patients with chronic myeloid leukemia. Blood 79:642-649

Yamasaki K, Taga T, Hirata Y, Yawata H, Kawanishi Y, Seed B, Taniguchi T, Hiranoto T, Kishimoto T (1988) Cloning and expression of the human interleukin-6 (BSF-2/IFNβ 2) receptor. Science 241:825-828

CYTOMETRIE TOUT TERRAIN or BUSH DNA CYTOMETRY

Spencer C. Brown
Cytométrie
Institut des Sciences Végétales
Centre National de la Recherche Scientifique
91198 Gif-sur-Yvette
France

This unconventional title derives partially from our subject matter being flow cytometry applied to plants, but principally from a proposition developed in the conclusion: namely, that an industrial application of cytometry in the diagnosis of plant ploidy levels will be vastly more economic when the cytometer can go to the plants - on a wheel-barrow if necessary - rather than the samples coming to the laboratory!

Before coming to that conclusion, we shall offer a succinct guide to the literature of plant flow cytometry, then with respect to DNA analysis cite the methods and tabulate the applications in research and in industry. Several needs for development are emphasized.

GENERAL

Our bibliography merely updates several broad reviews of plant flow cytometry which are still current, namely Brown and Bergounioux (1989), Galbraith (1989) and Fox and Galbraith (1990). These discuss the major constraints of handling plant material.

Few cell clusters in plant cultures are sufficiently fine - let alone individualized - to lend themselves to flow analysis. Even conventional quantitative fluorescence microscopy has been difficult on plant material. Fortunately, confocal microscopy appears to be breaching this lacune - with even modest spatial resolution, it may be a boon to quantitative plant microscopy in intact or living tissue.

NATO ASI Series, Vol. H 67
Flow Cytometry
Edited by A. Jacquemin-Sablon

As individualized cells, plant protoplasts (enzymatically-produced "cells without walls") have been analysed for secondary products such as indole alkaloids (Brown et al. 1984), for response to stress (Gantet et al. 1990) or toxins (Berglund et al. 1988), for uptake of plasmid (e.g. to quantify and so to optimize electroporation steps; Tagu et al. 1987) or virus, but especially to understand the processes of regeneration and proliferation (Bergounioux et al. 1992). Protoplasts have been sorted to establish cell-lines, for biochemical analysis (e.g. specialized cells synthesizing C10-monoterpenes; Matine, 1990), and to enrich heterofusions of different protoplast species in order to develop interspecific plants, regenerated from sorted protoplasts (Hammutt et al. 1990). Another source of living cells conducive to flow cytometry is, of course, pollen and the embryogenic microspores obtained *in vitro* (Deslauriers et al. 1991). These applications have typically stimulated techniques for sorting large particles up to 100 um diameter (Harkins and Galbraith, 1987). In all this work, natural cellular constituents such as chlorophyll and alkaloids are ubiquitous fluorochromes, to be exploited or circumvented.

Properties of chloroplasts (Petit et al. 1989; Xu et al. 1990), mitochondria (Petit, 1992) and vacuoles (Brown et al. 1984) have been studied. For instance, different classes of chloroplasts can be defined by light scattering signals, then analysed for galactose receptor sites on the chloroplast envelope and thylakoid membrane (Schröder and Petit 1990).

Plant chromosome sorting, reviewed by Conia and Muller (1989), is still in the developmental stage but with renewed interest from major laboratories. Current programs concern the sex chromosomes of *Melandrium album* (2n=24, X,Y) (Ye et al. 1991; Veuskens et al. 1992); cell cultures of tomato (Aramuganathan et al. 1991) and of wheat (Leitch AR, John Innes Inst., Norwich, UK, *pers. comm.*).

Perspective

The afore-mentioned reviews also address the perspectives, but in 1992 one must admit that the discipline of cytometry is yet dimly perceived by the community of plant biologists! Why?

On the one hand, cytometry was initially associated with cell sorting of fused plant protoplasts (reviewed in Lister 1989, and Hammutt et al. 1990). Given that little heterofusion protoplast enrichment is now necessary, "cell sorter" has obscured the conceptual (and practical) contribution of cytometric techniques, as if sorting were to be the sole aim. True, sorting is still very useful - with nuclei, chromosomes, occasionally with protoplasts, and generally during development of methods.

Secondly come the logistical problems: the equipment is relatively expensive; divers applications call for divers configurations, yet polyvalent installations are rare and dedicated machines have been adopted. Many "in-a-hurry-scientists" could benefit from cytometry in their work but the learning-threshold to adapt the technique and to understand the data appears excessive. In fact, common service facilities or networking can build up experience such that new plant projects turn rapidly, whereas such common units are less justified in routine fields of mammalian cell biology. Our own polyvalent equipment has been used for about 130 projects European-wide during the past 8 years.

Thirdly, it is only now that some of the concepts of computer-assisted multiparametric cytology and non-invasive techniques can find their appropriate tools. Micro-injection techniques are available for plant cells; confocal microscopy enables cellular microfluorometry, otherwise too often compromised by out-of-field signals; fluorescent probes have evolved. Valuable additions to our arsenal of cytometric methods (static, confocal or flow) with plants <u>would be</u>:

- vital single-cell reporter gene protocols (suggested by Galbraith seven years ago);
- calcium and other signal transduction assays (which would also open up the field to toxicology and plant-microorganism interactions);
- cell surface receptor assays;
- vital cell cycle staining and sorting (a potentially simple and precise manner of obtaining cycle-synchronised material);
- cell cycle marker antigens (to distinguish differentiating, quiescent and proliferating cells, whether at 2C, 4C or higher DNA levels);

and

- reliable chromosome sorting with agronomic species.

One major application will modify this panorama: simple machines using simple methods are finding extensive application for ploidy analysis in plant research and production laboratories. This offers the chance for more scientists to glimpse the method. However, this should be an *entrée:* it is important that subsequently the power and extent of the discipline should be understood. Nevertheless, the remainder of this chapter is given to this single area.

DNA ANALYSIS

Methods

We have recently resumed many of the cytometric assays of plant nucleic acids (Brown et al. 1991a):

- nuclear DNA for cell-cycle determination, for ploidy diagnosis, to identify heterokaryon or hybrid plants;
- DNA versus RNA (Bergounioux et al. 1988a; Bergounioux and Brown 1990);

and

- DNA versus bromo-deoxy-uridine incorporation.

To this list one should add:

- DNA versus protein (Sgorbati et al. 1989; Ulrich and Ulrich 1991);
- DNA versus nuclear antigens (Bergounioux et al. 1992);
- what may well be termed *high-fi DNA analysis* of Ulrich and Ulrich (1991). This is a longer but particularly tight method for low Coefficients of Variation (around 1%) and clean baselines as necessary when assessing aneuploid subpopulations.

Our review of methods (Brown et al. 1991a) included nine buffers and step-by-step procedures, appropriate for nuclei extracted from tissue by simple chopping (Galbraith et al. 1983), from protoplast lysis (Bergounioux and Brown 1990), and in squashes (Sgorbati et al. 1986). A trouble-shooting guide addressed questions such as choice of fluorochrome, adjustment of buffers and their reductants, statistical processing, etc.

Calibration and DNA references

The following paragraph is adapted from that review.

The unit of ploidy is **x**, one copy of the genetic complement. The **1C** value for any genotype is the DNA content of the unreplicated haploid chromosome complement; the number of chromosomes in the unreplicated haploid cell is **n**.

The fluorescence intensities indicated on a histogram are normally in arbitrary units, channel numbers derived from photomultiplier outputs. To assign ploidy values this scale must be calibrated with a reference.

The simplest calibration is to analyse regularly a reference plant of known ploidy (*e.g.* **2x**) of the same species, note the position of its G1 nuclei (**2C**), and characterize all other samples by the relative position of their G1 peak. Units are thus **C** values. It is then common practice to assign a "DNA ploidy index" to observed peaks, assuming that DNA level (**C** values) and genetic complements (**x** values) can be assimilated.

Using reference species of known nuclear DNA amount, one can calibrate the intensity scale in absolute units (pg) of DNA. Ploidy can then be deduced when screening plants of a species whose C-value is known in pg. Ideally several reference species should be used, avoiding extrapolations. These should be internal references, in the same buffer and receiving the same treatments as the sample.

It is sound to work with an internal reference (even when absolute DNA level is not needed) as this enables one to compensate for sources of variation such as the number of isolated nuclei per ml, temperature, browning, flow-rate effects in the cytometer, disalignment. Fluorescent polystyrene beads cannot monitor the principal source of variation, namely, inconsistencies in the isolation and staining procedures.

Direct assessment of the *ploidy index* is to count chromosomes in meristematic tissue thought to be indicative of the germinal cell line. Many indirect methods are used in practice: plant or pollen phenotype, stomatal cell size or chloroplast content, and nuclear DNA level. To acknowledge the indirect nature of this latter diagnosis, we use the term *DNA ploidy index*.

Preparative sorting of nuclei

Sorting nuclei (Perennes et al. 1990) is a simple and elegant way of obtaining "synchronised" material. As an example, only in host nuclei sorted from the S and early G2 position does one find the double-stranded replicative DNA of *Digitaria* streak virus (Accotto G-P, Instituto de Fitovirologia Applicata, CNR, Turin, *pers. comm.*). Nuclei can be sorted for subsequent assessment of mRNA levels according to cell-cycle phase (Bergounioux et al. 1992). Given the inconsistency of most synchronisation procedures, the demand of scientists for such sorting of plant nuclei is considerable.

Genome size

Another prime demand is for determination of genome size. Bennett and Smith (1976) and Bennett et al. (1982) collated C-values for almost 1000 plants, generally obtained by Feulgen micro-densitometry. Arumuganathan and Earle (1991) have published cytometric data concerning the absolute C-values for more than 100 species; strangely, European data have been systematically omitted (e.g. Ulrich et al. 1988; Brown and Bergounioux 1989). We have also analysed about 130 species. It is indeed urgent that the dispersed published data of C-values be collated in a base.

It is even more urgent that problems of quality control be addressed. Two important aspects follow.
- There is a need for uniform reference nuclei covering the range 0.1 pg to 100 pg DNA commonly found in higher plant and algal species, material readily available to the various laboratories, storable, and characterized for AT/GC composition.
- There is a need for coherence in the calculation procedures with notably the base-pair specific dyes, taking into consideration the curvilinear function relating relative base-pair composition and relative intensities of an unknown to a reference species (see below).

DNA base composition and fluorochrome intensity

It is commonly found that different stains give different DNA estimates. Indeed, Costich et al. (1991) use this in sex identification of dioecious plant species having heterochromosomes.

Is Hoechst fluorescence linearly proportional to AT base content? No! Godelle, comparing 5 species within the plant complex of *Crepis*, has validated a novel procedure for calculating base composition in whole genomes with base-

specific dyes and an internal reference species (Godelle et al., 1990; 1992). The commonly assumed linear relation between fluorescence intensity and base content often leads to large errors: a difference of 1% in GC% between an unknown and a reference will produce an error of about 6% in the 2C calculation with Hoechst. A curvilinear function is necessary, essentially due to the fact that bisbenzimide Hoechst needs five AT in a row to bind and to fluoresce. Similar precautions are necessary with, for instance, Mithramycin.
The advantage is that use of these base-specific stains with internal reference nuclei allows calculation of the base composition - but using curvilinear (Godelle et al. 1992) and not linear functions! Le Gall et al. (1992) give examples for 10 algal species.

Polysomaty complicates interpretation

In *Arabidopsis thaliana* cv. Bensheim only 15% of rosette leaf cells are diploid (Brown et al. 1991a): most of the somatic tissue is 4C, 8C, with cells up to 128C! This is a routine developmental process, and the Coefficients of Variation of the nuclear subpopulations are the same at each ploidy level. Polysomaty is commonplace in plant tissue (Sree Ramulu and Dijkhuis 1986; De Rocher et al. 1990): it refers to an organism having somatic cell lines with different ploidy levels (Langlet 1927). Indeed, two situations may arise from endomitosis: one where terminal endomitosis is associated with certain histological cell types (endopolyploidy) and one where the endomitosis establishes a proliferating somatic cell line (polysomaty).

The presence of multiple ploidy levels complicates interpretation of DNA histograms. For instance, the nuclei at 4C in a diploid plant may conceptually come from four situations: proliferating diploid cells at G2, diploid cells in quiescence or differentiated at 4C (this is clearly possible in plants: Bergounioux et al. 1988b), tetraploid cells in G1. In addition to advancing our fundamental understanding of cell-cycle entry and exit, it would be beneficial to have markers of these different states during ploidy assessment in complex tissues or in polyploid cell cultures.

Routine Applications

A recent French article (Brown et al. 1991b) reviewed the applications and problems of interphasic nuclear DNA analysis in plant research, breeding and production, with reference to the methods that it might replace in industry. Table 1 indicates several examples. Some of these will remain modest laboratory operations, but we would like to forecast that a massive adoption of the method is underway. Since 1983 when Galbraith reported his chopping method and Bergounioux showed DNA analysis of interspecific crosses, scientists have applied this approach (e.g. Petit et al. 1986; De Laat et al. 1987; Fahleson et al. 1989) through the 80s - as did discretely certain seed companies. But while cytometric DNA analyses strictly within France totalled only several thousand in 1989, we expect this figure to be several hundred thousand in a couple of years.

The demand is for simple, rapid, precise, robust, inexpensive tests of uniformity and of plant ploidy.

The offer needs to be equipment capable of bulk processing: automated storage (or fixation), nuclear extraction, filtering, staining, sampling, print-out with code, statistics, sample storage for review, etc. Throughput should be comparable to that of contemporary cytometers in immunology, namely about 90 samples/h. Without yet attaining such performance, present cytometric procedures compare favorably with previous industrial methods on economic and analytical criteria.

Moreover, additional factors have yet to come into play.

Turn-around time. An industrialist currently assessing this new option will tend to compare its unit cost with any present technique such as chromosome counts, phenotypic identification, size of stomatal cells, etc. (see Brown et al. 1991). But the true comparison should take into account turn-around time. If a breeder is holding 15000 plants in glass-houses during winter, or in cool-rooms during summer, what is the cost advantage of an automated method capable of screening the lot in two weeks rather than two months? The breeder will then typically put to use his same infrastructure for an additional lot.

Legal expertise. The assessment of seed quality includes ploidy uniformity. The acceptance or rejection of substantial contracts for seed production will increasingly call for competence in accepting or rejecting cytometric evidence of possible contamination.

Quality control. Plant production *in vitro* has been generalized and somatic embryo production is approaching the industrial stage. Here, errors or instability will be enormously costly, so quality control becomes vital. Ploidy is one small indicator of uniformity, particularly valid in somatic embryogenesis. For in-line or total quality control, flow cytometry will no doubt prove useful whereas any human-based assessment would inevitably tire at the odds of identifying the rare but critical non-uniform event.

Wheel-barrow or bush cytometry. What is the cost of subsampling plants, dispatching to the laboratory, coordinating the results with execution of subsequent culture treatments? Once the cytometer workstation has been optimised, the object must be to remove these unproductive transfers, namely, to take the cytometer to the plants and not the reverse. In the culture room, in the glass-house or in the field, if the sampling and assay can only take a few seconds, the sorting and treatment of material will be on the spot and in-line. Herein lies a real gain of productivity in plant production involving manipulation of ploidy levels.

Call it what you like, but we think that wheel-barrow cytometry or bush cytometry comes close, or of course *cytométrie tout terrain* for the French.

Table 1. Applications of plant nuclear DNA assessment
(adapted from Brown et al. 1991b)

Screening triploid plants after crossing a male-sterile diploid with a pollinating tetraploid e.g. in sugarbeet seed production, irregular ploidy levels indicate contamination from extraneous pollen.

Screening haploid plants e.g. cereal breeding calls increasingly for the production of haploid plants *via* anther culture. These will be treated with colchicine to rapidly produce fully homozygous diploid plants for subsequent selection. But spontaneous diploidisation occurs in some plants; these advantageous cases must be identified and left untreated.

Screening diploid plants when establishing genetic linkage e.g. in chromosome walking strategies, tetraploid transformants unnecessarily complicate the genetic analysis.

Identification and selection on genome size from regenerated plants following heterologous protoplast fusions e.g. selection of plants with the sum of two genomes (Chaput et al. 1990).

Seed conformity. That at least the interphasic nuclear DNA levels are uniform (Petit et al. 1986).

Plant conformity, notably in those derived by *in vitro* multiplication or somatic embryogenesis. During culture of undifferentiated tissue and in transformation procedures, ploidy instability and aneuploidy often arises (Sree Ramulu and Dijkhuis, 1986).

Search for novel ploidy levels in plants e.g. in the creation of ornamental cultivars.

Identification of tissue development stages appropriate as transformation targets (Sangwan et al. 1992) e.g. to avoid transforming endoreplicated cells which tend to give polyploid plants.

Assessment of genome size (C-value)

Calculation of the **base composition** of the genome (GC%)

Physiological studies of the **cell cycle**

Physiological studies of the **extent and control of polysomaty**

Ecological studies e.g. predator-host interactions

Speciation studies and population dynamics

Early screening of sex in dioecious plants e.g. when heterochromosomes exist, the base content or GC composition of male and female somatic nuclei may be sufficiently distinctive.

REFERENCES

Arumuganathan K, Earle ED (1991) Nuclear DNA content of some important plant species. Plant Mol Biol Rep 9:208-218

Arumuganathan K, Slattery JP, Tanksley SD, Earle ED (1991) Preparation and flow cytometric analysis of metaphase chromosomes of tomato. Theor Appl Genet 82:101-111

Bennett MD, Smith JB (1975) Nuclear DNA amounts in Angiosperms. Philos Trans R Soc, Ser B 274:227-274

Bennett MD, Smith JB, Heslop-Harrison JS (1982) Nuclear DNA amounts in Angiosperms. Proc R Soc Biol 216:179-199

Berglund DL, Strobel S, Sugawara F, Strobel GA (1988) Flow cytometry as a method for assaying the biological activity of phytotoxins. Plant Sci 56:183-188

Bergounioux C, Brown S (1989) Plant cell cycle analysis with isolated nuclei. In: Crissman HA, Darzynkiewicz Z (ed) Flow Cytometry, Academic Press, New York (Methods in Cell Biology, 33, 563)

Bergounioux C, Brown SC, Petit P (1992) Flow cytometry and plant protoplast cell biology. Physiol Plant, in press

Bergounioux C, Perennes C, Brown SC, Gadal P (1988a) Nuclear RNA quantification in protoplast cell-cycle phases. Cytometry 9:84-87

Bergounioux C, Perennes C, Brown SC, Sarda C, Gadal P (1988b) Relation between protoplast division, cell cycle stage and nuclear chromatin structure. Protoplasma 142:127-136

Brown SC, Bergounioux C (1989) Plant flow cytometry. In: Yen A (ed) Flow Cytometry: Advanced research and clinical applications. vol II. CRC Press, Boca Raton Florida, p 195

Brown, S, Bergounioux, C, Tallet, S, Marie, D (1991a) Flow cytometry of nuclei for ploidy and cell cycle analysis. In: Negrutiu I, Gharti-Chherti GB (ed) A laboratory guide for cellular and molecular plant biology. Birkhäuser Verlag, Basel, p 326

Brown SC, Devaux P, Marie D, Bergounioux C, Petit P (1991b) Cytométrie en flux: Application à l'analyse de la ploïdie chez les végétaux. Biofutur 105 (October):1-14 (Technoscope n°47)

Brown SC, Renaudin JP, Prevot C, Guern J (1984) Flow cytometry and sorting of plant protoplasts: Technical problems and physiological results from a study of pH and alkaloids in *Catharanthus roseus*. Phyiol Vég 22:541-554

Chaput MH, Sihachakr D, Ducreux G, Marie D, Barghi N (1990) Somatic hybrid plants produced by electrofusion between dihaploid potatoes: BF15 (H1), Aminca (H6) and Cardinal (H3). Plant Cell Rep 9:411-414

Conia J, Muller P (1989) Isolation, classification, and

flow cytometric sorting of plant chromosomes. In: Yen A (ed) Flow Cytometry: Advanced research and clinical applications. vol I. CRC Press, Boca Raton Florida, p 143
Costich DE, Meagher TR, Yurkow EJ (1991) A rapid means of sex identification in *Silene latifolia* by use of flow cytometry. Plant Mol Biol Rep 9:359-370
De Laat AM, Gohde W, Vogekzang MJ (1987) Determination of ploidy of single plants and plant populations by flow cytometry. Plant Breeding 99:303-307
De Rocher EJ, Harkins KR, Galbraith DW, Bohnert HJ (1990) Developmentally regulated systematic endopolyploidy in succulents with small genomes. Science 250:99-101
Deslauriers C, Powell AD, Fuchs K, Pauls KP (1991) Flow cytometric characterization and sorting of cultured *Brassica napus* microspores. Biochim Biophys Acta 1091:165-172
Fahleson J, Dixelius J, Sundberg E, Glimelius K (1988) Correlation between flow cytometric determination of nuclear DNA content and chromosome number in somatic hybrids within *Brassicaceae*. Plant Cell Rep 7:74-77
Fox MH, Galbraith DW (1990) Application of flow cytometry and sorting to higher plant systems. In: Melamed M, Lindmo T, Mendelsohn M (ed) Flow cytometry and sorting, 2nd edn. Wiley-Liss, New York, p 632
Galbraith DW (1989) Analysis of higher plants by flow cytometry and cell sorting. Int Rev Cytol 116:165-228
Galbraith DW, Harkins KR, Maddox JM, Ayres NM, Sharma DP, Firoozabady E (1983) Rapid flow cytometric analysis of the cell cycle in intact plant tissues. Science 220:1049-1051
Gantet P, Hubac C, Brown SC (1990) Flow cytometric fluorescences anisotropy of lipophilic probes in epidermal and mesophyll protoplasts from water-stressed *Lupinus albus* L. Plant Physiol 94:729-737
Godelle B, Cartier D, Marie D, Brown SC, Siljak-Yakovlev S (1990) Relationship between flow cytometric determination of nuclear DNA content and the amount of heterochromatin within the *Crepis praemorsa* complex. Biol Cell 70(1-2):28a
Godelle B, Cartier D, Marie D, Brown S, Siljak-Yakovlev S (1992) Heterochromatin study reveals non-linearity of fluorometry for genomic base composition. *pers. comm.*
Hammatt N, Lister A, Blackhall NW, Gartland J, Ghose TK, Gilmour DM, Power JB, Davet MR, Cocking EC (1990) Selection of plant heterokaryons from diverse origins by flow cytometry. Protoplasma 154:34-44
Harkins KR, Galbraith DW (1987) Factors governing the flow cytometric analysis and sorting of large biological particles. Cytometry 8:60-70
Langlet O (1927) Svensk Bot Tids 21:397-422
Le Gall Y, Brown S, Marie D, Mejjad M, Kloareg B (1992) Quantification of nuclear DNA and GC% in marine macroalgae by flow cytometry of isolated nuclei. J Phycology in press
Lister A (1989) Flow cytometry for selection of plant cells *in vitro*. In: Dix PJ (ed) Selection of plant cells, VCH Verlagsgesellschaft, Weinheim, chpt 3
Matine A (1990) Biosynthèse d'isoprenoïdes par des protoplastes et plastes isolés du fruit de calamondin. Université de Bordeaux I, Ph.D. thesis

Perennes C, Bergounioux C, Gadal P (1990) Direct assessment of hormone stimulated transcription in protoplasts and isolated nuclei of *Petunia hybrida*. Plant Cell Rep 8:684-686

Petit PX (1992) Flow cytometric analysis of Rhodamine 123 fluorescence during modulation of the membrane potential in plant mitochondria. Plant Physiol 98:279-286

Petit PX, Conia J, Brown SC, Bergounioux C (1986) Cytométrie en flux et biotechnologies végétales. Biofutur 5 (November):128-137

Petit PX, Diolez P, de Kouchkovsky Y (1989) Flow cytometric analysis of energy transducing organelles: mitochondria and chloroplasts. In: Yen A (ed) Flow Cytometry, Advanced research and clinical applications. vol I. CRC Press, Boca Raton Florida, p 271

Sangwan R, Bourgeois Y, Brown SC, Vasseur G, Sangwan-Norreel B (1992) Characterization of competent cells and early events of *Agrobacterium* mediated genetic transformation in *Arabidopsis thaliana*. Planta, in press

Schröder W, Petit PX (1990) Study of lectin-binding sites on the chloroplast envelope using flow cytometry analysis. Physiol Plant 79(2, part 2):A59

Sgorbati S, Levi M, Sparvoli E, Trezzi F, Lucchini G (1986) Static and flow cytometry of 4',6-diamidino-2-phenylindole (DAPI)-stained suspensions of nuclei released from fresh and fixed tissues of plants. Physiol Plantarum 68:471-476

Sgorbati S, Sparvoli E, Levi M, Chiatante D (1989) Bivariate cytofluorimetric analysis of nuclear protein and DNA relative to cell kinetics during germination of *Pisum sativum* seed. Physiol Plantarum 75:479-484

Sree Ramulu K, Dijkhuis P (1986) Flow cytometric analysis of polysomaty and *in vitro* genetic instability in potato. Plant Cell Rep 3:234-237

Tagu D, Bergounioux C, Perennes C, Brown SC, Muller P, Gadal P (1987) Analysis by flow cytometry of the primary events of *Petunia hybrida* protoplast transformation. Plant Sci 51:215-223

Ulrich I, Fritz B, Ulrich W (1988) Application of DNA fluorochromes for flow cytometric DNA analysis of plant protoplasts. Plant Sci 55:151-158

Ulrich I, Ulrich W (1991) High-resolution flow cytometry of nuclear DNA in higher plants. Protoplasma 165:212-215

Veuskens J, Marie D, Hinnisdeals S, Brown S (1992) Flow cytometry and sorting of plant chromosomes. In: Radbruch A (ed) EMBO Course Methods, Flow cytometry and cell sorting, Springer, Berlin Heidelberg New York (Monograph Series, Methods in biology and medicine), in press

Xu C, Auger J, Givindjee (1990) Chlorophyll *a* fluorescence measurements of isolated spinach thylakoids obtained by using single-laser-based flow cytometry. Cytometry 11:349-358

Ye D, Oliveira M, Veuskens J, Wu Y, Installe P, Hinnisdaels S, Truong AT, Brown SC, Mouras A, Negrutiu I (1991) Sex determination in the dioecious *Melandrium*. The X/Y chromosome systems allows complementary cloning strategies. Plant Sci 80:93-106

PART III

Nuclear Components: Form and Function

BROMODEOXURIDINE PROCEDURES FOR ANALYSIS OF DNA SYNTHESIS

Harry A. Crissman
Life Sciences Division
Los Alamos National Laboratory
LS-1, MS M888
Los Alamos, NM 87545

Significant advances have been made in studies on DNA synthesis, cell cycle traverse and cell proliferation through the use of radiolabeled DNA precursors. The use of tritiated thymidine to label cells actively synthesizing DNA enabled Howard and Pelc (1953) to subdivide the cell cycle in the G1, S, and G2/M phases. The capability for following the traverse of labeled S phase cells through the subsequent phases of the cell cycle has also dramatically improved investigations on cell kinetics and cell division patterns. Recently, 5-bromodeoxyuridine (BrdUrd), the base analog of thymidine, has been substituted for the radioactive precursor. The use of BrdUrd has several advantages in that (a) it can be more safely utilized in studies that would prohibit or limit the use of radiolabeled precursors, (b), methods for assaying BrdUrd, approaching the sensitivity for detection of triatiated thymidine, are currently available. One of the most widely used

NATO ASI Series, Vol. H 67
Flow Cytometry
Edited by A. Jacquemin-Sablon

cytological technique for detection of BrdUrd-substituted DNA is the immunofluorescent assay developed by Gratzner (1982). This method requires partial denaturation of DNA to the single stranded conformation, by heat or acid treatment, to provide access of the incorporated BrdUrd to the antibody, which is usually tagged with the green fluorochrome, fluorescein isothiocyanate (FITC). A modification of this procedure by Dolbeare et al. (1983) incorporates the red fluorescent counterstain, propidium iodide (PI), for the additional measurement of cellular DNA content. Using two-color, flow cytometric analysis, cells containing BrdUrd incorporated for pulse-labeling periods as short as 10 min are readily detected and their cell cycle position is easily assessed.

An alternative cytochemical method for detecting BrdUrd-substituted DNA was first demonstrated by Latt (1973) using the A-T base-binding fluorochrome, Hoechst 33258, that is quenched when bound to A-BrdUrd regions in double-stranded DNA. The reduction in the uv-excited, blue fluorescence of the Hoechst dye is proportional to the quantity of BrdUrd incorporated into DNA. Attempts to use the BrdUrd/Hoechst fluorescence-quenching technique in flow cytometry have been limited due to the lack of sensitive methods for quantitatively measuring the reduction in HO fluorescence when bound to BrdUrd. Some modifications of the technique including the combination of the red fluorescent DNA dye, ethidium bromide (EB), with Hoechst have improved BrdUrd detection. However, BrdUrd labeling periods of 6h (Ellwart and Dohmer, 1985) or longer (Bohmer RM, 1979; Bohmer and Ellwart, 1981; Kubbies and Rabinovitch, 1983) are usually required for cells to incorporate sufficient amounts of the base analog for sensitive detection and quantitation of HO quenching by the EB-HO combination.

Recently, we described a sensitive, flow cytometric method involving differential fluorescence analysis of two DNA-binding fluorochromes for detecting BrdUrd-labeled DNA in cells treated for 30 min or less (Crissman and Steinkamp, 1987). The technique combines two nonintercalating, non-competing DNA dyes: Hoechst 33342 (HO) and mithramycin (MI) (G-C binding) (Crissman and Tobey, 1974; Crissman et al., 1979), a dye whose fluorescence in the presence of BrdUrd remains stoichiometric to DNA content, under the conditions employed. Using dual-wavelength excitation, the blue (HO) and green-yellow (MI) fluorescence emissions are measured and a differential amplifier (Steinkamp and Stewart, 1986) subtracts the blue fluorescence from the green-yellow fluorescence signal amplitude on a cell-by-cell basis. The resulting difference signal is then amplified. Cells in S phase show significant BrdUrd/Hoechst quenching and produce greater differential fluorescence signals compared to signals obtained for cells in the G1 and G2+M phase which, except for minor differences in HO and MI stainability, show relatively small (near zero) fluorescence differences. The technique is easy and highly reproducible; it does not require denaturation of DNA and involves only a one-step staining procedure. Since the method is mild it minimizes cell loss and loss of other important cellular markers such as DNA and/or chromatin, RNA, and proteins, including cellular antigens. The sensitivity of the method is comparable to the immunofluorescent assay.

In this chapter we briefly describe the two procedures for flow cytometric analysis of BrdUrd content. Details on the application of the monoclonal antibody method (Dolbeare et al., 1990) and the differential fluorescence analysis approach (Crissman and Steinkamp, 1990) have recently become available.

Detection of BrdUrd with Monoclonal Antibodies

The introduction of monoclonal antibodies against BrdUrd by Gratzner (1982) provided an immunofluorescent approach for tagging cells actively synthesizing DNA. For such studies viable cells are routinely labeled with BrdUrd in culture at 37° C while protected from intense light since BrdUrd is light sensitive and can deteriorate. In addition, cells incorporating BrdUrd can suffer chromatin damage and lose viability during exposure to strong light. The length of the labeling period and the BrdUrd concentration varies depending on the cell cycle time and the rate of DNA synthesis. Rapidly growing L1210 cells (i.e., T_G = 12 h), for example, may require only a 30-min pulse with 10-30 μM BrdUrd, whereas more slow-growing cells may require pulses of 1-2 h to incorporate sufficient quantities of BrdUrd for detection. Excessively high concentrations of BrdUrd can induce a lethal block to cells in the G2 phase. Preliminary growth studies and FCM analyses of BrdUrd-treated and untreated cell populations are usually necessary to determine optimal conditions for BrdUrd labeling.

Simultaneous analysis of DNA content and the amount of incorporated BrdUrd significantly increased the cytokinetic applications of the antibody technique (Dolbeare et al., 1983). Analysis of the BrdUrd and DNA contents is achieved by measuring the FITC and PI fluorescence, respectively, following excitation of the two dyes with a single laser tuned to 488 nm. Although the emission of the dyes overlap somewhat, adequate fluorescence resolution is achieved by the use of appropriate color filters. As mentioned above, the BrdUrd labeling procedure requires partial, but not excessive, denaturing of DNA since subsequent staining by PI requires double-stranded DNA. Several methods for denaturing DNA include the use of high molarity hydrochloric acid (1.5-4.0 N HCL), elevated temperature in 50% formamide or in low ionic strength solution. Restriction endonucleases combined with exonuclease III have also been used (Dolbeare and Gray, 1988). The procedure involves several centrifugation steps which can induce significant cell loss due to cell clumping, particularly following the DNA denaturation step. Clumps may be disaggregated by a combination of mild vortexing and syringing the suspension gently through a 25-gauge needle. Cell adherence to the centrifuge tube can be decreased by siliconizing the tubes to reduce surface tension. There are a large number of commercial sources of monoclonal antibodies to BrdUrd (see Linscott's Directory of Immunological and Biological Reagents, Fifth Edition, Linscott's Directory, Mill Valley, CA). The choice of antibody source will be determined on the basis of the purity, affinity, and specificity of the antibody, all factors that influence the final sensitivity of the measurement.

A sample of cells that have not been pulsed with BrdUrd should be included in the preparation to serve as a negative control and also to assess background fluorescence. An autofluorescence standard can be a cell sample that has been processed through the protocol with exclusion of the antibody or PI labeling steps.

The primary application of the technique is in the quantification of phase fractions, phase durations, doubling time, labeling index, and growth fractions. One may use multiple sample points (Dolbeare et al., 1983) or single sample points (Begg et al., 1985) to obtain labeling index and potential doubling times.

The method has been applied to the analysis of dispersed solid tumors, bone marrow, spleen, liver, bladder epithelial, neural cells, dermal cells, cultured mammalian cells, and cell nuclei.

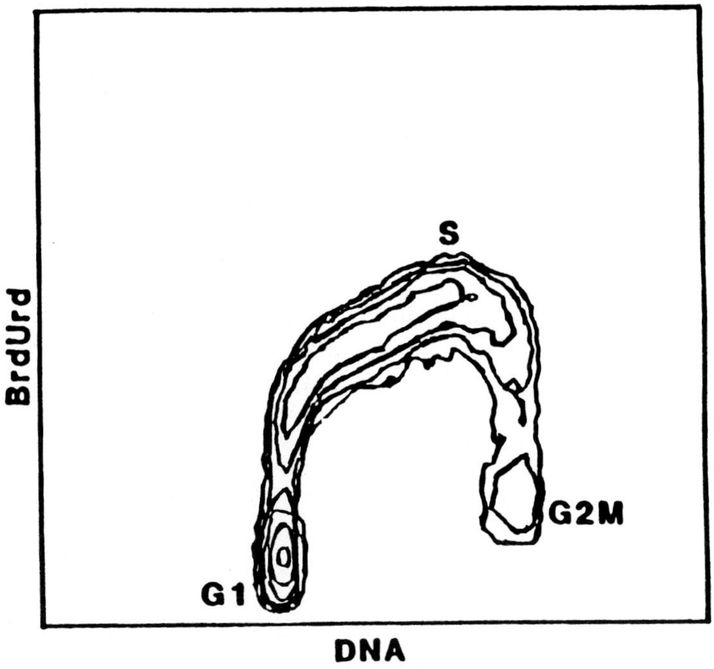

Figure 1. Bivariate histogram showing DNA content (PI fluorescence) and BrdUrd content (fluorescein fluorescence) for CHO cells treated for 30 min with 10 μM BrdUrd

Results

Figure 1 shows a bivariate DNA-BrdUrd contour histogram generated by CHO cells stained according to the protocol. This is the typical bivariate contour display generated by FCM analysis after a 30-min pulse of 10 μM BrdUrd. The G1 and G2 + M populations show only background green fluorescence. Cells in S phase have green fluorescence and produce the horseshoe-shaped pattern with mid-S-phase cells having the highest fluorescence. The fraction of cells with S-phase fluorescence divided by the total population will give the labeling index. Cells that have not been treated with BrdUrd but processed with the protocol will exhibit only background fluorescence.

Detection of BrdUrd by Differential Fluorescence Analysis

Quantitative FCM measurement of the HO quenching is accomplished by electronic cell-by-cell subtraction of the HO fluorescent signal from the signal of a second dye that is essentially unaffected by BrdUrd and thus remains proportional to DNA content. Mithramycin (MI) is used with HO since, under appropriate conditions, dye-dye interference and energy transfer between the two dyes are minimal and thus do not affect FCM BrdUrd analysis. Initially, staining experiments are performed on untreated (non-BrdUrd containing) cells to determine the concentrations of HO and MI that can be used in combination to obtain DNA content histograms with low coefficient of variation (CV) values that are comparable in numerical value for the two dyes.

Two lasers are used for FCM analysis, one operating in the uv (333.6-363.8 nm) and one tuned to 457.9 nm. The lasers are separated by 250 µm to provide sequential excitation and analysis of each fluorochrome. Initially the electronic gains are adjusted so that the G1 peaks of the FCM generated HO- and MI-DNA content histograms are in the same channel number. The fluorescence signals are then subtracted electronically (i.e., MI minus HO) on a cell-by-cell basis (Steinkamp and Stewart, 1986) to measure quantitatively the quenching of HO fluorescence. The adjustment of the G1 peak positions in the same channel sets the zero value for the subtraction process for all cells across the remainder of the cell cycle. The arbitrary assignment of zero values for G1 cells is based on the fact that these cells do not contain BrdUrd and thus do not exhibit HO quenching.

Results

Untreated Chinese hamster (line CHO) cells examined by this technique show only the slight differences in stainability between the two dyes that are detected with increased sensitivity and amplified in the MI-HO fluorescence difference vs DNA content (mithramycin) profile (Fig. 2B). In contrast, a population of CHO cells treated for 30 min with 30 μM BrdUrd shows a significant quenching of HO fluorescence in S phase as seen in Fig 2D. The magnitude of the MI-HO fluorescence differences, reflecting BrdUrd/Hoechst quenching, shown in Fig. 2D is most significant for cells in S-phase that have incorporated BrdUrd in the 30-min pulse-labeling period. The percentage of BrdUrd containing S-phase cells detected by this technique (i.e., relative fraction of cells in the boxed region in Fig. 2D) was 35%. This value is in good agreement with the flow

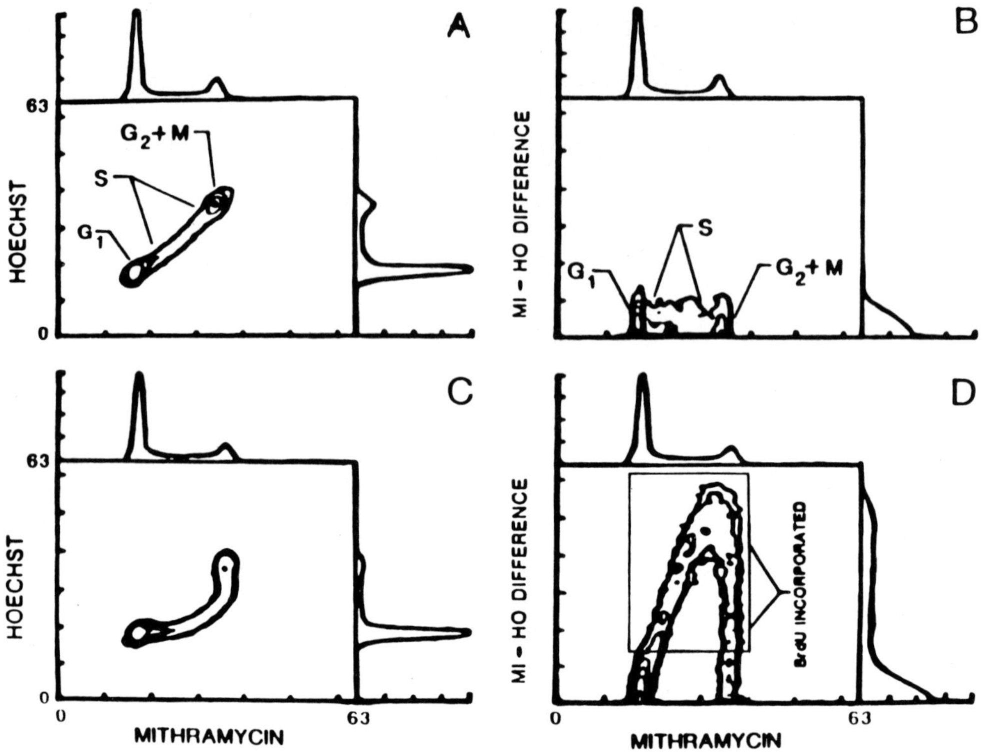

Figure 2. DNA content (HO and MI) and MI-HO signal difference frequency distribution histograms and the corresponding bivariate contour diagrams for untreated CHO cells (A and B) and for CHO cells treated in culture with 30 μM BrdUrd for 30 min (C and D). The x and y axes are linear relative units. The MI-HO difference signal amplitudes from G1, S, and G2+M phase control (untreated) cells are shown in the bivariate diagram (B). The difference signal values for untreated cells are near zero or slightly negative due to differences in stainability of DNA by the two dyes. In (D), the S-phase cells show positive MI minus HO fluorescence differences values, the magnitude of which varies with the amount of BrdUrd a cell has incorporated. The data counts from most G1 and G2+M phase cells are seen along the x axis. The difference amplifier gain was initially adjusted to give a maximum expansion of the fluorescence difference signal range from cells that had incorporated BrdUrd. The untreated cell population was analyzed at this same instrument gain setting.

cytometric (FCM) computer-fit S phase calculation of 37% and also with a 33% thymidine labeling index derived by standard autoradiography. The bivariate contour distribution for the BrdUrd-treated population (Fig. 2D) is also similar to distributions we obtain routinely for BrdUrd-treated CHO cells using the BrdUrd-antibody technique.

Acknowledgments

This work was supported by the Los Alamos National Flow Cytometry and Sorting Research Resource funded by the Division of Research Resources of NIH (Grant P41-RR01315) and the Department of Energy.

References

Begg AC, McNally NJ, Shrieve DC, Karcher HA (1985). A method to measure the duration of DNA synthesis and potential doubling time from a single sample. Cytometry 6:620-626

Bohmer RM (1979). Flow cytometric cell-cycle analysis using the quenching of 33258 Hoechst fluorescence by bromodeoxyuridine incorporation. Cell Tissue Kinet. 12:101-110

Bohmer RM, Ellwart J (1981). Combination of BUdR-quenched Hoechst fluorescence with DNA-specific ethidium bromide fluorescence for cell cycle analysis with a two-parameter flow cytometer. Cell Tissue Kinet. 14: 653-658

Crissman HA, Steinkamp JA (1987). A new method for rapid and sensitive detection of bromodeoxyuridine in DNA replicating cells. Exp. Cell Res. 173: 256-261

Crissman HA, Steinkamp JA (1990). Detection of bromodeoxyuridine-labeled cells by differential fluorescence analysis of DNA fluorochromes. In: Darzynkiewicz Z, Crissman HA (eds), Flow Cytometry. Methods in Cell Biology, Vol 33, San Diego: Academic Press, pp 199-206

Crissman HA, Stevenson AP, Kissane RJ, and Tobey RA (1979). Techniques for quantitative staining of cellular DNA for flow cytometric analysis. In: "Flow Cytometry and Sorting" (Melamed MR, Mullaney PF, and Mendelsohn ML, eds), New York: John Wiley & Sons, pp 243-261

Crissman HA, Tobey RA (1974). Cell cycle analysis in 20 minutes. Science 184: 1297-1298

Dolbeare F, Gratzner H, Pallavacini M, and Gray JW (1983). Flow cytometric measurement of total DNA content and incorporated bromodeoxyuridine. Proc Natl Acad Sci USA 80: 5573-5577

Dolbeare F, Gray JW (1988). Use of restriction endonucleases and exonuclease III to expose halogenated pyridines for immunochemical staining. Cytometry 9: 631-635

Dolbeare F, Kuo W-L, Beisker W, Vanderlaan M, Gray JW (1990). Using monoclonal antibodies in bromodeoxyuridine-DNA analysis. In: Darzynkiewicz Z, Crissman HA (eds), Flow Cytometry. Methods in Cell Biology, Vol 33, San Diego: Academic Press, pp 207-216

Ellwart J, Dohmer P (1985). Effect of 5-fluoro-2'-deoxyuridine (FdUrd) on 5-bromo-2 deoxyuridine (BrdUrd) incorporation into DNA measured with a monoclonal BrdUrd antibody and by BrdUrd/Hoechst quenching effect. Cytometry 6: 513-520

Gratzner H (1982). Monoclonal antibody against 5-bromo- and 5-iododeoxyuridine: A new reagent for detection of DNA replication. Science 218: 474-475

Howard A, Pelc SR (1953). Synthesis of deoxyribonucleic acid in normal and irradiated cells and its relation to chromosome breakage.

Heredity (suppl) 6: 261-273

Kubbies M, Rabinovitch PS (1983). Flow cytometric analysis of factors which influence the BrdUrd Hoechst quenching effect in cultivated human fibroblasts and lymphocytes. Cytometry 3: 276-281

Latt SA (1973). Microfluorometric detection of deoxyribonucleic acid replication in human metaphase chromosomes. Proc Natl Acad Sci USA 70: 3395-3399

Steinkamp JA, Stewart CC (1986). Dual laser, differential fluorescence correction method for reducing cellular background autofluorescence. Cytometry 7: 566-574

FLOW CYTOGENETICS: FUNDAMENTALS AND NEW DEVELOPMENTS

L. Scott Cram, J. Fawcett, and L. L. Deaven
Life Sciences Division
Center for Human Genome Studies
University of California, Los Alamos National Laboratory
Los Alamos, NM 87545

I. INTRODUCTION

The ability to rapidly and quantitatively analyze and separate single chromosomes with a flow cytometer has made a significant impact on several biomedical research disciplines. The analysis and sorting of single specific chromosomes in a flow cytometer is called flow cytogenetics. One area where flow cytogenetics has served an enabling role is the human genome program. To reduce the complexity of a genome, chromosome sorting offers the advantage of sorting individual chromosomes for the cloning and mapping of separate distinct components. By dividing up a cells genetic material into the 24 different types of chromosomes and producing chromosome specific libraries from these sorted populations, the task of mapping the human genome has been greatly simplified.

The advantage of chromosome specific libraries is even more significant now that yeast artificial chromosome (YAC) libraries containing large genomic inserts are being constructed from flow sorted chromosomes (McCormick, 1992). These libraries have an average insert size of 200 kb and are extremely valuable for identifying and constructing contiguous elements derived from smaller cosmid libraries. In addition to sorting, the ability to rapidly and quantitatively analyze thousands of chromosomes per second has proven to be an important complement to cytogenetic analysis. The throughput, high sensitivity, and quantitative aspects of flow cytogenetic analysis has made it possible to process of large numbers of samples for cytogenetic analysis.

Opportunities for further developments and new applications in this field are numerous and will emerge from the challenges involved in overcoming current limitations. For example, intracellular chromosome identity is lost in the present chromosome isolation process. Solutions to this problem have been proposed

NATO ASI Series, Vol. H 67
Flow Cytometry
Edited by A. Jacquemin-Sablon

(Cram, 1985) and one implemented (Poletaev, 1991). Improved resolution of individual chromosome types, new probes with better specificity, and a basic understanding of the binding properties and characteristics of presently used probes would expand both clinical and research applications of flow cytogenetics.

The purpose of this brief chapter is to provide 1) an introduction to flow cytogenetics and 2) resource information for additional and more detailed information for those interested in either using or further developing the technology. Several recent reviews are available and provide in-depth information on all aspects of flow cytogenetics, this information will by and large, not be repeated here (Gray, 1989c, Melamed, 1990, Darzynkiewicz, 1990). Examples of common types of flow cytogenetic data are included along with an explanation of their interpretation. In addition, recent developments are discussed as examples of what one can anticipate for the future. Further developmental needs are discussed in hopes of stimulating new ideas.

The field continues to expand as evidenced by the development of new protocols designed to meet the requirements of new cloning vectors, new chromosome labeling protocols, and chromosomes from new species are being analyzed, identified, and sorted.

II. CURRENT TECHNOLOGIES

Univariate Analysis. Remarkable resolution of distinct chromosome types can be achieved with univariate (single parameter) analysis. For some applications the use of a single "DNA specific" fluorochrome resolves most if not all chromosome types. For species such as Chinese hamster, propidium iodide stained chromosomes can be completely resolved into twelve populations, one peak in the flow karyotype for each chromosome type. Figure 1 illustrates typical results obtained with a commercial flow cytometer. Procedures used for obtaining these results are available from several references; cell culture (Lozes, 1989), chromosome isolation and staining (Cram, 1990a, Cram, 1990b, Trask, 1989, Trask 1990), sample handling (Gray, 1989c, van den Engh, 1985), and instrument modifications and alignment (Bartholdi, 1987a, Gray, 1989a).

Figure 1. Chromosomes isolated from cloned Chinese hamster embryo cell sat passage 16 were stained with propidium iodide and analyzed on a flow cytometer. Each peak is identified by its chromosome assignment.

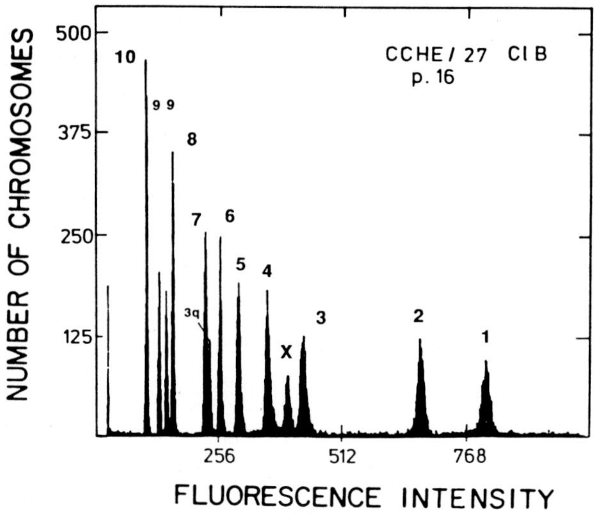

Each peak in figure 1 is a single chromosome type. Chinese hamster chromosome 1 contains the most DNA and is therefore the brightest and is in the most fluorescent peak to the far right. Peak area is proportional to the number of chromosomes of that type and peak position is proportional to DNA content. In the distribution shown, the integrated number of events in chromosome peaks 1,2, 3, and 4, for example, are the same because the cell line is near diploid (two homologues of each chromosome type). The two number 9 chromosomes (the two homologues) have a slightly different DNA content, one homologue having been inherited from each parent (Gray, 1989b). Peak height and peak width changes across the distribution as the result of the data recording technique. There is no biological significance associated with this change in shape.

Numerical chromosome aberrations (trisomy for example) are reflected in peak area and structural aberrations (non-reciprocal translocations, deletions, etc.) can result in a variety of changes in the flow karyotype. Any net change in DNA content by more than a couple of percent will in principle, result in a new peak. In the case of a marker chromosome which ends up with the same DNA content as another chromosome, a new peak would not appear. In the initial stages of karyotype instability only a fraction of the cells will typically contain a marker chromosome. A flow karyotype in which only a fraction of the cells contain a marker chromosome (chromosomes with a net gain or loss of DNA) will exhibit a new peak with an area proportional to the number of cells containing the marker. A new peak will also be present with an increase or loss of fluorescence in proportion to the marker chromosomes gain or loss of DNA.

Univariate flow karyotypes have been used for the analysis of chromosomes isolated from a broad range of species and cell types including fibroblasts, peripheral lymphocytes, cultured tumor cells, and primary cells (Lozes, 1989). Applications have included the analysis of karyotype instability associated with spontaneous neoplastic progression (Bartholdi, 1987b), detection of chromosome polymorphisms (Ray, 1984), analysis of radiation damage to chromosomes (Aten, 1989), and the detection of random chromosome changes as a measure of genetic change (Green, 1989). Chromosome sorting from univariate flow karyotypes has also been used for identifying peaks, gene mapping and for all the applications described below.

Bivariate Analysis. The addition of a second fluorochrome with different binding specificity adds an independent variable and results in increased resolution. Bivariate or two fluorochrome staining of human chromosomes, and to a lesser degree other species as well, greatly enhances the resolving power of flow cytogenetics. Figure 2 illustrates the resolution of normal human chromosomes stained with Hoechst 33258 (Ho258) and Chromomycin A3 (CA3). Ho258 preferentially binds to adenine-thymine (AT) rich regions of DNA and CA3 preferentially binds to guanine-cytosine (GC) rich regions (Langlois, 1989). The amount of binding of each fluorochrome is measured for each chromosome as the chromosome is sequentially excited by two laser beams, the first tuned to excite Ho258 (ultraviolet) and the second to excite CA3 (457 nm). The correlated value from each of the two signals is plotted as a bivariate distribution. The number of accumulated events is recorded as an isocontour where a contour represents events of equal number (Figure 2). The interpretation of numerical and structural changes is analogous to the situation with univariate analysis except now the one dimensional peaks are two dimensional isocontours. Alternatively, the data could be displayed as three dimensional distributions with the number of events plotted on the Z-axis, the plane perpendicular to the plane of the paper.

Figure 2. Human chromosomes isolated from a normal, euploid tissue culture cells and stained with Hoechst 33258 and Chromomycin A3 and analyzed on an EPICS V flow cytometer.

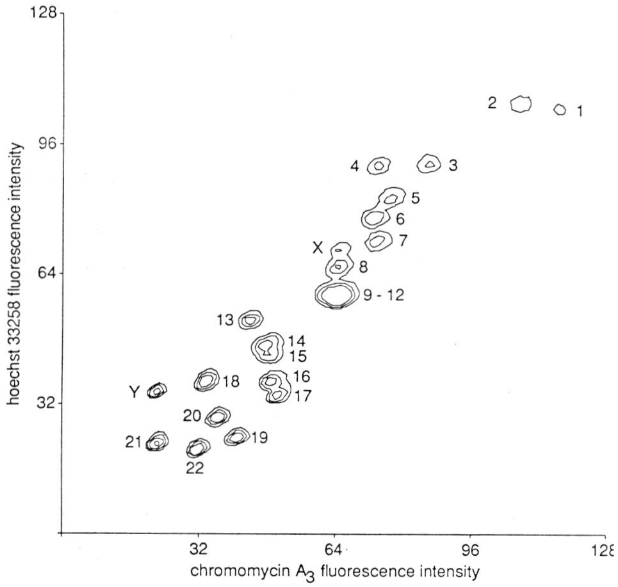

Fluorescent debris and chromosome fragments combine to form a underlying continuum on top of which resides individual chromosome peaks. Improved buffers and isolation procedures have largely eliminated the relatively small amount of contamination due to these background counts. Bivariate distributions provide greatly improved resolution for human chromosomes but the advantages are less pronounced in other species. Some human chromosome types have very nearly the same (AT)/(GC) ratio (chromosomes 9,10,11,12) and fall within the same peak. A large variation in Hoechst binding alone separates chromosomes 13 through 17. Human heterogeneity and chromosome polymorphisms limit the clinical utility of bivariate chromosome analysis. For example, skewed peak shape can be the result of variation in homologue DNA content or indicative of a marker chromosome (Moore, 1989, Gray, 1989b).

Chromosome Sorting. This is one of the few areas in which the sorting capabilities of flow cytometers are heavily exercised. Sorting is used to identify peaks by banding of the sorted material (Young, 1989), for construction of chromosome specific libraries (Van Dilla, 1989), and for gene mapping (Lebo, 1989). Sorting rates are a function of analysis rate and the fractional representation of the population being sorted. Sorting a single human chromosome from a euploid cell line [22 (X,Y)] consists of sorting one chromosome out of every 24 chromosomes. Typical chromosome analysis rates are 1,200 chromosomes per second, which corresponds to a sort rate of about 50 chromosomes per second. A useful benchmark for sorting is the recovery of one million chromosomes per day when sorting from a normal human cell line (Albright, 1991).

Chromosome sorting purity can be determined by parametric analysis of one-dimensional projections of the bivariate flow karyotype and/or by fluorescence in-situ hybridization. Hybridization procedures are preferred since sorted chromosomes are counted directly. The procedure is straightforward when analyzing chromosomes sorted from somatic cell hybrids (REFERENCE). Typical purity values range between 85 to 95% when sorting human chromosomes from somatic cell hybrids. Chromosome recovery, the number of chromosomes actually recovered versus the number of chromosomes sorted by the instrument, is typically 80% to 90%.

Figure 3 illustrates the resolution one can achieve using a somatic cell hybrid as a source of a single human chromosome. CY-18 is a mouse cell line containing a single human chromosome (REFERENCE). Chromosome 16 was sorted from this cell line with a purity of 96% as determined by fluorescence in-situ hybridization and 97% as determined by parametric data analysis of the bivariate data.

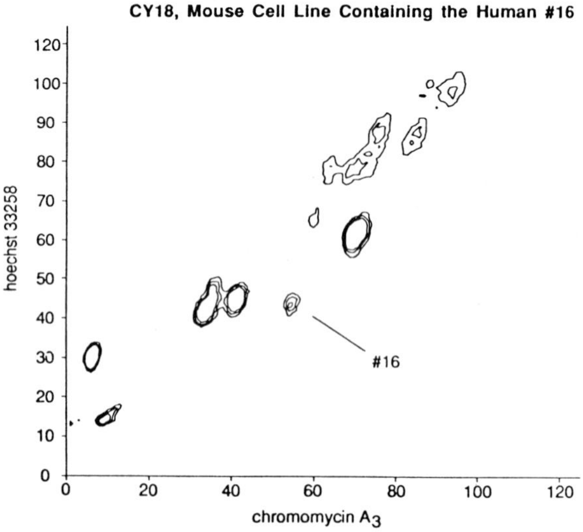

Figure 3. Bivariate flow karyotype of the CY-18 somatic cell containing the human 16 chromosome. Chromosomes were stained with Hoechst 33258 and Chromomycin A3 and analyzed on an EPICS V flow cytometer.

Sorting human chromosomes from somatic cell hybrids offers several advantages. The fractional representation of a human chromosome relative to Chinese hamster chromosomes is higher, a single homologue is sorted, and contamination in the final product is from a different species and can be easily distinguished from material of human origin. Somatic cell hybrids which contain intact, normal human chromosomes in a high percentage of cells (>90%) are required for most applications. Extensive screening of cell lines reported to contain single or a low number of human chromosomes is often required to find a suitable somatic cell hybrid for a particular chromosome.

III. NEW DEVELOPMENTS AND PROCEDURES

Additional species are being added to the list of those from which chromosomes have been isolated for flow cytometric analysis and sorting. Two recent examples are mouse and porcine; both species are of importance for different reasons. Mouse univariate flow karyotypes consist of five peaks which contain the 21 different types of chromosomes [19(X,Y] (Cram, 1989). Recent bivariate analysis indicates many more peaks can be resolved and sorted, see Figure 4. Thirteen populations have been resolved and sorting for confirmation of peak identity is underway (Cram, 1992). Alternative approaches for resolving the mouse karyotype have been demonstrated using mouse cell lines with multiple Robertsonian translocations (Baron, 1984, Baron, 1986). Several groups are involved in porcine chromosome sorting for the Pig Gene Mapping Project (Haley, 1990). The pig karyotype consists of 38 chromosomes; 18 autosomal pairs, X and Y. Roughly 17 peaks plus the Y chromosome are resolved using Ho258 and CA3 for bivariate flow karyotype analysis. Several of the peaks have been identified and confirmed (Schmitz, 1991).

A new and improved procedure for labeling chromosomes in suspension with antibodies has recently been described. Levy and colleagues report the use of anti-kinetochore staining to improve the resolution of Indian muntjac chromosomes (Levy, 1991). Extension of this technique using antibodies specific for a limited number of chromosomes in combination with other fluorochrome stains will offer new approaches to resolving additional chromosomes and for rapid screening of specific aberrations.

Figure 4. Bivariate flow karyotype of chromosomes isolated from the *Mus spretus* cell line CL 5A. Positive identification of the individual peaks is in progress.

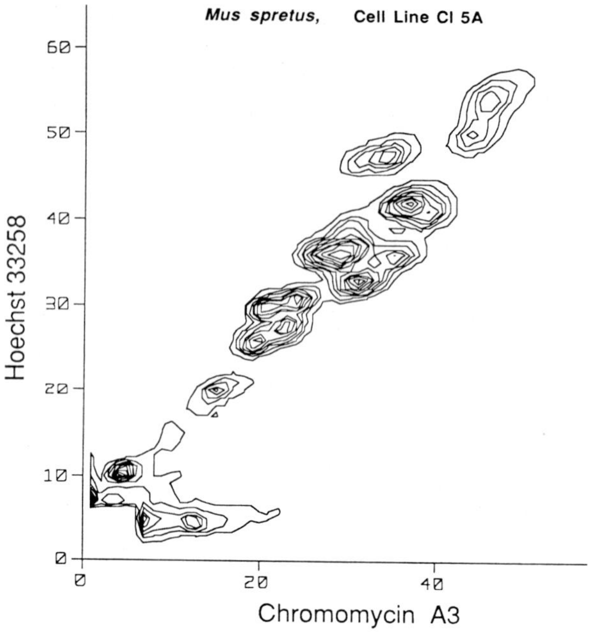

Prospects for flow cytogenetic research applications include a wide variety of requirements for sorted chromosomes not only extending current procedures but the sorting of break point regions and other specialized requirements. Clinical applications are presently limited to confirmatory work (McConnell, 1991), however with specific DNA probes and the means to accomplish in-situ hybridization on chromosomes in suspension, rapid rare event flow karyotype analysis would become a significant technology for clinical analysis - just as flow karyotype sorting has for the research community.

REFERENCES

Albright KL, Cram LS, Martin JC (1991) Separation techniques used to prepare highly purified chromosome populations: sedimentation, centrifugation, and flow sorting. In: Kompala DS, Todd P (eds) American chemical society symposium series cell separation science and technology vol 464. American Chemical Society p 73

Aten JA (1989) Relation between radiation-induced flow karyotype changes analyses by Fourier analysis and chromosome aberrations. In: Gray JW (ed) Flow Cytogenetics Academic Press, San Diego, p 151

Baron B, Metezeau P, Kelly F, Bernheim A, Berger R, Guenet JL, Goldberg ME (1984) Flow cytometry isolation and improved visualization of sorted mouse chromosomes. Exp Cell Res 152:220-230

Baron B, Metezeau P, Hatat D, Roberts C, Goldberg ME, Bishop C (1986) Cloning of DNA libraries from mouse Y chromosomes purified by flow cytometry. Som Cell Mol Gen 12:289-295

Bartholdi MF, Meyne J, Albright K, Luedemann M, Campbell E, Chritton D, Cram LS (1987a) Chromosome sorting by flow cytometry. In: Gottesman (ed) Molecular genetics of mammalian cells. Academic Press, Orlando, p 252

Bartholdi MF, Ray FA, Cram LS, Kraemer PM (1987b) Karyotype Instability of Chinese hamster cells during in vivo tumor progression. Som Cell and Mol Gen 13: 1-10.

Cram LS (1985) Karyotype instability and chromosome aneuploidy, Gordon Research Conference on Genetic Toxicology and Bioassays.

Cram LS, Bartholdi MF, Ray FA, Cassidy M, Kraemer PM (1989) Univariate flow karyotype analysis. In: Gray JW (ed) Flow Cytogenetics Academic Press, San Diego, p 113

Cram LS, Ray FA, Bartholdi MF (1990a) Univariate analysis of metaphase chromosomes using the hypotonic potassium chloride - propidium iodide protocol. In: Darzynkiewicz Z, Crissman, H (eds) Methods in cell biology, vol 33, flow cytometry. Academic Press, San Diego, p369

Cram LS, Campbell M, Fawcett, JL, Deaven, LL (1990b) Polyamine buffer for bivariate human flow cytogenetic analysis and sorting. In: Darzynkiewicz Z, Crissman H (eds) Methods in cell biology, vol 33, flow cytometry. Academic Press, San Diego, p377

Cram LS, Chapman V, Gray JW (1992) personal communication.

Darzynkiewicz Z, Crissman H (eds) Methods in cell biology, 33, flow cytometry (1990). Academic Press, San Diego.

Gray JW, van den Engh GJ, Trask BJ (1989a) Instrumentation for chromosome analysis and sorting. In: Gray JW (ed) Flow Cytogenetics Academic Press, San Diego, p 17

Gray JW, van den Engh GJ (1989b) Bivariate flow karyotyping. In: Gray JW (ed) Flow Cytogenetics Academic Press, San Diego, p 137

Gray JW (ed) (1989c) Flow Cytogenetics. Academic Press, San Diego

Green KD, Fantes JA, Evans JH (1989) Detection of randomly occurring aberrant chromosomes as a measure of genetic change. In: Gray JW (ed) Flow Cytogenetics Academic Press, San Diego, p 161

Haley C, Archibald A (1990) The pig gene mapping project, Commission of the european communities, BRIDGE programme, DG XII-F-2, Rue de la Loi 200, B-1049, Brussels.

Langlois RG (1989) DNA stains as cytochemical probes for chromosomes. In: Gray JW (ed) Flow Cytogenetics Academic Press, San Diego, p 61

Lebo RV (1989) Gene mapping strategies and bivariate flow cytogenetics. In: Gray JW (ed) Flow Cytogenetics Academic Press, San Diego, p 225

Levy, HP, Schultz RA, Ordonez JV, Cohen MM (1991) Anti-kinetochore staining for single laser bivariate flow sorting of Indian muntjac chromosomes. Cytometry 12:695-700

Lozes C (1989) Cell culture for chromosome isolation. In: Gray JW (ed) Flow Cytogenetics Academic Press, San Diego, p 35

Melamed MR, Lindmo T, Mendelsohn ML (eds) (1990) Flow cytometry and sorting, 2nd ed. Wiley-Liss, New York

McConnell TS, Fawcett JJ, Baczek NA, Cram LS (1991) Chromosome analysis by flow cytometry. App Cytogenet 17:1-4

McCormick MK, Campbell E, Deaven L, Moyzis R (1992) Non-chimeric yeast artificial chromosome libraries from flow sorted human chromosomes 16 and 21. Proc Natl Acad Sci (in press).

Moore DH (1989) Methods for estimating components of multipeaked flow histograms. In: Gray JW (ed) Flow Cytogenetics, Academic Press, San Diego, p83

Poletaev A., personal communication.

Ray FA, Bartholdi MF, Kraemer PM, Cram LS (1984) Chromosome polymorphism involving heterochromatic blocks in Chinese hamster chromosome 9. Cytogenet and Cell Genet 38:257-263

Schmitz A, Chardon P, Chaput B, Frelat G, Vaiman M (1991) TNF alpha gene localization by PCR on swine flow sorted chromosomes. Cytometry Suppl 5: 29

Trask B (1989) Chromosome isolation procedures. In: Gray JW (ed) Flow Cytogenetics, Academic Press, San Diego, p 43

Trask B, van den Engh G (1990) Chromosome and nuclei isolation with the $MgSO_4$ procedure. In: Darzynkiewicz Z, Crissman, H (eds) Methods in cell biology, vol 33, flow cytometry. Academic Press, San Diego, p363

van den Engh G, Trask BJ, Gray JW, Langlois RG, Yu LC (1985) Preparation and bivariate analysis of suspensions of human chromosomes. Cytometry 6: 92-100

Van Dilla MA, Deaven LL, Albright KL, Allen NA, Bartholdi MF, Brown NC, Campbell EW, Carrano AV, Christensen M, Clark LM, Cram LS, Dean PN, de Jong P, Fawcett JJ, Juscoe JC, Gray JW, Hildebrand CE, Jackson PJ, Jett JH, Kolla S, Longmire JL, Lozes CR, Luedemann ML, McNinch JS, Mendelsohn JL, Meyne J, Meincke LJ, Moyzis RK, Mullikiin J, Munk AC, Perlman J, Pederson L, Peters, DC, Silva AJ, Trask BJ, van den Engh GJ (1989) The national laboratory gene library project. In: Gray JW (ed) Flow Cytogenetics, Academic Press, San Diego, p 257

Young BD (1989) Human chromosome analysis by flow cytometry. In: Gray JW (ed) Flow Cytogenetics, Academic Press, San Diego, p 1

In Situ Hybridization

Jan A. Bayer
Department of Molecular Pathology
TNO Institute for Applied Radiobiology and Immunology
Lange Kleiweg 151
2288 GJ RIJSWIJK
The Netherlands

Ger van den Engh
Lawrence Livermore National Laboratory
Biomedical Sciences Division (L-452)
P.O. Box 5507
Livermore, CA 94550
USA

General

Hybridization of a nucleic acid probe to a nucleic acid target within cytological preparations permits a high degree of resolution as to the spatial localization of this target sequence. This localization can be useful in a number of ways for answering biological questions. For example, *in situ* hybridization (ISH) to the DNA of condensed chromosomes can be used to map the chromosomal localization of particular sequences (e.g. genes). Hybridization to the DNA of interphase nuclei can be used to study the functional or spatial organization of specific sequences within the diffuse chromatin that characterises this stage of the cell cycle, or to study numerical aberrations ("interphase cytogenetics"). Furthermore, ISH to cellular RNA allows a very precise analysis of the tissue distribution of any RNA species of interest. In addition, ISH makes it possible to study the RNA of individual cells unaffected by the RNA of other cells in the tissue. At present, it is possible to detect RNAs that are present in a very small subset of cells only. Such RNAs might never be detectable in RNA extracted from whole tissue because of the dilution by other RNA species from the majority of the cells which

NATO ASI Series, Vol. H 67
Flow Cytometry
Edited by A. Jacquemin-Sablon

do not contain the RNA of interest. Thus ISH may be useful not only to show where an RNA is localized but, in some cases, it may even be the best way to show the RNA exists at all.

A Historic Overview

The technique of *in situ* hybridization was introduced by Gall and Pardue and by John, in 1969, using radioactively labeled nucleic acid probes (Gall and Pardue, 1969 and John et al., 1969). At that time it was called *in situ* hybridization (ISH), to emphasize the difference with the biochemical hybridization method, introduced in 1961 by Hall and Spiegelman (Hall and Spiegelman, 1961). Since analysis by ISH in reality is not performed *in situ* , but on microscopic preparations, be it glass slides or cell suspensions, the name hybridocytochemistry would be more correct for this elegant method. This term however is rarely used in the literature.

Using radioactive isotopes, the insulin gene was mapped to chromosome 11 by Harper in 1981 (Harper et al., 1981). The first serious report on non-radioactive ISH came from Rudkin and Stollar in 1977 (Rudkin and Stollar, 1977). They used a rabbit-derived antiserum with specificity towards RNA•DNA hybrid molecules to detect *in situ* hybrids. Bound antibody was visualized using secondary fluorochrome labeled anti-rabbit immunoglobulins. This technique was further explored by Van Prooijen-Knegt (Van Prooijen-Knegt et al., 1982), but the difficulties in obtaining good anti-DNA•RNA antibodies prevented the widespread use of this method. A direct approach was used by Bauman (Bauman et al., 1980), who applied fluorochrome-labeled RNA as a probe for the detection of for instance 5S ribosomal RNA genes in Drosophila melanogaster.

The most important labeling techniques for ISH came with the introduction of enzymatically or chemically modified nucleic acid probes. In 1981, Langer reported the first enzymatic synthesis of biotin-labeled polynucleotides (Langer et al., 1981). In 1982 the first applications were reported by using this technique for mapping genes on Drosophila chromosomes (Langer-Safer et al., 1982) and centromeric DNA on mammalian chromosomes (Manuelidis et al., 1982). The detection of the hybridized biotinylated probe is accomplished by anti-biotin antibodies, or, more often, by employing the biotin/avidin detection system. In either case visualization is achieved through coupled fluorochromes (FITC, TRITC, AMCA etc.) or enzymes (peroxidase or alkaline phosphatase). Chemical modification of probe DNA was first described by Tchen (Tchen et al., 1984) and Landegent (Landegent et al., 1984). They used 2-acetylaminofluorene (AAF) to modify the guanine residues in the nucleic acids. This technique was further refined by Landegent (Landegent et al., 1985), who reported the first detection of a single copy sequence by non-isotopic ISH. Using AAF-modified DNA probes and anti-AAF antibodies, they were able to localize part of the human thyroglobulin gene. The most recent developments have been dedicated to new techniques for the labeling of nucleic acid probes. Hopman modified cytosine residues with mercury acetate and detected these with haptenized sulfhydryl containing ligands (Hopman et al., 1986[a] and 1986[b]). Digoxigenin- or direct FITC-labeling of probes can be achieved by enzymatical introduction of digoxigenin- or FITC carrying nucleotides (Boehringer Mannheim), and also chemical sulfonation is one of the possibilities (Chemiprobe, FMC Bioproducts). All of these new labeling procedures imply their simultaneous use in multiple color hybridization experiments (Nederlof et al., 1989).

Nucleic Acid Hybridization; the Concept

A DNA molecule in its native state is a double stranded molecule. Each strand consists of a sequence of the bases adenine (A), guanine (G), thymidine (T) and cytosine (C), and two opposite strands are "bound" to each other by way of interstrand base pairing. Because base pairing can only occur by formation of hydrogen bonds between an A- and a T-base, or between a G- and a C-base (Fig. 1), the two strands are called "complementary". A G-base in one strand is hydrogen bonded to a C-base of the opposite strand by a triple hydrogen bond, whereas an A-base is bonded to a T-base by a double hydrogen bond only. This basepairing governs DNA replication during S-phase of the cell cycle. Because these interstrand hydrogen bonds are relatively weak, the individual strands of a double stranded DNA molecule

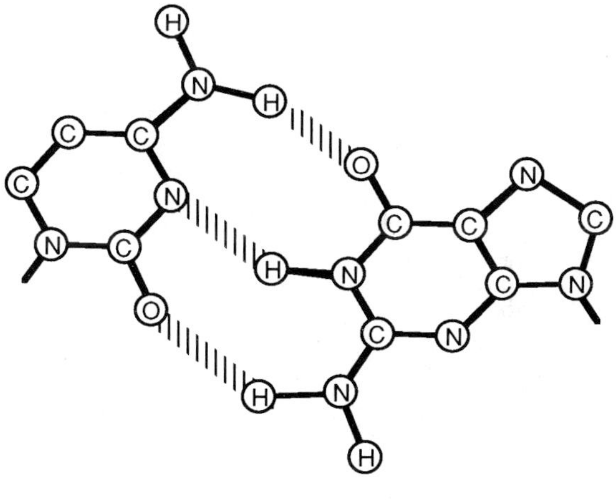

Figure 1.
Model of a guanine-cytosine base pair.

will seperate relatively easy with a moderate increase in temperature. The temperature (Tm) at which this strand separation (called "melting" or "denaturation") occurs depends on the relative content of GC base-pairs, because these GC base-pairs are held together more strongly by three hydrogen bonds. Also the Tm depends on the length of the hybrid molecule and physical parameters such as salt-strength and concentration of formamide.

The same characteristics hold for pieces of nucleic acid that are molecularly cloned into a bacterial plasmid or a bacteriophage. Upon cloning of a DNA fragment of interest, a whole range of techniques for *in vitro* labeling become eligible, e.g. nick translation, random primer extension and Polymerase Chain Reaction. Through labeling and subsequent melting of the "probe", a modified or tagged single stranded DNA or RNA is obtained, that is complementary to the sequence under investigation (e.g. a gene). Upon bringing together the denatured nucleic acid target and the single stranded probe under conditions favourable for duplex formation, a hybrid molecule can be formed between probe and target. This process is called "hybridization" (Fig. 2). The reader is referred to "Nucleic Acid Hybridization" for a complete overview (Hames and Higgins, 1985).

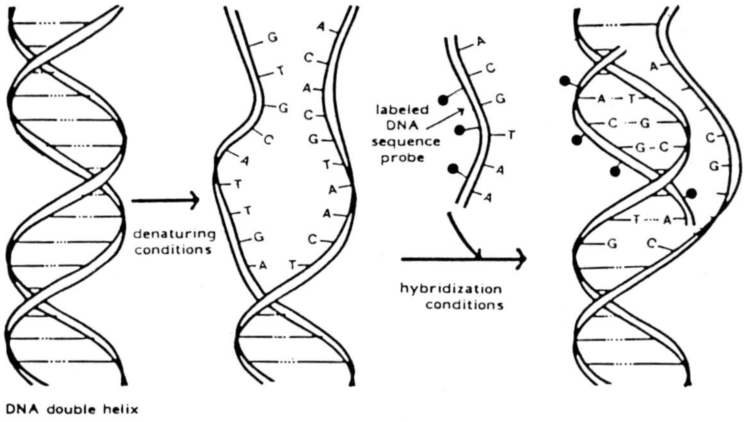

Figure 2.
Schematical representation of the formation of a hybrid DNA molecule.

Non Radioactive *In Situ* Hybridization

Non radioactive *in situ* hybridization can be performed using any of the probe modification methods mentioned above. The use of biotinylated nucleic acid probes however has become a "golden standard". Incorporation of biotin (but also digoxigenin) into nucleic acids is possible using enzymatic methods like nick translation. For biotin, also a photochemical introduction is possible, by using photobiotin. Sulfonation, mercuration and AAF modifications are chemically achieved, but have not found as widespread use.

After hybridization, detection of bound probe can be performed using either of two possible detection routes. One can detect bound probe with fluorochrome-tagged reagents, but also enzyme-linked reagents are being used (schematically depicted in Fig. 3). Examples of the latter possibility are horseradish peroxydase- or alkaline phosphatase-linked streptavidin. These can be used in the detection of hybridized biotinylated probes. Upon supplying the substrates for these respective enzymes, a visible non-soluble reaction product is deposited at the place of formation. The less diffusion of the enzyme product, the better the spatial resolution. These detection schemes are predominantly used in *in situ* hybridization on tissue sections. Especially paraffin embedded materials are not very suitable for fluorescence detection schemes, because of substantial autofluorescence. The advantage of these colorimetric detection procedures is that results can be observed with a standard light-microscope.

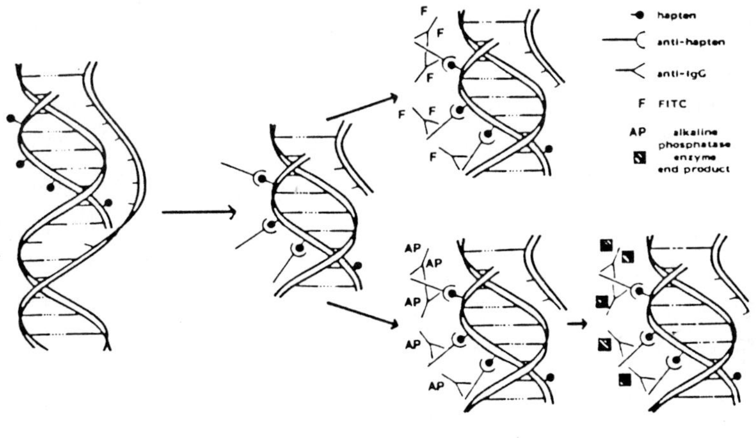

Figure 3
Detection of target DNA sequences using fluorescent or enzymatic labeling. In both schemes a small hapten (e.g. biotin) was bound to the probe and detected using anti-hapten molecules.

Fluorescence detection schemes are used very frequently at present. Both the detection of DNA targets in metaphase and interphase, as well as RNA targets in cells deposited on slides has been accomplished. A lot of the procedures used in radioactive *in situ* hybridization can be readily adapted for use in non radioactive *in situ* hybridization. These procedures have been thouroughly studied and optimized (Singer et al., 1986). Special (cell sorter-) grades of avidin-FITC for the detection of biotinylated probes have especially gained a lot of interest since the development of elegant strategies for amplification of primary signals. The use of biotinylated anti-avidin antibodies was described by Pinkel (Pinkel et al., 1986).

Flow Cytometric Fluorescence *In Situ* Hybridization (FC-FISH)

The earliest applications in flow cytometry aimed at the detection of nucleic acids by way of specific staining protocols. Procedures for staining nuclear DNA and cellular RNA were extensively used, and predominantly employed more general chemical stains. At later times, more or less sequence dependent stains were used.Very well known are the fluorescent dyes Hoechst 33258 and Chromomycin A3. Hoechst 33258 stains DNA predominantly at regions with a relative high number of AT-base pairs, whilst Chromomycin A3 stains more GC-rich areas. Using these stains, flow karyotyping and sorting of individual chromosomes has become an important application. The detection however of unique sequence elements is impossible using these types of dyes. For this purpose only nucleic acid "probes" can be used. In order to be able to analyze *in situ* hybridization by flow cytometry the hybridization must be performed in suspension (Gray, 1989). The methods needed for the detection of specific DNA sequences (e.g. alphoid repeats or even genes) are different from the methods for the detection of specific mRNA species. The first detection of repetitive DNA sequences in fixed nuclei in suspension using flow cytometry was described by Trask (Trask et al., 1985). Van Dekken described the flow cytometric detection of a highly repeated alpha satellite sequence located on the centromeric region of the human chromosome 1 (Van Dekken et al., 1990). The method proved sensitive enough to detect polymorphism of this region between individuals. Also the simultaneous detection of a probe for the centromeric region of chromosome 1 and a probe for the Y-chromosome, using three-laser flow cytometry, was accomplished.

The detection of *in situ* hybridization of isolated metaphase chromosomes was very cumbersome because of detrimental effects of ISH protocols on chromosome integrity. A report by Dudin for the first time described successful application of such a procedure (Dudin et al., 1987).

The detection of specific RNA species in cells in suspension has relied on the development of a fixation protocol that *i*) preserves cellular integrity, *ii*) immobilizes the RNA molecules in the cellular matrix, but, somewhat contradictory, *iii*) does not impair diffusion of probe and detection reagents through the cells and *iiii*) inactivates the ever-present RNase activity. In the case of RNA *in situ* hybridization, the denaturation of the target can be omitted since RNA is a single stranded molecule.

A procedure was developed that meets these requirements and the detection of 28S ribosomal RNA (Bauman and Bentvelzen, 1988) and β-globin RNA (Bayer and Bauman, 1990) were described. The procedure is based on a formalin/EtOH fixation. We prefer to use single strand RNA probes that can be produced by way of *in vitro* transcription. The generated transcripts can be either complementary or identical to the RNA target. The transcript (or probe) that is identical to the RNA target serves the purpose of a non-hybridizing control probe that is of exactly the same composition as the complementary (hybridizing) probe in terms of overall GC-content and regional GC-distribution. In this way, these strand specific RNA probes can give the ultimate assurance that the desired hybrids are indeed formed. These controls, we feel, are very important in RNA *in situ* hybridization. Until now, the sensitivity of this method for the flow cytometric detection of gene expression is such that only higher copy number mRNA species can be detected (±1,000 copies per cell).

References

Bauman JGJ, Wiegant J, Borst P and van Duijn P (1980). A new method for fluorescence microscopical localization of specific DNA sequences by in situ hybridization of fluorochrome labeled RNA. Exp. Cell Res. 138: 485-490.

Bauman JGJ and Bentvelzen P (1988) Flow cyometric detection of ribosomal RNA in suspended cells by fluorescent in situ hybridization. Cytometry 9: 517-524.

Bayer JA and Bauman JGJ (1990) Flow cytometric detection of β-globin mRNA in murine haemopoietic tissues using fluorescent in situ hybridization. Cytometry 11: 132-143.

Dudin G, Cremer T, Schardin M, Hausmann M, Bier F and Cremer C (1987) A method for nucleic acid hybridization to isolated chromosomes in suspension. Hum. Genet. 76: 290-292.

Gray J (Ed) (1989) Flow Cytogenetics. Acad. Press Ltd., New York, pp 275-301.

Gall JG and Pardue ML (1969) Molecular hybridization of radioactive DNA to the DNA of cytological preparations. Proc. Natl. Acad. Sci. USA 64: 600-604.

Hall BD and Spiegelman S (1961) Sequence complementarity of T2-DNA and T2-specific RNA. Proc. Natl. Acad. Sci. USA 47: 137-146.

Hames BD and Higgins SJ (eds) (1985) Nucleic acid hybridization IRL Press.

Harper ME, Ullrich A and Saunders GF (1981) Localization of the human insulin gene to the distal end of the short arm of chromosome 11. Proc. Natl. Acad. Sci. USA 78: 4458-4460.

Hopman AHN, Wiegant J and van Duijn P (1986) A new hybridocytochemical method based on mercurated nucleic acid probes and sulfhydryl-hapten ligands. I. Stability of the mercury-sulfhydryl bond and influence of the ligand structure on immunocytochemical detection of the hapten. Histochem. 84: 169-178.

Hopman AHN, Wiegant J and van Duijn P (1986) A new hybridocytochemical method based on mercurated nucleic acid probes and sulfhydryl-hapten ligands. II. Effect of variations in ligand structure on the in situ detection of mercurated probes. Histochem 84, 179-185.

John H, Birnsteil ML and Jones KW (1969) RNA-DNA hybrids at cytological levels. Nature 223: 582-587.

Langer PR, Waldrop AA and Ward DC (1981) Enzymatic synthesis of biotin-labeled polynucleotides: Novel nucleic acid affinity probes. Proc. Natl. Acad. Sci. USA 78: 6633-6637.

Langer-Safer PR, Levine M and Ward DC (1982) Immunological method for mapping genes on Drosophila polytene chromosomes. Proc. Natl. Acad. Sci. USA 79: 4381-4385.

Landegent JE, Jansen in de Wal N, Baan RA, Hoeijmakers JHJ and van der Ploeg M (1984) 2-Acetylaminofluorene-modified probes for the indirect hybridocytochemical detection of specific nucleic acid sequences. Exp. Cell Res. 153: 61-72.

Landegent JE, Jansen in de Wal N, van Ommen GJB, Baas F, de Vijlder JJM, van Duijn P and van der Ploeg M (1985) Chromosomal localization of a unique gene by non-autoradiographic in situ hybridization. Nature 317: 175-177.

Manuelides L, Langer-Safer PR and Ward DC (1982) Sensitive, high-resolution mapping of satellite DNA using biotin-labeled DNA probes. J. Cell Biol. 95: 619-625.

Nederlof PM, Robinson D, Abuknesha R, Wiegant J, Hopman AHN, Tanke HJ and Raap AK (1989) Three-color fluorescence in situ hybridization for the simultaneous detection of multiple nucleic acid sequences. Cytometry 10: 20-27.

Pinkel D, Straume T and Gray JW (1986) Cytogenetic analysis using quantitative, high sensitivity, fluorescence hybridization. Proc. Natl. Acad. Sci. USA 83: 2934-2938.

Rudkin GT and Stollar BD (1977) High resolution detection of DNA-RNA hybrids in situ by indirect immunofluorescence. Nature 265: 472-473.

Singer RH, Lawrence JB and Villnave C (1986) Optimization of in situ hybridization using isotopic and non-isotopic methods. BioTechniques 4: 230-243.

Tchen P, Fuchs RPP, Sage E and Leng M (1984) Chemically modofied nucleic acids as immunodetectable probes in hybridization experiments. Proc. Natl. Acad. Sci. USA 81: 3466-3470.

Trask BJ, van den Engh G, Landegent J, Jansen in de Wal N and van der Ploeg M (1985) Detection of DNA sequences in nuclei in suspension by in situ hybridization and dual beam flow cytometry. Science 230: 1401-1403.

van Dekken H, Arkesteijn GJA, Visser JWM and Bauman JGJ (1990) Flow cytometric quantification of human chromosome specific repetitive DNA sequences by single and bicolor fluorescent in situ hybridization to lymphocyte interphase nuclei. Cytometry 11: 153-164.

Van Prooijen-Knegt AC, van Hoek JFM, Bauman JGJ, van Duijn P, Wool IG and van der Ploeg M (1982) In situ hybridization of DNA sequences in human metaphase chromosomes visualized by an indirect fluorescent immunocytochemical procedure. Exp. Cell Res. 141: 397-407.

THE MOLECULAR BASIS OF MAMMALIAN CELL GROWTH CONTROL

Gerard I. Evan
ICRF Laboratories,
44, Lincolns Inn Fields,
London WC2A 3PX,
Great Britain.

Introduction

Multicellular organisms have a unique problem with regulation of proliferation of their component cells. Failure of even a large proportion of cells to divide within a tissue is generally of little consequence because of substantial cellular redundancy: i.e. there are always other cells present than can renew the affected tissue. In contrast, tight control of proliferation is an absolute requirement for multicellularity because unrestrained proliferation of even one cell and its progeny will be lethal: this is the disease we call cancer. Recently, there have been many spectacular advances in our understanding of the molecular processes that regulate cell proliferation. Concomittantly with this advance has also arisen a confusion over what regulation of cell proliferation entails. Is regulation exerted at the level of control of the cell cycle, or at the point when a cell decides whether or not to proliferate, quiesce or differentiate, or at the level of the whole tissue or organism by balancing cell gain and cell loss? In this review I will outline the major features of cell growth regulation as they pertain to mammalian cells and attempt to provide a more intergated framework for further discussion.

Tiers of control

Mammalian cell proliferation comprises three linked processes:

1. The cell cycle. This is the process whereby DNA synthesis and cell division alternate and it is ensured that one process finished before the other starts. Clearly, the cell cycle and its machinery is likely to be substantially conserved amongst all cells, whether from a unicellular or a multicellular organism. For this reason, many of the recent advances in our undertsanding of the cell cycle have come from studies on the unicellular budding yeast Saccharomyces *cerevisiae* and the fission yeast Schizosaccharomyces *pombe*.

NATO ASI Series, Vol. H 67
Flow Cytometry
Edited by A. Jacquemin-Sablon

2. The regulation of proliferation. This is arguably the key growth control point in normal mammalian cells. It is presumably damaged in many tumour cells and likely to be absent or extensively modified and adapted in unicellular eularyotes. Accordingly, normal untransformed mammalian cells are the only effective model in which to study this process.

3. Control of tissue mass. The homoestatic control of tissue mass and tissue architecture relies on the net gain or loss of cells from tissues. This is a dynamic process that includes those mechanisms leading to cell gain (i.e. cell cycle and proliferation control), cell stasis (via quiescence and differentiation) and cell loss (shedding, cell death).

In seeking to understand how these three processes are interlinked at the molecular level it is first necessary to examine what is known of the molecular machinery mediating each one. I will therefore begin by discussing the cell cycle, control of proliferation and regulation of tissue mass in turn and try to show how each might be intergrated with the others.

The cell cycle

The cell cycle is, as already outlined, necessarily common to all cells and, for this reason, much of the molecular machinery that regulates is also highly conserved through evolution. Current understanding of the regulation of the cell cycle comes from the study of temperature-sensitive (ts) cell division cycle (called "*cdc*" mutants in *pombe* and "*CDC*" in *cerevisiae*) mutants in yeast which are unable to proceed through the cell cycle above a permissive temperature (Bartlett and Nurse., 1990). The first of these to be characterised in S. *pombe* was *cdc2* (equivalent to *CDC28* in *cerevisiae*). *cdc2* mutants arrest at both the G1/S and G2M boundaries both of these cell cycle checkpoints share a common gene product. *cdc2* encodes a 34 kDa serine/threonine protein kinase, $p34^{cdc2}$, that is substantially conserved throughout eukaryotic evolution and that phosphorylates a wealth of intracellular target proteins. Early studies showed, however, that the levels of $p34^{cdc2}$ are invariant thoroughout the cell cycle whereas its kinase activity is detected only at the G1/S and G2/M boundaries. It is now known that $p34^{cdc2}$ is active only when associated with members of a family of protein called cyclins. Cyclins had initially been identified as protein that demonstrate a cell cycle-dependent pattern of expression in sea urchins. Thus, $p34^{cdc2}$ activity is cell-cycle-dependent because it is regulated by a cell cycle-dependent cyclin subunit. The question then becomes: what regulates the cyclins? This is best illustrated by examining the cell cycle transition that is currently best understood, G2/M.

Control of the G2/M transition

G2/M cyclins are the A and B cyclins. Several A and B cyclins are known but it is unclear whether each has a different specificity or function or whether their numbers simply reflect inherent redundancy. Both A and B cyclins accumulate throughout S and G2 phases of the cell cycle, with A accumulateing slightly after B (Pines., 1992). As both are synthesised they complex with available $p34^{cdc2}$ but the complex is immediately inactivated by phosphorylation of $p34^{cdc2}$ at positions Tyr15 by the mitotic inhibitory kinase $p107^{wee1}$ and $p66^{mik1}$. $p34^{cdc2}$ -cyclin A/B is thus held in abeyance until the end of G2 at which $p34^{cdc2}$ is dephosphorylated and activated via the $p180^{cdc25}$ phosphatase. Active $p34^{cdc2}$ -cyclin A/B then initiates mitosis following which cyclins A and B are rapidly degraded. Indeed, evidence suggests that cyclin A and B degradation is necessary for complete exit from M. What activates $p180^{cdc25}$? The answer is presently unclear but some evidence suggests that it, itself, is in part regulated by cyclin B. One of the paradigms that therefore emerges from our knowledge of the cell cycle is that certain components are influenced by factors that are themselves sensitive to the components they regulate. Regulation is therefore truly cyclic!

Control of the G1/S transition

Can what we know of regulation of the G2/M transition teach us anything about the G1/S transition? The answer appears to be yes. Just as there are G2/M cyclins there are G1 and S cyclins. In S. *cerevisiae* these have been identified as the CLN 1, 2 and 3 which are some 25% homologous with cyclins A and B. CLN1 and 2 are unstable and their levels are regulated principally at the level of transcription. They possess a cell cycle control sequence (the Swi4/6-dependent cell cycle box or SCB) which, together with another enhancer sequence (the Mlu1 cell-cycle box or MCB) is present in a number of genes whose expression is regulated within G1 and S phase (see below). CLN3, on the other hand, is a stable protein. Expression of CLN1 1nd 2 is repressed by mating factor and this is thought to be responsible for mating factor-induced growth arrest. G1 cyclins in S. *pombe* have proven more elusive. To date two candidates, the *puc1*+ and *cig1*+ gene products are known.

In mammalian cells, comittment to proceed through the cell cycle maps to a "restriction point" (often called "R") late in G1, typically about two hours before the onset of DNA synthesis. R is commonly thought to represent the G1/S regulatory transition step analogous to that at G2/M. Three G1/S cyclins are known, C, D and E (Pines., 1991, Pines., 1992). C peaks early in G1, E just before S. Both are expressed in a cyclic manner. In contrast, expression of the D cyclins, of which three are known) appears to be dependent upon mitogen availability, at least in macrophages in which they were first characterised (Matsushime, et al., 1991). D cyclins may therefore represent cdc2 controlling subunits whose expression is not linked to cell cycle position but to the decision of whether or not to enter cycle in the first place. D cyclins may thus act as a bridge between cell cycle and proliferation control.

Regulation of gene expression in the cell cycle

A defined pattern of gene expression is clearly necessary during cell cycle progression. For example, the onset of DNA synthesis requires expression of proteins needed for DNA replication such as thymidylate kinase, thymidylate synthase, DNA polymerase and dihydrofolate reductase; histones and some other chromatin components must be synthesised during S phase. In *S. cerevisiae,* genes for many of these G1/S-specific proteins contain the SCB or MCB boxes described above in their regulatory regions. The SCB box (5'-CACGAAA-3') is bound by the product of the *SWI4* gene which, in association with the *SWI6* product, causes activation of relevant genes (Andrews., 1992). *SWI6* associates with a different protein to bind the MCB box (5'-ACGCGT-3')and induce analogous activation. The S.*pombe* counterpart of *SWI6* is the product of the *cdc10* gene. In mammalian cells, the analogous transcription factor controlling G1/S genes is called E2F. E2F is itself the centre for much interest because it is found associated with cyclin A as cells proceed from G1 to S in a complex that also includes a 107 kDa nuclear protein (Mudryj, et al., 1991). This p107 has homology with the product of the retinoblastoma gene, $p105^{RB}$, with which E2F is associated in G1 (Bagchi, et al., 1991, Chellappan, et al., 1991, Chittenden, et al., 1991). Current dogma asserts that E2F is inactive when complexed, and it is displacement of $p105^{RB}$ by p107 that leads to E2F activation and thence to activation of appropriate genes. This simplistic model is certain to be further modified over the next few months or years but provides a basis for modelling cell cycle control.

Factors restraining cell cycle progression

Given the dangers of unrestrained cell proliferation in multicellular organisms, it might be expected that cell cycle progression is modulated not only by positive factors driving cell proliferation but also by factors that restrain it. At least two potential candidate endogenous growth suppressors are known. One, the product of the retinoblastoma locus $p105^{RB}$ has already been mentioned above. The other, p53, was first identified as a protein that complexed with the large T antigen of the SV40 DNA tumour virus (Levine., 1992). $p105^{RB}$ undergoes a cyclic phosphorylation and dephosphorylation during the cell cycle.

At about the time of cell cycle commitment, p105^{RB} is phosphorylated, a process that is presumed to inactivate it (Buchkovich, et al., 1989, Chen, et al., 1989). The function of active p105^{RB} is thus presumed to be to prevent entry into S phase. p105^{RB} is involved in formation of complexes with the E2F transcription factor as well as with a number of other transcription factors such as the nuclear products of the oncogenes c-*myc* (Rustgi, et al., 1991), c-*fos* and c-*jun*. p53 may also be involved in controlling entry into S phase although the mechanism is unclear.

Cell cycle and cancer

Perturbation of cell cycle components is associated with neoplasia in a variety of ways. The cyclin A gene is the frequent site of intergration of the hepatitis B virus, associated with hepatocellular carcinoma. Viral insertion typically induces deregulated expression of nearby genes and presumably deregulated cyclin A expression contributes to tumour formation in this instance (Hunter and Pines., 1991). Both the *bcl-1* gene, translocated to the immunoglobulin gene heavy chain locus in many B cell lymphomas, and the *Prad1* locus, translocated to the parathyroid hormone gene in parathyroid tumours, encode D type cyclins. Translocations of this kind also cause deregulated expression of attendant genes, suggesting that damage to control of cyclin D expression is also carcinogenic (Hunter and Pines., 1991).

Both the retinoblastoma protein p105^{RB} and p53 are intimately involved in human cancers. The retinoblastoma gene RB is functionally inactivated in human retinoblastoma as well as in a variety of other tumours (Weinberg., 1992) suggesting that its growth-limiting properties act as an effective bulwark against spontaneous carcinogenesis. p105^{RB} is also involved in complexes with the large T antigen of SV40, the E1A protein of adenovirus and the E7 protein of human papilloma virus, all tumour viruses. Current thinking is that p105^{RB} is functionally inactivated in such complexes and that, by promoting host cell growth, this facilitates attendant viral proliferation. p53, which is functionally inactivated in some 60% of all human tumours, also forms complexes with the tumour antigens of DNA tumour viruses; specifically SV40 large T, adenovirus E1B and Human papilloma virus E6 (Levine., 1992). Once again, sequestration and concomittant inactivation of p53 is thought to confer an advantage on the virus because by promoting host growth the virus promotes its own.

Thus, controlled expression and activation of cell cycle regulatory components is an essential prerequerisite for effective control of cell proliferation. With the possible exception of the D cyclins, the factors regulating the cell cycle are not in the main involved in determining whether or not a cell is in cycle or out of cycle (quiescent). Yet, the transition form quiescence to prolferation is, as discussed, perhaps the major control transition in mammalian cells. We will therefore next examine what is known of the molecular events that accompany exit from G0 and entry into cycle - the regulation of proliferation.

The regulation of proliferation

Although the importance of positive and negative factors mediating transition through cell cycle checkpoints in cellular growth control is clear, many, perhaps most cells in an adult human are not in cycle at all. Rather, they are in a state of growth quiescence. In many cases, quiescence arises as a result of terminal differentiation and irreversible loss of proliferative capacity. In other cases, however, quiescent cells retain full proliferative capacity and can respond rapidly to mitogenic stimulation by entry into the cell cycle and eventual division. Such is the case in, for example, fibroblasts, lymphocytes and hepatocytes. Quiescent cells are not usually arrested at either of the cell cycle checkpoints. Instead, they reside in a non-cycling state often termed G0. The transition from G0 into G1 is, therefore, the major point at which mammalian cell proliferation is regulated.

Immediate early genes

When a quiescent fibroblast or lymphocyte is stimulated with mitogens *in vitro* there is an abrupt induction of some 40-100 genes (Almendral, et al., 1988, Lord, et al., 1990, Mohn, et al., 1991), the so-called "immediate early growth response genes." These genes encode proteins thought to be involved in enabling the cell to engage in growth. Some are cytokines, some elements of signal transduction pathways, some structural and cytoskeletal proteins, and others transcription factors. Substantial attention has focussed on this latter group, the transcription factors, because they presumably mediate genetic changes that accompany proliferation. Amongst them are the products of the c-*fos*, c-*jun* and c-*myc* genes, all known to have oncogenic properties if deregulated and all sequence-specific DNA-binding proteins. The overt proto-oncogenic properties of the c-*fos*, c-*jun* and c-*myc* genes suggests that they have crucial roles in regulating cell proliferation and the decision whether to proilferate or arrest. I will therfore discuss them in detail, although this should not imply that other immediate early genes are irrelevant to growth control.

Fos and Jun

Both the c-*fos*, c-*jun* genes were first identified as viral oncogenes in the FBJ and FBR murine osteosarcoma viruses and in the ASV17 avian sarcoma virus respectively. Both c-*fos* and c-*jun* are members of a gene families that encodes a group of transcription factors which includes c-*jun*, *jun*-B, *jun*-D, c-*fos*, *fos*-B and *fra*-1 (Abate and Curran., 1990, Curran and Franza., 1989). The various *fos* and *jun* proteins function as a variety of homo- and hetero-dimers which form through a characteristic dimerisation motif called a leucine zipper which they all possess (Jones., 1990). The Fos-Jun dimers constitute the AP-1 transcription factor activity originally isolated from nuclear extracts by its specificity for the common TGACTCA motif and which confers inducibility by the phorbol ester TPA (Curran and Franza., 1989). The TGACTCA is found upstream of a variety of genes activated during the G0/G1 transition.

Although their oncogenic properties strongly imply that both Fos and Jun proteins are intimately involved in cell proliferation, it is clear that they can function in other contexts also. For example, both are expressed in non-mitotic CNS neurones in response to synaptic stimulation (Sagar, et al., 1988, Williams, et al., 1990a, Williams, et al., 1990b, Wisden, et al., 1990). Thus, their function appears to act as general switches to initiate a variety of cellular responses: what they do depends upon the context in which they are expressed.

Myc

The c-myc proto-oncogene was originally isolated by its homology to the transforming genes of the avian myelocytic leukaemia viruses. C-*myc* is one member of a family of *myc* genes in man which includes N-*myc*, L-*myc*, B-*myc*, and S-*myc* (Ingvarsson., 1990), although only c-*myc* appears to be expressed in adult cells. In untransformed cells, c-*myc* expression is linked with growth state. It is absent in quiescent cells but rapidly induced by mitogenic growth factors (Kelly, et al., 1983). For this reason, it has frequently been grouped with other immediate early growth response genes such as c-*fos* and c-*jun*. However, whereas most other immediate early genes are expressed transiently and only at the G0/G1 interface (Almendral, et al., 1988) or in response to other adventitious stimuli, c-*myc* appears to be expressed at a constant rate throughout the cell cycle in proliferating cells (Hann, et al., 1985, Rabbitts, et al., 1985, Thompson, et al., 1985, Waters, et al., 1991) but in a manner that is mitogen dependent (Dean, et al., 1986, Waters, et al., 1991). As both the c-*myc* mRNA and protein have extremely short half-lives, continuous expression of c-*myc* can only occur as a result of continuous synthesis. This suggests that c-*myc* exerts a continuous function throughout cell proliferation.

Almost all tumour cells express steady state levels of c-*myc* mRNA and protein which are substantially higher than those seen in untransformed proliferating fibroblasts (Moore, et al., 1987). Although the significance of this elevated expression is unknown, altered regulation of the c-*myc* gene, in particular by amplification and chromosomal translocation, has been implicated in the genesis of a number of human and non-human tumours (Alitalo, et al., 1987). Interestingly, in many tumour cells that express high levels of c-*myc* protein there is no obvious genetic lesion to account for the over-expression. However, because c-*myc* expression appears to be autoregulating (Penn, et al., 1990b), over-expression of c-*myc* in tumour cells is unlikely to be simply a concomittent effect of some other oncogenic lesion. Growth factor deprivation does not, in general, lead either to growth arrest of tumour cells or to repression of c-*myc* expression as is the case in untransformed fibroblasts (Campisi, et al., 1984). C-*myc* expression alone will drive quiescent cells into cycle in the absence of mitogens (Eilers, et al., 1991), and keep them there (Evan, et al., 1992) and (Littlewood *et al.*, manuscript submitted). Moreover, specific anti-sense inhibition of c-*myc* expression effectively blocks entry of cells into cycle in response to mitogens (Heikkila, et al., 1987). Myc is therefore both necessary *and* sufficient for cell proliferation. Clearly, this is a potentially very dangerous state of affairs because any mutation that deregulates c-*myc* expression will in principle be oncogenic. Recently, this paradox has been resolved by the discovery that Myc is a potent inducer of programmed cell death (Evan, et al., 1992) - see below.

The c-*myc* gene product, Myc, is a sequence-specific DNA binding protein (Blackwell, et al., 1990) that forms an active complex with a heterologous partner called Max (Blackwood and Eisenman., 1991, Littlewood, et al., 1992). Dimerisation is mediated by a C-terminal domain, part of which shares structural homology with the leucine zipper present in Fos and Jun proteins described above. In addition, however, both Myc and Max possess another mutually interactive domain, a Helix-Loop-Helix (HLH) motif, that further stabilises dimerisation. Structures for the Myc-Max interaction have been modelled (Vinson and Garcia., 1992). The Myc-Max dimer is almost certainly a transcription factor that modulates expression of specific genes that then mediate the observed effects of Myc expression - mitogenesis (Eilers, et al., 1991 , Evan, et al., 1992), transformation (Land, et al., 1983, Stone, et al., 1987), autosuppression (Penn, et al., 1990a, Penn, et al., 1990b), inhibition of differentiation (Freytag., 1988) and apoptosis (Evan, et al., 1992).

Role of growth response genes

From this brief survey of the characteristics of the c-*fos*, c-*jun* and c-*myc* genes and their proteins it is clear that they fall into two categories. Fos and Jun exemplify a class of transcription factor that mediates a transition from one state (eg quiescence) to another (eg proliferation) but is not required to maintain the new state. Such factors are also involved in a variety of other molecular switching events which are not associated with cell proliferation. I suggest that Fos and Jun are examples of generic "toggle switches" that flick a cell from one stable genetic state into another. In contrast, c-*myc* exemplifies a switch that must be continuously engaged in order to have its effect. I liken this to an "accelerator" which must be continuously activated in oredr to maintain a response. However, both Fos/Jun and Myc classes operate through their own target genes and it is only by determining the nature and function of these target genes that we will uncover the biological role of these molecular switches.

Control of tissues - the problem of integration

Tissues are dynamic structures whose architecture and size depends upon the net behaviour of large numbers of cells working in concert. How is such integration achieved. In principle, there are four possible fates for any one cell: quiescence, proliferation, differentiation and death. I have discussed what is known about quiescence and proliferation, but what of differentiation and death? In many ways, differentiation can be seen as the genetic obverse of proliferation in that as cells differentiate they almsot always lose their capacity for self-renewal. Thus growth and differentiation are necessarily and inversely linked. In this scheme, the onset of differentiation, mediated by interactions with specific cytokines, automatically leads to the cessation of proliferation. A common fate of the differentiated cells is eventual death. This can occur by a variety of mechanisms, for example physical loss (eg. shedding of skin cells or loss of colonic epithelium into the gut lumen), by accident (eg. random poisoning, lethal mutation etc.) and also by programmed cell death. One of the mechanisms of programmed cell death is the process of apoptosis - an energy-dependent active process involving a specific pathway of chromatin condensation and fragmentation, cellular vesicularisation and activation of surface markers signalling phagocytosis (Bursch, et al., 1990, Cohen, et al., 1992, Harmon, et al., 1991, McConkey, et al., 1990, Williams., 1991, Wyllie., 1987). Apoptosis is involved in the shaping of tissues in birth embryo and adult and is also one of the patways of tumour cell destruction in response to chemo- and radio-therapy (Cotter, et al., 1990, Lennon, et al., 1990).

Apoptosis

The genetics of apoptosis is best understood in the nematode Caenorhabditis *elegans* tin which it has been shown to depend on the interplay of specific cell death (ced) genes - in particular *ced 3* and *ced 4*, which induce death, and *ced 9* which blocks it (Ellis and Horvitz., 1991). In mammalian cells the pathway is far less well defined. The gene *bcl*-2, first identified as the site of reciprocal translocations in human follicular lymphomas, is able to block apoptosis (Hockenbery, et al., 1990, Hockenbery, et al., 1991, Korsmeyer, et al., 1990, Sentman, et al., 1991) and may be a major regulator of the pathway in mammals. Recently, the c-*myc* gene has been shown to be a potent inducer of apoptosis (Evan, et al., 1992) whose lethal effects are inhibited by BCL-2 expression (Fanidi, Harrington and Evan, paper in press). The c-*myc* gene thus appears to sit in a pivotal position as a regulator not only of proliferation, but also of cell death. What might be the sense in having one intracellular component regulating two such contradictory processes? The answer seems to be that c-*myc* is, as outlined above, both necessary *and* sufficient for cell proliferation. Thus any mutation deregulating c-*myc* expression would be oncogenic and lethal. If, however, in addition to regulating genes mediating proliferation, Myc also activates genes mediating apoptosis, then all cells expressing Myc would necessarily be primed for programmed cell death. Successful proliferation would then presumably occur only if apoptosis were actively inhibited, perhaps by activation of complementary signal transduction pathways. The dual properties of Myc therefore serve to integrate the different tiers of control of cell proliferation and thereby give rise to a mutlicellular organism that can maintain its tissue architecture yet retain proliferative capacity in its cell necessary for repair and growth.

References

Abate C and Curran T (1990) Encounters with *fos* and *jun* on the road to AP-1. Sem. Cancer Biol. - Transcription factors, differentiation and cancer **1**:19-26

Alitalo K, Koskinen P, Makela TP, Saksela K, Sistonen L and Winqvist R (1987) Myc oncogenes: activation and amplification. Biochem. Biophys.Acta **907**:1-32

Almendral JM, Sommer D, MacDonald-Bravo H, Burckhardt J, Perera J and Bravo R (1988) Complexity of the early genetic response to growth factors in mouse fibroblasts. Mol. Cell. Biol. **8**:2140-2148

Andrews B (1992) Dialogue with the cell cycle. Nature **355**:393-394

Bagchi S, Weinmann R and Raychaudhuri P (1991) The retinoblastoma protein copurifies with E2F-I, an E1A-regulated inhibitor of the transcription factor E2F. Cell **65**:1063-72

Bartlett R and Nurse P (1990) Yeast as a model system for understanding the control of DNA replication in Eukaryotes. Bioessays **12**:457-63

Blackwell TK, Kretzner L, Blackwood EM, Eisenman RN and Weintraub H (1990) Sequence-specific DNA binding by the c-Myc protein. Science **250**:1149-51

Blackwood EM and Eisenman RN (1991) Max: a helix-loop-helix zipper protein that forms a sequence-specific DNA-binding complex with Myc. Science **251**:1211-7

Buchkovich K, Duffy LA and Harlow E (1989) The retinoblastoma protein is phosphorylated during specific phases of the cell cycle. Cell **58**:1097-1105

Bursch W, Kleine L and Tenniswood M (1990) The biochemistry of cell death by apoptosis. Biochem Cell Biol **68**:1071-4

Campisi J, Gray HE, Pardee AB, Dean M and Sonenshein GE (1984) Cell-cycle control of c-myc but not c-ras expression is lost following chemical transformation. Cell **36**:241-247

Chellappan SP, Hiebert S, Mudryj M, Horowitz JM and Nevins JR (1991) The E2F transcription factor is a cellular target for the RB protein. Cell **65**:1053-61

Chen P-L, Scully P, Shew J-Y, Wang JYJ and Lee W-H (1989) Phosphorylation of the retinoblastoma gene product is modulated during the cell cycle and cellular differentiation. **58**:1193-1198

Chittenden T, Livingston DM and Kaelin WJ (1991) The T/E1A-binding domain of the retinoblastoma product can interact selectively with a sequence-specific DNA-binding protein. Cell **65**:1073-82

Cohen J, Duke R, Fadok V and KS S (1992) Apoptosis and programmed cell death in immunity. Ann Rev Immunol **10**:267-293

Cotter TG, Lennon SV, Glynn JG and Martin SJ (1990) Cell death via apoptosis and its relationship to growth, development and differentiation of both tumour and normal cells. Anticancer Res. **10**:1153-9

Curran T and Franza BR (1989) *Fos* and *jun:* the AP-1 connection. Cell **55**:395-397

Dean M, Levine RA, Ran W, Kindy MS, Sonenshein GE and Campisi J (1986) Regulation of c-*myc* transcription and mRNA abundance by serum growth factors and cell contact. J. Biol. Chem. **261**:9161-6

Eilers M, Schirm S and Bishop JM (1991) The MYC protein activates transcription of the alpha-prothymosin gene. EMBO J. **10**:133-41

Ellis RE and Horvitz HR (1991) Two C. elegans genes control the programmed deaths of specific cells in the pharynx. Development **112**:591-603

Evan G, Wyllie A, Gilbert C, Littlewood T, Land H, Brooks M, CM W, Penn L and Hancock D (1992) Induction of apoptosis in fibroblasts by c-*myc* protein. Cell **63**:119-125

Freytag SO (1988) Enforced expression of the c-*myc* oncogene inhibits cell differentiation by precluding entry into a distinct predifferentiation state in G0/G1. Mol. Cell. Biol. **8**:1614-1624

Hann SR, Thompson CB and Eisenman RN (1985) C-*myc* oncogene protein synthesis is independent of the cell cycle in human and avian cells. Nature **314**:366-369

Harmon BV, Takano YS, Winterford CM and Gobe GC (1991) The role of apoptosis in the response of cells and tumours to mild hyperthermia. Int J Radiat Biol **59**:489-501

Heikkila R, Schwab G, Wickstrom E, Loke SL, Pluznik DH, Watt R and Neckers LM (1987) A c-*myc* antisense oligodeoxynucleotide inhibits entry into S phase but not progress from G0 to G1. Nature **328**:445-449

Hockenbery D, Nunez G, Milliman C, Schreiber RD and Korsmeyer SJ (1990) *Bcl-2* is an inner mitochondrial membrane protein that blocks programmed cell death. Nature **348**:334-6

Hockenbery DM, Zutter M, Hickey W, Nahm M and Korsmeyer SJ (1991) BCL2 protein is topographically restricted in tissues characterized by apoptotic cell death. Proc Natl Acad Sci U S A **88**:6961-5

Hunter T and Pines J (1991) Cyclins and cancer. Cell **66**:1071-4

Ingvarsson S (1990) The myc gene family proteins and their role in transformation and differentiation. Semin Cancer Biol **1**:359-69

Jones N (1990) Transcriptional regulation by dimerization: two sides to an incestuous relationship. Cell **61**:9-11

Kelly K, Cochran BH, Stiles CD and Leder P (1983) Cell specific regulation of the c-*myc* gene by lymphocyte mitogens and platelet-derived growth factor. Cell **35**:603-610

Korsmeyer SJ, McDonnell TJ, Nunez G, Hockenbery D and Young R (1990) Bcl-2: B cell life, death and neoplasia. Curr Top Microbiol Immunol **166**:203-7

Land H, Parada LF and Weinberg RA (1983) Tumorigenic conversion of primary embryo fibroblasts requires at least two cooperating oncogenes. Nature **304**:596-602

Lennon SV, Martin SJ and Cotter TG (1990) Induction of apoptosis (programmed cell death) in tumour cell lines by widely diverging stimuli. Biochem. Soc. Trans. **18**:343-5

Levine A (1992) The p53 tumour suppressor gene and product. Cancer Surveys **12**:59-79

Littlewood T, Amati B, Land H and Evan G (1992) Max and c-Myc/Max DNA binding activities in cell extracts. Oncogene **In Press**:

Lord KA, Hoffman-Liebermann B and Liebermann DA (1990) Complexity of the immediate early respnse of myeloid cells to terminal differentiation and growth arrest includes ICAM-1, *Jun*-B and histone variants. **5**:387-396

Matsushime H, Roussel MF, Ashmun RA and Sherr CJ (1991) Human D-type cyclin. Cell **65**:701-13

McConkey DJ, Orrenius S and Jondal M (1990) Cellular signalling in programmed cell death (apoptosis). Immunol Today **11**:120-1

Mohn KL, Laz TM, Hsu JC, Melby AE, Bravo R and Taub R (1991) The immediate-early growth response in regenerating liver and insulin-stimulated H-35 cells: comparison with serum-stimulated 3T3 cells and identification of 41 novel immediate-early genes. Mol Cell Biol **11**:381-90

Moore JP, Hancock DC, Littlewood TD and Evan GI (1987) A sensitive and quantitative enzyme-linked immunosorbence assay for the c-*myc* and N-*myc* oncoproteins. Oncogene Res. **2**:65-80

Mudryj M, Devoto SH, Hiebert SW, Hunter T, Pines J and Nevins J (1991) Cell cycle regulation of the E2F transcription factor involves an interaction with cyclin A. Cell **65**:1243-53

Penn L, Brooks M, Laufer E, Littlewood T, Morgenstern J, Evan G, Lee W and Land H (1990a) Domains of human c-*myc* protein required for autosuppression and cooperation with *ras* oncogenes are overlapping. Mol. Cell. Biol. **10**:4961-4966

Penn LJZ, Brooks MW, Laufer EM and Land H (1990b) Negative autoregulation of c-*myc* transcription. EMBO J. **9**:1113-1121

Pines J (1991) Cyclins: wheels within wheels. Cell Growth & Diff. **2**:305-310

Pines J (1992) Cell proliferation and control. Curr. Opin. Cell Biol. **4**:144-147

Rabbitts PH, Watson JV, Lamond A, Forster A, Stinson MA, Evan G, Fischer W, Atherton E, Sheppard R and Rabbitts TH (1985) Metabolism of c-*myc* gene products: c-*myc* mRNA and protein expression in the cell cycle. EMBO J. **4**:2009-2015

Rustgi AK, Dyson N and Bernards R (1991) Amino-terminal domains of c-myc and N-myc proteins mediate binding to the retinoblastoma gene product. Nature **352**:541-4

Sagar SM, Sharp FF and Curran T (1988) Expression of c-*fos* protein in brain: metabolic mapping at the cellular level. Science **240**:1328-1331

Sentman CL, Shutter JR, Hockenbery D, Kanagawa O and Korsmeyer SJ (1991) bcl-2 inhibits multiple forms of apoptosis but not negative selection in thymocytes. Cell **67**:879-88

Stone J, de Lange T, Ramsay G, Jakobvits E, Bishop JM, Varmus H and Lee W (1987) Definition of regions in human c-*myc* that are involved in transformation and nuclear localization. Mol. Cell. Biol. **7**:1697-1709

Thompson CB, Challoner PB, Neiman PE and Groudine M (1985) Levels of c-*myc* oncogene mRNA are invariant throughout the cell cycle. Nature **314**:363-366

Vinson C and Garcia K (1992) Molecular model for DNA recognition by the family of basic-helix-loop-helix-zipper proteins. New Biologist **4**:396-403

Waters C, Littlewood T, Hancock D, Moore J and Evan G (1991) c-*myc* protein expression in untransformed fibroblasts. Oncogene **6**:101-109

Weinberg R (1992) The retinoblastoma gene and gene product. Cancer Surveys **12**:43-57

Williams GT (1991) Programmed cell death: apoptosis and oncogenesis. Cell **65**:1097-8

Williams S, Evan G and Hunt S (1990a) Spinal c-*fos* induction by sensory stimulation in neonatal rats. Neuroscience Lett. **9**:309-314

Williams S, Evan GI and Hunt SP (1990b) Changing patterns of c-fos induction in spinal neurons following thermal cutaneous stimulation in the rat. Neuroscience **36**:73-81

Wisden W, Errington M, Williams S, Dunnett S, Waters C, Hancock D, Evan G, Bliss T and Hunt S (1990) Differential expression of immediate early genes in the hippocampus and spinal cord. Neuron **4**:603-614

Wyllie AH (1987) Apoptosis: cell death in tissue regulation. J. Path. **153**:313-316

FLOW CYTOMETRIC ANALYSIS of NUCLEAR ASSOCIATED ANTIGENS

James V. Watson
MRC Clinical Oncology Unit,
The Medical School
Hills Road,
CAMBRIDGE, CB2 2QH,
England.

The nucleus contains a number of proteins as well as the most obvious constituent DNA, and is surrounded by a membrane. It is beginning to be realised that nuclear organization is highly complex and the nuclear proteins which are involved in the organisation can be divided into three broad categories namely, structural proteins, enzymes and proteins involved in gene regulation.

Structural components include the histones and nuclear matrix proteins. The latter are responsible for the architecture of the nucleus as a whole during interphase and of the chromosomes during mitosis. Histones are responsible for DNA packing and together with DNA they constitute chromatin. There are five histones. Four of these make up the cylindrical nucleosome core around which is wrapped two turns of double stranded DNA. The "input" and "output" DNA strands to the nucleosome are tacked into place by the fifth histone. DNA between each nucleosome is termed linker DNA and the whole can be described as "beads on a string".

The nuclear enzymes include DNA polymerases, ligases, restriction enzymes, damage recognition and repair enzymes, transcriptases, topoisomerases and DNAases. Polymerases, as their name implies, are responsible for linking together constituents of DNA during synthesis. Restriction enzymes cut DNA at very specific base sequences. They are probably involved in physiological geonomic rearrangements of the types which occur at the immunoglobulin locus with the $C\mu$ to $S\mu$ heavy chain switch as spleen B-cells convert from IgM expressors to IgG secretors. The ligases join DNA strands together at specific sites and their function

NATO ASI Series, Vol. H 67
Flow Cytometry
Edited by A. Jacquemin-Sablon

is the converse of restriction enzymes. DNA damage recognition/repair enzymes are not clearly understood. Thymine dymers, induced by UV irradiation, are repaired in normal cells and there may well be a component where damage has to be recognised before it can be repaired. The repair process involves DNAases which digest out a stretch of bases either side of the damaged site before reconstruction takes place on the complementary strand. It is, however, contentious as to whether specific enzymes exist which recognise the damage before it can be repaired. Transcription of the genetic code contained within the DNA into messenger RNA which subsequently is translated into protein in the ribosomes is carried out by transcriptases which are DNA-binding proteins. Topoisomerases are DNA strand passing enzymes and are of two types, I and II, which respectively pass single and double stranded DNA. They are responsible for the topological integrity of DNA and are involved in winding and unwinding super-helical DNA by strand breakage and rejoining.

A number of nuclear proteins are involved with both reception and/or transduction of gene regulatory signals from the cytoplasm. Considerable interest is centered around growth regulatory proteins which include various steroid receptors and the protein products of a number of the proto-oncogenes.

PRINCIPLES of NUCLEAR ANTIGEN QUANTITATION

Flow cytometric measurement of nuclear proteins relies upon the availability of suitable antibodies and some means of getting the antibody to the nucleus. Such techniques have been developed using fixation or membrane permeablization to quantitate protein and DNA simultaneously (Watson, Sikora and Evan, 1985; Clevenger, Bauer and Epstein, 1985; Bauer et al, 1986; Elias-Jones et al, 1986; Dent et al, 1989) but as yet no satisfactory method has emerged for viable cells. All of these techniques have used fluorescein (green fluorescence) staining of the proteins and propidium iodide (red fluorescence) staining of DNA with an argon laser tuned to the 488nm line for excitation.

Immunofluorescence staining can be either direct or indirect. In the former, the antibody to the antigenic determinant is directly fluorescenated. In the latter, the antibody is not conjugated and a second fluorescenated antibody is used to probe the first. Indirect immunofluorescence staining is preferable to direct for three reasons. Firstly, direct fluorescenation of some monoclonal antibodies interferes with their specific binding. Secondly, it is not possible to carry out controls to determine the degree of non-specific binding or antibody trapping with the direct method. Finally, a degree of signal amplification is obtained with the indirect method as a number of second antibody molecules, each conjugated with fluorochromes, will bind to the first antibody.

A number of antibodies for probing nuclear antigens have been produced and some are now commercially available. They include antibodies to p53 (Crawford et al, 1981), $p62^{c-myc}$ (Evan, Lewis, Ramsey and Bishop, 1985), $p55^{c-fos}$ (Evan et al, 1986) and topoisomerase II (Liu, 1983). Some antibodies recognise nuclear antigens in stimulated cells, 6-B1012/N (Reeve, personnal communicatin) and Ki-67 (Gerdes, Schwab, Lemke and Stein, 1983) and others recognise preferentially cell cycle dependent nuclear antigens. These include specific regions in condensed chromatin (MAB 244-7, mW 34Kd), interchromatin granules (MAB 780-3, mW 105Kd and 41Kd) and euchromatin (MAb 58-15, mW 36Kd) (Epstein and Clevenger, 1985; Clevenger, Bauer and Epstein, 1985). This list is by no means exhaustive and the number of antibodies which recognise nuclear components is increasing very rapidly.

CELL PREPARATION

Cell preparation techniques must satisfy three criteria namely, epitope and DNA preservation plus the capacity to measure both parameters after preparation and to date five methods are available. It must be pointed out however that all of these techniques are very new, are still being evaluated and they must be regarded as being within the experimental domain. The five methods eluded to are freeze-thaw, fixation with or without detergent, non-ionic detergent treatment, lysolecithin permeabilization and extraction of nuclei from archival wax embedded material.

1) **Freeze-thaw technique** A freeze-thaw technique, adapted from the method described by Ganesan et al (1981), has been developed in our laboratories (Smith et al, 1985) which can be used to study nuclear associated antigens. It can be used for both monolayers and suspension cultures but it was primarily designed for the former. Monolayers are washed twice in phosphate buffered saline (pH 7.2), drained well and overlayed with low salt (LS) buffer. This contains 10mM NaCl, 10mM EDTA, 1.0mg ml^{-1} bovine serum albumen in 10mM Tris-HCl buffer pH 8.0. The quantity of buffer added to each flask depends on flask size and should be just sufficient to completely cover the monolayer. Flasks of $75cm^2$ require between 1.5 and 2.0 mls and 150 cm^2 flasks need about 3.0 mls. The flask is then frozen rapidly in a horizontal position by immersing the lower half in a dry ice/methanol bath for 30-45 seconds. The frozen monolayer is then thawed rapidly in a water bath at 37°C. Most cells detach from the plastic surface after the first freeze-thaw cycle but some will still be adherent and many will be in "sheets" if the monolayer was relatively dense. A second freezing is then carried out after which the frozen cells can be kept at -70°C for up to 6 months (maximum time investigated to date) if necessary. A single cell suspension suitable for immunofluorescence and propidium iodide staining can then be obtained by gentle syringing or pipetting after the second thawing. Inevitably, there

are some clumps using this method so filtering through 35μm nylon mesh is usually necessary.

Suspension cultures are treated similarly except that the pellet is resuspended in LS buffer after centrifugation and the medium has been poured off. Thereafter, the resuspended cells are then subjected to two cycles of freeze-thaw as above.

This method was designed to avoid both detergent treatment, which might interfere with subsequent antibody reaction, and fixation which can modify or destroy some epitopes rendering those samples useless. Moreover, it also enables samples to be stored. It depends on mechanical sheer stress to puncture the external membrane due to ice crystal formation during rapid freezing and most epitopes should be preserved.

2) **Fixation** A number of fixatives including paraformaldehyde, methanol, ethanol and acetone in various combinations have been used by both Clevenger, Bauer and Epstein (1985) and Elias-Jones et al, (1986) in assays for nuclear associated antigens. The best fixative for a particular antigen and assay system has to be determined experimentally. There is insufficient knowledge at present to make any worthwhile generalizations. Clevinger et al (1985) found that fixation with paraformaldehyde followed by treatment with triton X-100 gave the optimum results with Raji cells, HeLa cells and peripheral lymphocytes. A pellet of 1-3x10^6 cells was resuspended in 10ml of 0.5% paraformaldehyde (Polysciences, Warrington, PA, USA) in PBS for 10 minutes. After centrifugation at 200G for 10 mins the fixed cells were resuspended in 0.1% triton X-100 (Sigma, St. Louis, MO, USA) in PBS for 3 minutes as a washing step, recentrifuged and resuspended in 0.1% triton X-100/PBS in which all further washes and handing were carried out. Elias-Jones et al (1986) found that fixation in 50% methanol/PBS for cervical brush biopsy specimens gave satisfactory results.

3) **Non-ionic detergent treatment** Low concentrations of non-ionic

detergents (MP-40 or Triton X-100) can be used to permeabalize the external membrane which allows access of antibodies to the nucleus. Concentrations within the range 0.1% to 0.5% in PBS are used (Hunt, Pini and Evan, 1987; Karn et al,1989). Binding of some antibodies can be inhibited using non-ionic detergents thus, pilot studies using a range of concentration and conventional microscopy should be undertaken.

4) **Lysolecithin** Lysolecithin is a naturally occurring lipid with profound actions on the external cell membrane. It probably replaces phospholipids in the membrane bilayer (Weltzien, 1979) and at high concentrations this leads to lysis. At low concentrations, which should to be determined experimentally for each cell type, lysolethicin can be used to permeabalize cells sufficiently to allow access of large molecules whilst maintaining the gross integrity of the membrane. Scroff et al (1984) introduced this permeabalization technique in flow cytometry but, it had been used previously in transfection studies (Miller, Castellot and Pardee, 1979). Scroff et al (1984) used concentrations of $50 \mu g\ ml^{-1}$ which allowed access of anti-IgM antibodies to chronic lymphatic leukaemia cells and anti-intermediate filament antibodies to peripheral blood lymphocytes. Dent et al (1989) have used lysolecithin very effectively for analysis of intracytoplasmic and nuclear associated antigens.

5) **Nuclear extraction from archival material** This technique was introduced by Hedley et al (1983), and with very little modification it can be used to assay both DNA and nuclear associated antigens from paraffin wax embedded biopsies (Watson, Sikora and Evan, 1985; Bauer et al, 1986). Briefly, $40 \mu m$ sections are cut on a microtome and these are dewaxed and rehydrated by sequential treatment with washes in xylene then 100% ethyl alcohol followed by rehydration with washes in decreasing concentrations of alcohol.

Everyone has their own minor modifications of this technique and our dewaxing procedure is carried out with a minimum of 2 washes in xylene. Approximately 5 mls of xylene are used for each thick section measuring about 1.5 x 2.0 cm. The section(s) is placed in a glass test tube (this is important as many plastics dissolve in xylene), 5 mls xylene are added and the test tube is aggitated. By looking oblequely through the test tube it is possible to see the wax dissolving in the solution by the refractive index changes which occur as dissolving takes place. After about 5 mins the xylene can be poured off leaving the tissue section adherent to the test tube wall by surface tension. A further 5 mls xylene are then added as a washing procedure and this too is similarly poured off after a further 5 minutes. Water free 100% ethanol is then added, the test tube is aggitated for 1 to 2 minutes, the supernatant is poured off and a further 5 mls 100% ethanol are added for 5 minutes with intermittent aggitation. During this step the tissue section will change from being translucent to opaque. The supernatant is again poured off and a further aliquot of ethanol is added for 1 - 2 minutes. 1.0 ml is then removed with a pipette and replaced by 1.0 ml distilled water added drop-wise with inversion mixing. This gradually reduces the alcohol concentration to 80%. If the dewaxing process has not been complete the supernatant will become cloudy during this initial rehydrating step, in which case the alcohol is removed, replaced by 100% ethanol then by xylene to complete the dewaxing. If all xylene and wax have been removed the supernatant will remain absolutely clear during initial rehydration which can then proceed by sequentially pipetting off aliquots of supernatant and replacing this with equal volumes of distilled water thus gradually reducing the alcohol concentration. When the latter has reached about 10% it can all be removed and the section is washed in distilled water, then resuspended in phosphate buffered saline (PBS), pH 7.2.

The tissue is finally incubated at 37°C in 5ml hydrochloric acid (pH 1.9) containing pepsin (EC 3.4.23.1) which must be made up fresh as required. A number of pepsins are available from various manufacturers, don't get the cheap stuff, it doesn't work. The preparation should be chromatographically purified, crystallized and lyophilized. This is the more expensive variety and Sigma (St. Louis, MO, USA) do a good one. The stock bottle must be kept at or below 0°C and it goes off very quickly if you get water in it. The pepsin concentration required for cytoplasmic digestion to release nuclei varies with tissue type but this usually will be between 0.01 mg ml^{-1} and 1 mg ml^{-1}. This is a very wide range and the optimum concentration can only be determined by carrying out pilot experiments. Most sections required a digestion time between 45 and 60 minutes with frequent aggitation and mixing after which the solution will appear colloidal. One to four sections, depending on the size of the biopsy are usually adequate to obtain sufficient nuclei. A small minority of sections fail to digest adequately even after prolonged incubation. These tend to be sections which are difficult to cut and contain very dense sclerotic fibrous tissue and calcification. Following partial digestion to release nuclei the material is filtered through a 35μm nylon mesh, centrifuged at 200G and resuspended in PBS at a concentration between $5x10^5$ and 10^6 nuclei ml^{-1}.

A certain amount of luck is needed when using this technique for nuclear associated antigens. Pepsin has a wide specificity but it cleaves proteins preferentially at leucine and phenylalanine residues in hydrophobic domains. It also cleaves at tryptophan and methionone residues again in hydrophobic domains but bonds to tyrosine are refractory to digestion. With antibodies raised to native protein there is no way of knowing in advance if the epitope recognised by the antibody contains residues which are digested by pepsin. If it does you will not see anything, and that is just bad luck. However, monoclonal antibodies can be raised to synthetic peptides corresponding to the primary amino acid sequence of the protein and are generally raised to hydrophillic domains (Evan, Lewis, Ramsey and Bishop, 1984). Thus, by choosing for antibody production a peptide with amino acid sequences which are deficient in both leucine and phenylalanine but which are tyrosine rich it is possible to minimize the chance that pepsin will digest the epitope. However, it is still possible for the epitope to be lost if bonds either side are cleaved and the epitope floats away into the sunset. So again, some luck is needed.

STAINING

Irrespective of the preparation method the staining proceedures essentially are identical and are illustrated in detail with a synthetic peptide induced antibody directed to the protein product of the c-myc gene. Work has also been carried out in our laboratories to assay the protein products of the c-fos and c-myb genes, topoisomerase II, p53, KI-67 and a cell cycle dependent antigen recognised by MoAb 6-B1012/N. Clevenger, Bauer and Epstein (1985) and Bauer et al (1986) have reported similar assays for specific regions in condensed chromatin, interchromatin granules and euchromatin.

Aliquots of 1.0 ml of the cell/nuclear suspension are placed into six 1.5 ml tubes and spun down in an Eppendorf centrifuge. The supernatants are carefully removed and four of the nuclear pellets are resuspended in 20μl of diluted mouse anti p62^{c-myc} monoclonal antibody (MYC 1-6E10, stock concentration 2mg ml^{-1}) and two pellets are resuspended in PBS to act as controls. We now use antibody dilutions of 1:3.16, 1:10, 1:31.6 and 1:100 as the optimum MYC 1-6E10 concentration tends to vary in material from different wax blocks. Following incubation for 45 minutes all samples are centrifuged and supernatants are removed. The four samples treated with the MYC 1-6E10 antibody and one of the controls are then suspended in 0.5 ml PBS as a washing step, centrifuged and the supernatants are removed.

The pellet is then incubated with 10μl of fluorescein isothiocyanate conjugated rabbit anti-mouse Ig (Dako Immunoglobulins, Denmark) diluted 1:50. After one hour all samples are suspended in 0.5 ml of a solution containing ribonuclease (Sigma, St. Louis, MO, USA) and propidium iodide, PI (Calbiochem, La Jolla, CA, USA), both at 0.05mg ml^{-1}, which counterstains DNA red against the green fluorescence from the fluorescein staining of the $p62^{c-myc}$. Thus, one control contains only PI stained nuclei, the other is stained with PI and 2nd antibody (fluorescence control) and the remaining four samples are stained with the varying dilutions of anti-$p62^{c-myc}$ plus fluorescent 2nd antibody and PI.

It is conventional wisdom to carry out some form of specificity control for antibody binding. Irrelevant antibodies, which are not expected to bind to the determinant of interest, are frequently used for this purpose. Our practice has been to use anti-β2 microglobulin, anti-fibronectin and anti-HLA (which do not recognise nuclear structures) as well as X-63 myeloma protein. These have never given a significant signal above the "background" of the 2nd antibody alone control and are of doubtful usefulness. With synthetic peptide induced antibodies it is possible to perform more meaningful specificity controls by blocking the antibody prior to the assay with the peptide used as the immunogen. Furthermore, if two or more such antibodies are available it is possible to carry out cross-blocking assays (Watson, et al, 1987). In this type of control both the "relevant" and "irrelevant" peptides are used to block each antibody where the irrelevant peptide for each antibody should not block subsequent binding whereas the relevant peptide should.

EXAMPLES

An example of results obtained with the freeze-thaw technique for exponentially growing 3T3 cells stained for $p62^{c-myc}$ using the MYC 1-6E10 synthetic peptide induced antibody (abscissa) versus DNA content (ordinate) is given in figure 1A. This shows that the protein content is independent of cell cycle phase. In figure 1B the antibody was treated with the peptide used as the immunogen before the assay which obliterates subsequent binding.

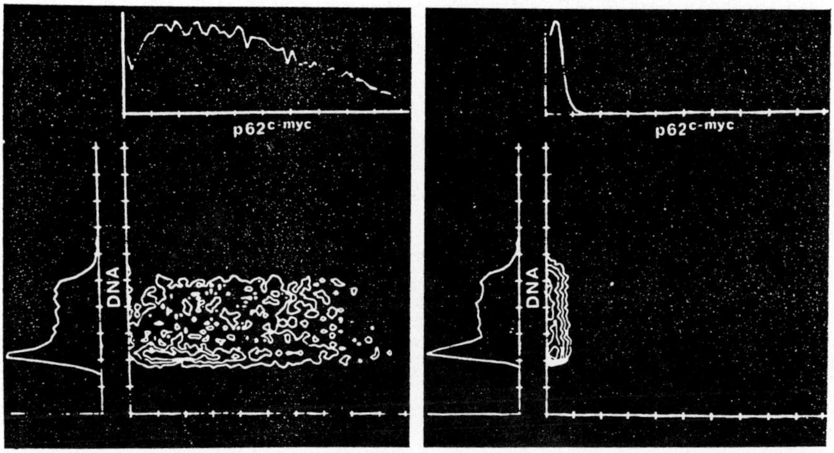

Figure 1. 3T3 cells stained for p62c-myc (ordinate) and DNA (abscissa) in the left panel. Pre-incubation of the antibody with the immunogenic peptide abolishes the p62c-myc fluorescence signal, right panel.

The green signal (abscissa) in figure 1B was not significantly different from the fluorescence control data obtained by treating with 2nd antibody alone (Rabbitts et al, 1985).

Figure 2 shows data from Clevenger, Bauer and Epstein (1985) which used paraformaldehyde/triton for MAB 780 directed towards interchromatin granules. The log of the MAb 780 associated fluorescence is plotted on the ordinate versus DNA on the abscissa. The arrowed population in the contour map is intensely stained and represents mitotic cells.

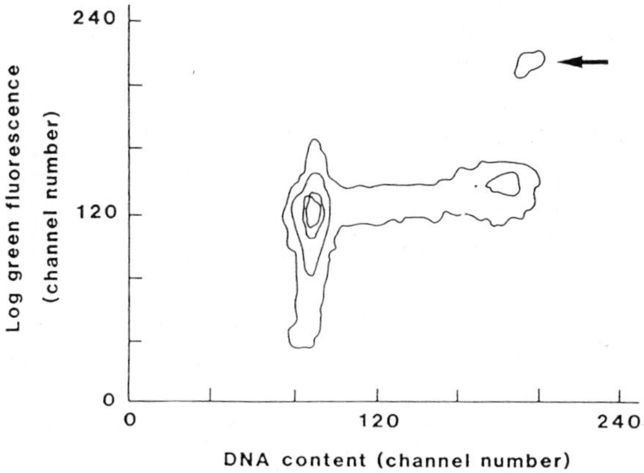

Figure 2

Data from a dysplastic cervical smear (Elias-Jones et al, 1986) stained for DNA (ordinate) versus $p62^{c-myc}$ (abscissa) using 50% methanol fixation is given in figure 3. The left and middle panels show the PBS (propidium iodide, PI, only) and fluorescence control data (PI plus second antibody only). Note the increased green signal in the right panel where both antibodies were used.

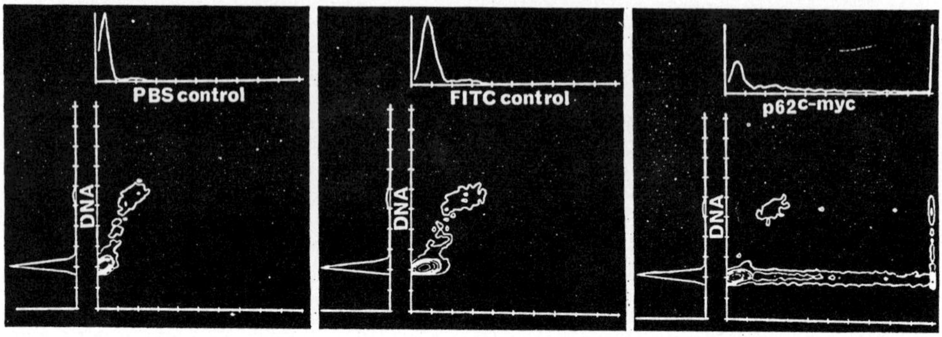

Figure 3) DNA (ordinate) versus $p62^{c-myc}$ (abscissa) using 50% methanol fixation of cells from a dysplastic cervical smear.

Figure 4 shown results obtained from nuclei extracted from paraffin embedded biopsies of a colonic polyp (left panel) and a colonic carcinoma (right panel) from the same patient stained for $P62^{c-myc}$ and DNA. Both the polyp and the carcinoma exhibit aneuploid components and in each of these the $p62^{c-myc}$ signal is higher than that of the diploid component. However, the $p62^{c-myc}$ signal is higher in the carcinoma than in the polyp, both of which are aneuploid.

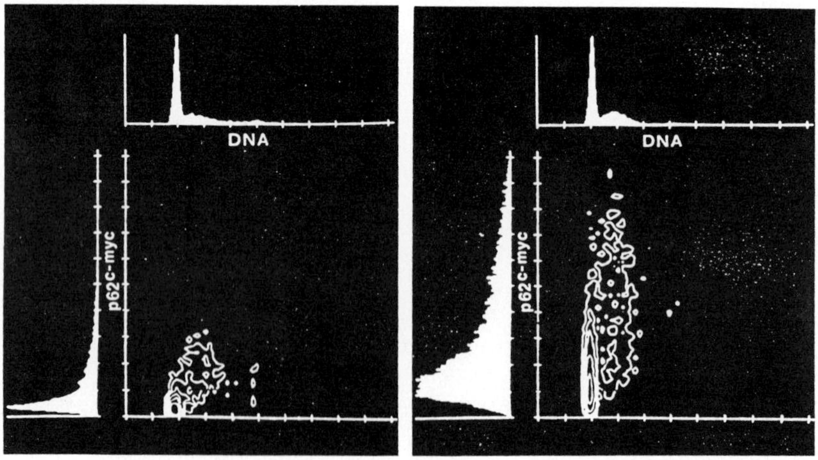

Figure 4

FLOW CYTOMETRY CONSIDERATIONS

a) Instrument configuration

Some nuclear antigens (specifically p62^{c-myc} and p55^{c-fos}) are present in very small quantities. Ideally, therefore, a high light collection efficiency flow chamber should be used (Watson, 1985). As well as red and green fluorescence, quantitating DNA and protein respectively, both forward and 90^0 light scatter should be collected.

The red and green signals should be separated by a dichroic beam splitter which reflects light below and transmits above 580nm. The green and red photomultipliers additionally should be guarded by a 515-560nm band pass and a 630nm long pass filter respectively. Even with this degree of optical filtration about 0.01% of the PI stained DNA signal will enter the green photomultiplier which is an advantage in this assay. The instrument is set up on the PI only stained control so that the diploid G1 peak is placed in about channel 200 on the DNA (red) axis and between channels 20 and 50 on the p62^{c-myc} (green) axis. This enables the instrument to be set up identically for each run by using similarly extracted and stained nuclei from a specimen of normal colonic mucosa.

It is also very helpful to use a crossed cylindrical lens pair which focusses the beam to a sheet of light across the cell stream to give a partial slit scan illumination effect and this has two advantages. Firstly, a very high light flux can be obtained at the focal point which increases fluorescence excitation efficiency and secondly, the pulse of light from each nucleus can be digitized into height, width (time of flight through the beam) and area for each detector. This has implications for quality control.

b) Data analysis and quality control

Material extracted from wax blocks inevitably contains a variable quantity of debris, clumps and nuclear fragments as well as intact single nuclei. Samples prepared by the freeze-thaw technique also contain a variable quantity of debris. The latter and large clumps can be gated out on a combination of forward and 90^0 scatter signals. Fragments of nuclei can be identified and gated out on a combination of 90^0 scatter and the shape of the pulse from the red (DNA) photomultiplier which should be set to trigger the system. The latter involves analysis of the pulse height, width and area under each pulse (PAW analysis) where the ratio of width*height/area should be constant to within narrow limits for a given pulse size (Watson, Sikora and Evan, 1985). Figure 5 illustrates the pulse shape analysis in a data set taken from a colon carcinoma sample where the tissue section thickness was 20μm.

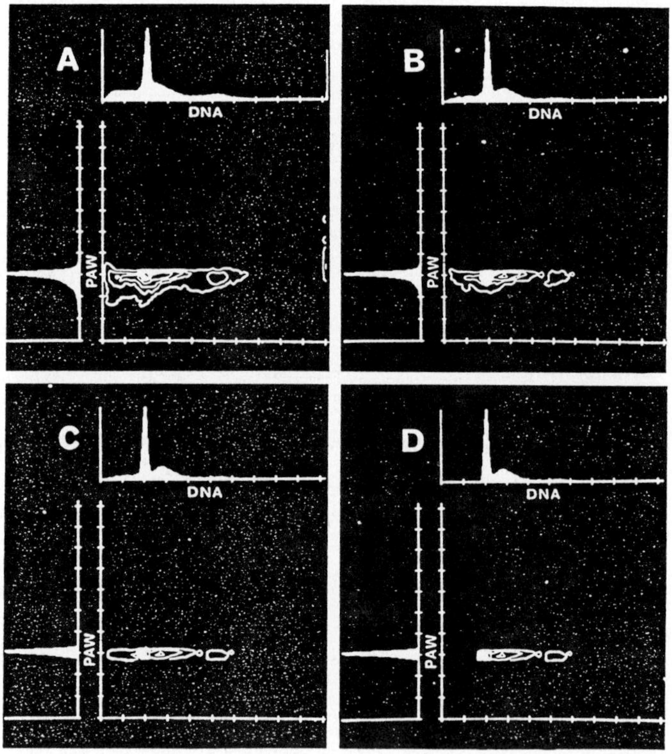

Figure 5. Pulse shape analysis where the product of pulse height*width/area of the red (PI) signal for each cell (PAW) is plotted on the ordinate against area under each pulse (abscissa). Panel A is the non gated raw data. Panel B shows the same data set after gating on forward and 90^0 scatter to exclude small debris and large clumps. Pulse shape analysis has been additionally applied in panel C. Panel D shows the final result where the mean and SD of the diploid peak in panel C were computed and any signals less than 2.5 SDs below the mean were excluded.

The product of pulse height*width/area of the red (PI) signal for each cell is plotted on the ordinate against area under the pulse (abscissa). Panel A is the non gated raw data where the 2-dimensional frequency is plotted as a contour map with the mono-dimensional histograms adjacent to the respective axes. Clearly, the derivative (ordinate) is essentially constant but there is a negative skew due to pulses with abnormal shapes. Panel B shows the same data set after gating on forward and 90^0 scatter to exclude small debris and large clumps. Many of the signals above 600 on the abscissa have now been gated out as have some of those at less than 180. These is also a dip appearing to the immediate right of the major spike which is the diploid DNA peak. Pulse shape analysis has been additionally applied in panel C. This was effected by computing the mean and variance (hence standard deviation, SD) of the Y-axis distribution in panel B. Only those pulses falling within $\pm$ 1.5 SDs of the mean of the Y-axis distribution of panel B were included in the display shown in panel C. There is now very clearly a bimodal DNA distribution (abscissa) with the diploid spike at about 200 and an aneuploid component peaking at about 290 which has a much wider distribution. There is still, however, a component being scored at less than about 180 on the abscissa. Panel D shows the final result in which the mean and SD of the diploid peak in panel C were computed and any signals less than 2.5 SDs below the mean were excluded. The original data set in panel A contained 20000 events; the final data set in panel D contained 6,500 events. Thus, only about 33% of the original raw data set was composed of isolated single nuclei which conformed to a specific size and shape distribution. In a number of experiments with various types of samples using wax block material with thicknesses between 10μm and 40μm we have found a direct relationship between section thickness and the proportion of events triggering the red detector which were compatible with whole nuclei. In each preparation examined the proportion was compatible with fluorescence microscope observations and, as expected, the proportion of single nuclei decreased as the section thickness decreased. The computer algorithms predict between 15 and 25% of events to be whole nuclei in 10μm sections, this rises to between 25 and 40% with 20μm sections and to between 50 and 60% for 40μm thicknesses.

c) Base line correction

All fluorescence emission contains a spectrum of wavelengths and no filter system can absolutely separate two different emission spectra. This gives rise to to some breakthrough of the red signal into the green detector hence, there is a background signal on the green channel which is proportional to DNA content. Thus, the base-line for any additional green fluorescence due to fluorescein is not parallel to the X-axis but angled to it. Compensation can be applied in software after data collection but this method of base line correction is a second best proceedure as a degree of resolution is lost on the immunofluorescence axis. It is far better to carry out an analogue subtraction within the electronics before the analogue-to-digital conversion as this preserves the resolution.

d) Subset identification

Forward and 90^0 light scatter signals can be used to identify subsets within heterogeneous samples. This is illustrated with a paraffin embedded ovarian cancer biopsy stained for p62^{c-myc} which contains an aneuploid component (Watson, et al, 1987). Figure 6 shows the p62^{c-myc} fluorescence (ordinate) versus DNA (abscissa) as contour plots and both a diploid and an aneuploid component is present. Also, there may be two subsets within the diploid component based on their p62^{c-myc} signal.

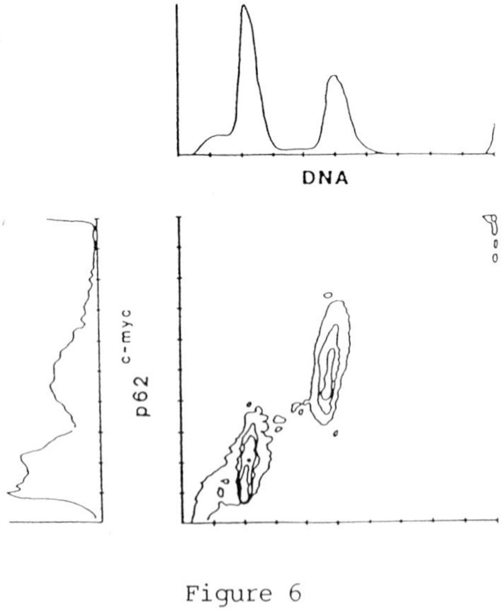

Figure 6

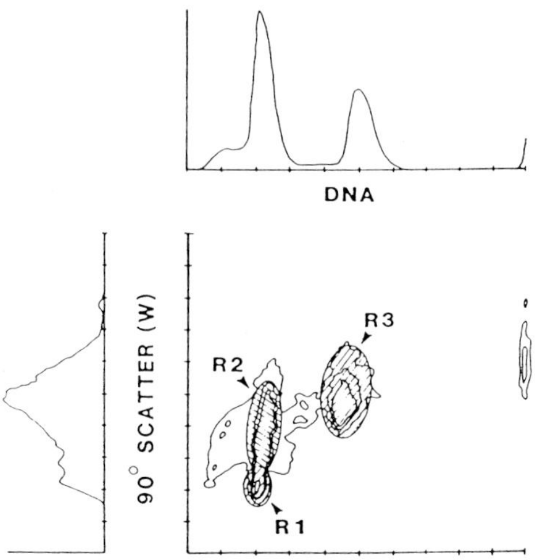

Figure 7

Figure 7 shows 90^0 scatter pulse width plotted against DNA for this same sample which suggests that there may be two subsets in the first DNA peak (channel 200) defined by their scatter characteristics. Elliptical gated regions, R1, R2 and R3, were set on the three putative subsets and the histograms of the 5 extracted parameters were generated for all regions in a single pass through the processed data file. These data are shown in figure 8 where the columns correspond to regions R1, R2 and R3 respectively and the rows (from the top) show DNA, oncoprotein, 90^0 scatter pulse width, forward scatter pulse width and 90^0 scatter pulse area respectively. Clearly, R3 is a discrete subset and from inspection of the data in columns 1 and 2 we can see R1 and R2 are also discrete entities. Subset R1 shows tighter forward scatter compared with subset R2 (row 4), it exhibits less total 90^0 scatter (rows 5), slightly less DNA fluorescence (row 1) and lower oncoprotein content (row 2). Nuclei is subset R1 were small (rows 3 and 5, figure 8), very uniform (row 4, figure 8) and the coefficient of variation of the DNA peak was in the range 3%-5% (row 1, figure 8). These findings suggest that these could be lymphocytes. Parallel experiments using formalin fixed pepsin digested peripheral blood lymphocytes separated using Ficol strongly support, but do not prove, this interpretation. Cells comprising subset R2 have somewhat fusiform nuclei corresponding to stromal cells. The $p62^{c-myc}$ content (ordinate) versus DNA (abscissa) is shown for each subset in figure 9.

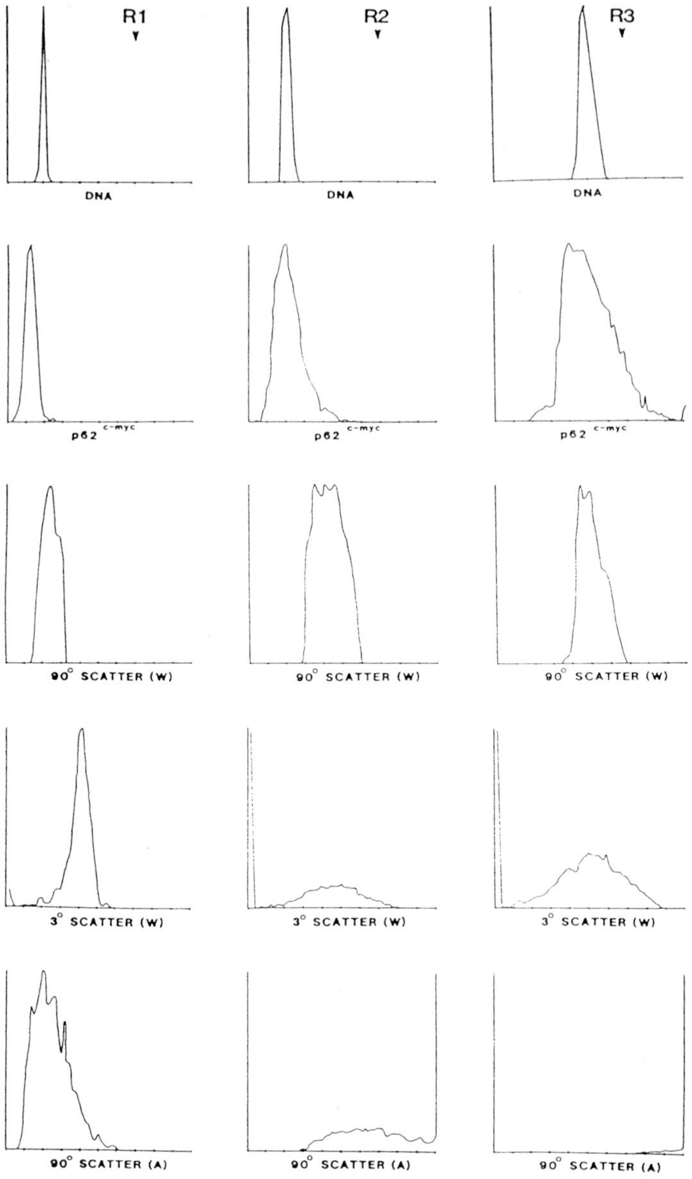

Figure 8) DNA, oncoprotein, 90^0 scatter pulse width, forward scatter pulse width and 90^0 scatter pulse area shown in the rows respectively for regions R1, R2 and R3 (columns) for the data in figures 6 and 7.

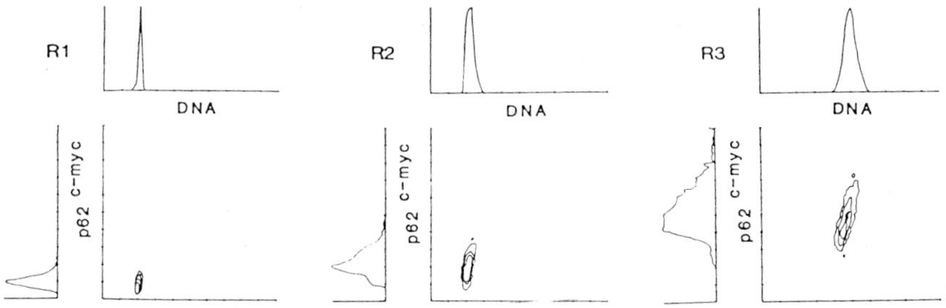

Figure 9) p62^{c-myc} content (ordinate) versus DNA (abscissa) for each of the 3 subsets defined in figure 7.

REFERENCES

Bauer, D.B., Clevenger, C.V., Endow, R.K., Murad, T., Epstein, A.L. and Scarpelli, D.G. (1986) "Simultaneous nuclear antigen and DNA content quantitation using paraffin-embedded colonic tissue and multiparameter flow cytometry" CANCER RES. 46; 2428-2434

Clevenger, C.V., Bauer, D.B. and Epstein, A.L. (1985) "A method for simultaneous nuclear immunofluorescence and DNA content quantitation using monoclonal antibodies and flow cytometry" CYTOMETRY 6; 208-214

Dent, G.A., Leglise, M.C., Pryzwansky, K.B. and Ross, D.W. (1989) "Simultaneous paired analysis by flow cytometry of surface markers, cytoplasmic antigens, or oncogene expression with DNA content" CYTOMETRY 10; 192-198

Elias-Jones, J., Hendy-Ibbs, P., Cox, H. and Watson, J.V. (1986) "Cervical brush biopsy specimens suitable for DNA and oncoprotein analysis using flow cytometry" J. CLIN. PATH., 39; 577-581

Epstein, A.L. and Clevenger, C.V. (1985) "Identification of nuclear antigens in human cells by immunofluorescence, immunoelectron microscopy, and immunocytochemical methods using monoclonal antibodies" In RECENT ADVANCES in NON-HISTONE CHROMOSOMAL PROTEIN RESEARCH, Vol 1, Bekhor, I. (Ed), CRC Press, Boca Raton Press, FL pp117-137.

Evan, G.I. and Hancock, D.C. (1985) "Studies on the interaction of the human c-myc protein with cell nuclei: p62$_{c-myc}$ as a member of a discrete subset of nuclear proteins" CELL 43; 253-261

Evan, G.I., Lewis, G.K., Ramsay, G. and Bishop, J.M. (1985) "Isolation of monoclonal antibodies specific for human and mouse proto-oncogene products" MOL. CELL. BIOL. 5; 3610-3616

Ganesan, A.K., Smith, C.A. and van Zeeland, A.A. (1980) in "DNA repair: a laboratory manual of research techniques" Vol 1, page 89. Eds, E.C. Friedberg and P.C. Hanawalt (New York: Marcel Dekker).

Gerdes, J. Schwab, U., Lemke, H. and Stein, H. (1983) "Production of a mouse monoclonal antibody reactive with a human nuclear antigen associated with cell proliferation" INT. J. CANCER 31; 13-20

Hedley, D.W., Friedlander, M.I., Taylor, I.W., Rugg, C.A. and Musgrove, E.A. (1983) "Method for analysis of cellular DNA content of paraffin-embedded pathalogical material using flow cytometry" J. HISTOCHEM. and CYTOCHEM. 31, 1333-1335

Hunt, S.P., Pini, A. and Evan, G. (1987) "Induction of c-fos-like protein in spinal cord neurones following sensory stimulation" NATURE 328; 632-634

Liu, L.F. (1983) "DNA topoisomerases - enzymes that can catalyse the breaking and rejoining of DNA" CRC CRIT. REV. BIOCHEM. 15; 1-24

Rabbitts, P.H., Watson, J.V., Lamond, A., Fischer, W., Forester, A., Stinton, M.A., Evan, G.I., Atherton, E., Sheppard, R.C. and Rabbitts, T.H. (1985) "Metabolism of c-myc gene products: c-myc mRNA and protein expression in the cell cycle" EMBO J. 4; 2009-2015

Smith, P.J., Nakeff, A. and Watson, J.V., (1985) "Flow-cytometric detection of changes in the fluorescence emission spectrum of a vital DNA-specific dye in human tumour cells" EXP. CELL RES. 159; 37-46

Watson, J.V. (1985) "A method for improving light collection by 600% from square cross section flow cytometry chambers" BRIT. J. CANCER 51; 433-435

Watson, J.V., Curling, O.M., Munn, C.F. and Hudson, C.N. (1987) "Oncogene expression in ovarian cancer: A pilot study of c-myc oncoprotein in serous papillary ovarian cancer" GYNAECOL. ONCOLOGY 28; 137-150

Watson, J.V., Sikora, K.E. and Evan, G.I. (1985) "A simultaneous flow cytometric assay for c-myc oncoprotein and cellular DNA in nuclei from paraffin embedded material" J. IMMUNOL. METHODS 83; 179-192

CHROMATIN CHANGES ACCOMPANYING APOPTOSIS DETECTED BY FLOW CYTOMETRY

Zbigniew Darzynkiewicz
Wojciech Gorczyca
Michel A. Hotz
Piotr Lassota
Silvia Bruno
Frank Traganos

The Cancer Research Institute
New York Medical College
100 Grasslands Road
Elmsford, NY 10523 USA

INTRODUCTION

Apoptosis is a term used to describe the mode of cell death, often triggered by diverse environmental agents, which has a well defined, morphological, biochemical and molecular pattern (Arends *et al.*, 1990; Kerr and Harmon, 1991; Kerr *et al.*,1972; Wyllie *et al.*, 1980; 1984). During this process, a sequence of specific events, the most characteristic of which is the activation of an endogenous endonuclease having affinity for internucleosomal (spacer) DNA regions, takes place, culminating in cell disintegration. The description of apoptosis, its association with cell differentiation, and the mechanisms involved is the subject of several reviews (Arends *et al.*, 1990; Dive and Hickman, 1991; Tomei and Cope, 1991; Wyllie *et al.*, 1980; 1984). Although the term "*programmed cell death*" is often used interchangeably with apoptosis, the mechanisms involved in "programmed cell death" and apoptosis, despite similarities, are not identical, and the terms are not synonymous (Lockshin and Zakeri, 1991). In contrast to apoptosis, or programmed cell death, *necrosis* is a nonspecific mode of cell death, often caused by excessively high concentrations of toxic agents (Kerr and Harmon, 1991; Lennon *et al.*, 1991; Schrek *et al.*, 1984; Tomei and Cope, 1991). Necrosis is characterized by rapid poisoning of the cell and immediate loss of integrity of the plasma membrane.

NATO ASI Series, Vol. H 67
Flow Cytometry
Edited by A. Jacquemin-Sablon

The observations that ionizing radiation or various classes of antitumor drugs induce apoptosis of tumor cells, and that efficacy of some antitumor regimens may correlate with the response of target cells by apoptosis (reviews, Bertrand *et al.*, 1991; Dive and Hickman, 1991; Kaufmann, 1989; Krishan and Frei, 1975; Szende *et al.*, 1989; Tritton, 1991), generated an interest in apoptosis in the field of oncology. If the mechanisms of this process can be better understood, it may be possible to modulate it towards a more selective tumor cell kill, which can open new strategies for improving chemotherapy. Furthermore, the ability to monitor early signs of apoptosis in samples from patient tumors may be predictive of the outcome of the treatment protocol.

So far, the methods to distinguish apoptotic cells by flow cytometry were based primarily on the phenomenon of decreased DNA stainability of these cells with different fluorochromes (Afanas'ev *et al.*, 1986; Compton *et al.*, 1988, Cowden and Curtis, 1981; Ojeda *et al.*, 1990; Swat *et al.*, 1991; Telrod *et al.*, 1991; 1992; Umansky *et al.*, 1981; Walker *et al.*, 1991). Because biochemical studies indicated that despite partial DNA degradation (nicking of the nucleosomal linker sections), the overall DNA content of apoptotic cells remains unchanged (Arends *et al.*, 1990), the lowered DNA stainability observed in cytometric studies was explained as being due to a change in chromatin structure, or DNA conformation, resulting in a decrease in DNA accessibility to the dyes.

We have recently observed that exposure of human myelogenous leukemic HL-60 or KG1 cells to the DNA topoisomerase I inhibitor camptothecin (CAM) or topoisomerase II inhibitors teniposide (TN) or *m*-AMSA, resulted in rapid (seen after 2 h) apoptosis of these cells (Del Bino and Darzynkiewicz, 1991; Del Bino *et al.*, 1991). Likewise, the same inhibitors, or a glucocorticosteroid hormone, prednisolone (PRED), equally rapidly caused apoptosis of rat thymocytes (Bruno *et al.*, 1992). In both cell systems, apoptosis was cell cycle phase specific; namely, only S phase cells of myelogenous leukemias (Del Bino *et al.*, 1991) and only G_0 thymocytes (Bruno *et al.*, 1992) were affected. In contrast, still another inhibitor of topoisomerase II, fostriecin (FST), induced apoptosis of HL-60 cells indiscriminately, regardless of the phase of the cell cycle (Hotz *et al.*, 1992). Changes in chromatin of the affected cells were presently explored by measuring the sensitivity of DNA in situ to denaturation (Darzynkiewicz, 1990 a; Darzynkiewicz, 1990 c). This data, which extends our previous observations (Hotz *et al.*, 1992), shows that the sensitivity of DNA to denaturation is markedly increased in cells undergoing apoptosis and necrosis, and that a technique based on this phenomenon can be used to identify such cells by flow cytometry.

MATERIALS AND METHODS

Cells: The HL-60 cell line was kindly provided by Dr. Harry A. Crissman of the Los Alamos National Laboratory (Los Alamos, NM). The cultures were maintained in RPMI 1640 medium (GIBCO, Grand Island, NY) supplemented with 10% fetal calf serum, 100 units/ml of penicillin, 100 μg/ml of streptomycin, and 2 mM L-glutamine, as described (Del Bino *et al.*, 1991). The cells were split every third day to ensure that the cell density in cultures did not exceed $5x10^5$ cells/ml.

Thymus glands were rapidly dissected from 2 day old Sprague-Dawley rats of either sex. A suspension of thymocytes was obtained by mincing the glands in phosphate buffered saline (PBS). After filtering through a 53 μm nylon mesh, the cells were rinsed and suspended in RPMI 1640 medium containing fetal calf serum and antibiotics, as above, at a density of $5x10^6$ cells/ml (Bruno *et al.*, 1992). Cell viability was assayed using the trypan blue exclusion test.

Chemicals: CAM, and prednisolone 21-acetate (PRED) were purchased from Sigma Chem. Co. (St. Louis, MO). TN (VM-26) was kindly provided by Mr. Henry H. Holava of Bristol Meyers Co., Wallingford, CT. FST, was a gift from Dr. Werbel (Parke-Davis, Ann Arbor, MI). Stock solutions of CAM, TN and genistein (GEN) were prepared at concentrations of 3 and 50 mM and 1 mg/ml, respectively, in dimethyl sulfoxide, and kept at -20°C. A stock solution of PRED (5 mM) was made in ethanol. Control cultures were treated with equivalent concentrations of dimethyl sulfoxide or ethanol alone.

Cell Staining and Fluorescence Measurements: Cellular DNA and protein content were measured by flow cytometry following cell staining with 1.0 μg/ml of DAPI and 10 μg/ml of sulforhodamine 101 (Eastman Kodak, Rochester, NY) dissolved in 10 mM PIPES buffer (Calbiochem, La Jolla, CA) containing 0.1 M NaCl, 2 mM $MgCl_2$, and 0.1% Triton X-100 (Sigma), pH 6.8, at 0-5°C, as described before (Del Bino *et al.*, 1991). Simultaneous staining of cellular DNA and RNA was accomplished using the metachromatic fluorochrome acridine orange (AO; Polysciences Inc., Warrington, PA), as described in earlier publications (Darzynkiewicz *et al.*, 1976; Darzynkiewicz, 1990 b). The sensitivity of DNA to denaturation was also measured using AO as a metachromatic dye to differentially stain denatured *versus* double stranded DNA, with minor modifications, as described previously (Darzynkiewicz, 1990 c). Briefly, the cells were fixed in 80% ethanol, and following rehydration in buffered saline, treated with RNase A, then with 0.1 N HCl, and subsequently stained with AO (20 μM), at pH 2.6. Under these conditions, the condensed denatured DNA-AO complexes produce red luminescence, whereas intercalation of AO to nondenatured DNA results in green fluorescence following excitation with blue light (Darzynkiewicz and Kapuscinski, 1990).

The fluorescence of individual cells stained with DAPI and sulforhodamine 101 was measured with an ICP-22 flow cytometer (Ortho Diagnostics, Westwood, MA) using a combination of dichroic mirrors and filters as described before (Del Bino *et al.*, 1991). The data were stored in and analyzed using a Compaq Deskpro 386 computer. The Phoenix Flow Systems (San Diego, CA) software package was used for data accumulation and analysis of the cell cycle distribution. The luminescence of cells stained with AO according to the protocol designed to differentially stain DNA and RNA (Darzynkiewicz *et al.*, 1976), was measured with the FACScan flow cytometer (Becton Dickinson, San Jose, CA). Luminescence of cells stained with AO to reveal denatured *versus* nondenatured DNA was measured with a Cytofluorograf 30S (Ortho Diagnostics Instruments, Westwood, MA), using Phoenix Systems software. All experiments were repeated at least twice.

Agarose Gel Electrophoresis of DNA: The cells, washed with phosphate buffered saline, were resuspended in 50 mM Tris/HCl, 0.15 M NaCl, 5 mM EDTA buffer at pH 8.0 and frozen in liquid nitrogen. DNA was then isolated from the thawed cells and the total nucleic acid content determined spectrophotometrically by measuring absorption at 260 nm (A_{260}). All samples had A_{260}/A_{280} >2. Approximately 2 μg of DNA from each sample was analyzed on 1.2% vertical agarose gels (Maniatis, *et al.*, 1982). Electrophoresis was performed at 2.5 V/cm and the DNA was visualized under UV light after staining with ethidium bromide (Polysciences). Other details of DNA analysis were presented before (Del Bino *et al.*, 1991).

RESULTS

Incubation of thymocytes with PRED (P), CAM (C) or TN (T) results in degradation of DNA in these cells (Fig. 1). The products of degradation have a molecular weight equivalent to that of single nucleosomes and oligonucleosomes, which produces a "ladder-like" sequence of bands on the gels. A similar pattern can be observed for DNA extracted from HL-60 cells treated with CAM.

Treatment of HL-60 cells with GEN induces apoptosis in a significant proportion of HL-60 cells (Traganos *et al.*, 1992), and as is evident from the data shown in Fig. 2, the stainability of DNA in apoptotic cells is diminished. On DNA frequency histograms, the population of apoptotic cells is located to the left of the G_1 peak. In numerous experiments, we observed that the extent of the decrease of DNA stainability varied markedly depending on the cell treatment following their fixation. Thus, when the cells were fixed in cold ethanol (-20°C), centrifuged, immediately stained with ice-cold DAPI or propidium, and their fluorescence measured within 1-5 minutes, the difference in DNA stainability of normal and apoptotic cells was minimal: the intensity of fluorescence of apoptotic cells was

Fig. 1. Agarose gel electrophoresis of DNA extracted from human myelogenous leukemic HL-60 cells (A) or rat thymocytes (B).

A; HL-60 cells, untreated, growing exponentially (Control, Ct), or treated for 4 h with 0.15 μM camptothecin (C). The three lowest bands on the M lane represent DNA of M equivalent of 2322, 2027 and 564 base pairs, respectively. For details, see Del Bino *et al.* (1991).

B; thymocytes were incubated in cultures either in the absence of any drugs (Control, Ct), or in the presence of 5 μM prednisolone (P), 0.15 μM camptothecin (C), or 1 μM teniposide (T) for 6 h. The three lowest bands on the molecular weight marker lane (M) represent DNA of MW equivalent of 4631, 2322 and 2027 base pairs, respectively. For details, see Bruno *et al.* (1992).

The typical "ladder" pattern of DNA extracted from the drug treated cells indicates the presence of DNA fragments of the size equivalent to single nucleosomes or oligonucleosomes. This pattern is characteristic of the DNA of apoptotic cells.

very close to that of G_1 cells. However, when, after removal of fixative and prior to fluorescence measurement, the cells were kept suspended in PBS at 24-37°C for periods longer than 5 min, a decrease in DNA fluorescence of apoptotic cells was clearly evident. The data shown in Fig. 2 illustrates the changes in DNA stainability with DAPI, in apoptotic cells stained and measured either immediately after removal of fixative, or kept at 37°C for 20 min prior to staining. The decrease in DNA-associated fluorescence of apoptotic cells correlated with the duration and extent of rinsing prior to measurement. Extensive rinsing, especially at elevated temperatures, (e.g. over 30 min at 37°C) results in a greater than 80% loss in DAPI or propidium iodide stainability of DNA of apoptotic cells (Lassota *et al.*, in preparation).

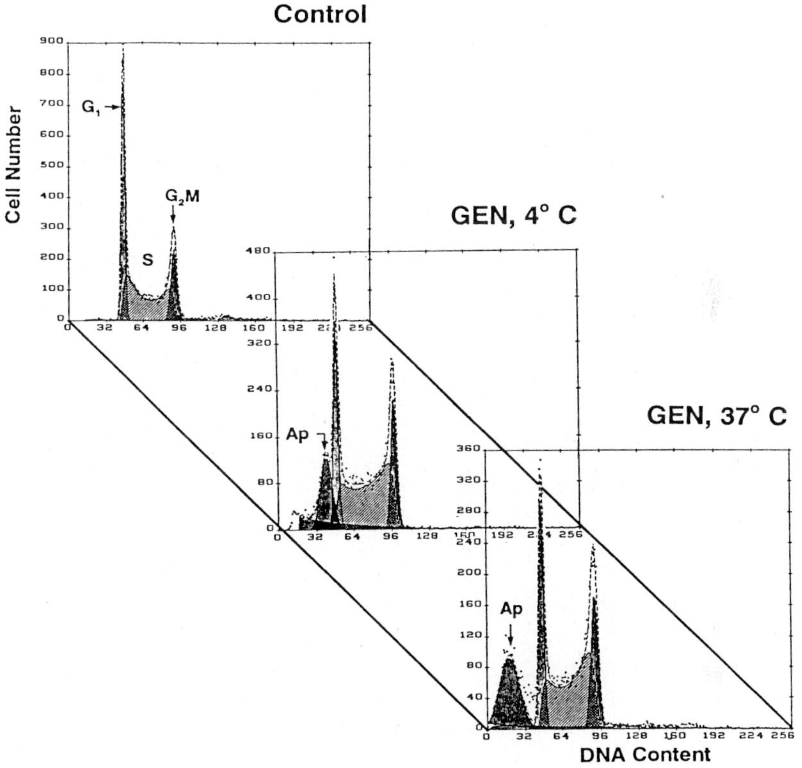

Fig. 2. DNA frequency histograms of HL-60 cells, untreated (Control) or treated with 100 μg/ml of Genistein for 8 h and then either stained immediately after fixation, at 4°C, or kept at 37°C for 20 min.

Cells were either stained immediately following fixation, at 0-4°C with DAPI and measured by flow cytometry (Control, and GEN, 4°C), or were, after fixation, suspended in Hanks' buffered saline and maintained at 37°C for 20 min prior to staining and measurement (GEN, 37°C). Genistein induces apoptosis of HL-60 cells; note that the position of apoptotic cells (Ap) is shifted to lower DNA content values when the cells are maintained in suspension at 37°C prior to measurement, suggestive of DNA extraction from these cells.

Fig. 3 shows changes in the sensitivity of DNA *in situ* to denaturation in cells which undergo apoptosis. In this experiment, the apoptosis of HL-60 cells was induced by CAM. The data shows the changes in DNA stainability with AO, following its partial denaturation with acid. CAM selectively causes apoptosis of

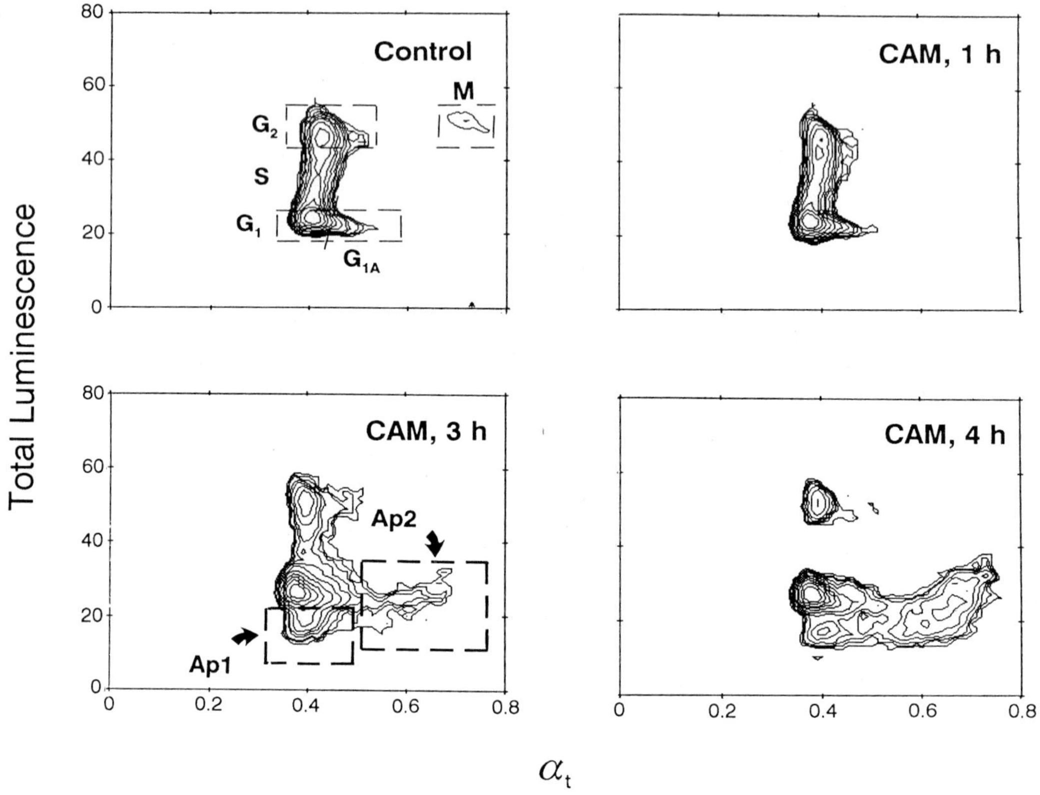

Fig. 3. Increased sensitivity of DNA *in situ* to denaturation, in apoptotic HL-60 cells, as a result of their treatment with camptothecin.

HL-60 cells, untreated (Control) or treated with camptothecin for 1, 3 and 4 h, were fixed, incubated with RNase A, treated with 0.1 M HCl and stained with acridine orange to differentially stain denatured and double stranded DNA. This staining makes it possible to discriminate, in control cultures, subpopulations of early G_1, G_{1A} cells, as well as the cells in S, G_2, and M. The decrease in total luminescence, seen in cells of the Ap1 population in cultures treated with camptothecin, reflects a decrease in DNA content in these cells, whereas the increase in α_t (Ap2) represents the proportion of denatured DNA, per cell. Note selective loss of S-phase cells as a result of treatment with CAM, coinciding with an increase in number of cells in regions Ap1 and Ap2.

S phase cells (Del Bino *et al.*, 1991), and it is evident that, indeed, only S-phase cells were affected by the treatment. The first changes were seen 3 h after addition of the drug, manifesting as a loss of S-phase cells which was paralleled by the appearance of a cell population with lower total luminescence (Ap1) and a another population, with increased α_t values (Ap2).

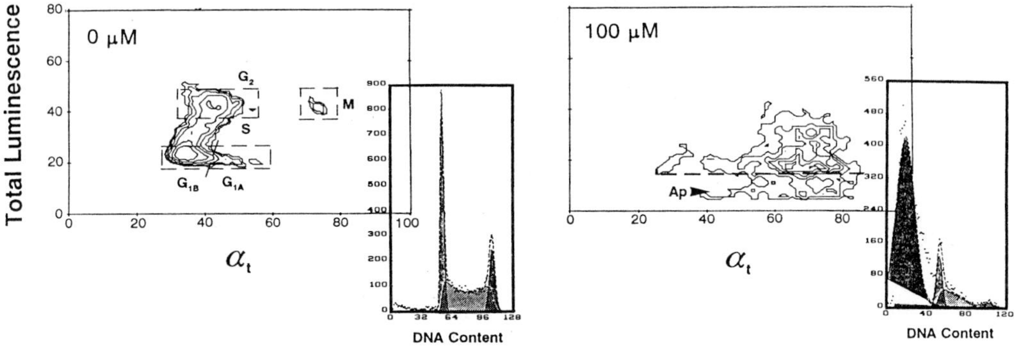

Fig. 4. Increased sensitivity of DNA *in situ* to denaturation in HL-60 cells undergoing apoptosis after treatment with fostriecin.

DNA in HL-60 cells, untreated (0 μM) or treated with 100 μM fostriecin for 24 h (100 μM) was subjected to acid denaturation and stained with acridine orange as described in Material and Methods. Note that unlike in the case of camptothecin (Fig. 3), the increase in α_t and a decrease in total luminescence (below the level of G_1 cells, broken line) in cultures treated with fostriecin, affect all cells, regardless of the phase of the cycle. DNA frequency histograms (after staining with DAPI) from the respective cultures are included in the insets.

A more detailed analysis of the data shown in Fig. 3 suggests that the decrease in total luminescence precedes the increase in α_t. For instance, 3 h after addition of CAM, 7% of the cells lose total luminescence (i.e. fall below that of G_1 cells) though their α_t values remain unchanged (Ap1 population) while 13% of the cells show an increase in α_t (Ap2). Treatment with CAM for 4 h results in 6% cells in the Ap1 and 36% cells in the Ap2 population. This sequence, the appearance first of cells in the Ap1 compartment then in the Ap2 compartment, was observed in repeated experiments (not shown).

Fig. 4 shows changes in the sensitivity of DNA *in situ* to denaturation, in HL-60 cells after their exposure to FST. Somewhat different conditions of acid denaturation were used in this experiment (Hotz *et al.*, 1992), compared to that illustrated in Fig. 3, which may explain the higher cell heterogeneity with respect to their α_t values, in the respective controls (untreated cells). Generally, however, the pattern of cell distribution with respect to total luminescence and α_t values is similar, and the same populations can be discriminated in both experiments. In contrast to CAM, which is selective for S phase cells, FST induces apoptosis of cells in all phases of the cycle (Hotz *et al.*, 1992). Indeed, as is evident from the data shown in Fig. 4, nearly all cells had markedly increased α_t values after exposure to the drug. A decrease in total cell luminescence, with a large number of cells falling below the luminescence values of the G_1 cell population, was also apparent.

The changes in stainability with AO of rat thymocytes, treated with PRED, CAM or TN are shown in Fig. 5. In contrast to the data presented in Figs. 3 and 4, the thymocytes were stained with AO without pretreatment with RNase, and at pH 6.8, i.e. according to the procedure designed to differentially stain DNA and RNA (Darzynkiewicz, 1990 c). This data illustrates changes in the cell cycle distribution and appearance of apoptotic cells as a result of exposure of thymocytes to these drugs. Based on cellular DNA content, it was possible to discriminate cells in $G_0 + G_1$, S and $G_2 + M$, whereas differences in RNA content made it possible to distinguish G_0 from G_1 cells (Fig 5). The apoptotic cells, appearing in large numbers in cultures treated with PRED, CAM or TN, were characterized by a decrease in green fluorescence. These cells, as is evident from the continuity of the contours in Fig. 5, originated from the G_0 population. The apoptotic cells in these contour maps are heterogenous with respect to their green and red luminescence values. Their characteristic position on the contour map indicates that the green and red luminescence values of individual cells are negatively correlated. Thus, for the apoptotic cells which originate from the G_0 population, the loss of green fluorescence is paralleled by a proportional increase in red luminescence. This would indicate that, in these cells, which (typical of the G_0 state) have minimal RNA content, the observed changes in stainability with AO actually represent a change in DNA stainability which is indicative of DNA denaturation. Thus, the progression of apoptosis in thymocytes is associated with a progressive increase in DNA denaturation, detectable by this assay. In this particular case of apoptotic cells, since there is an absence of any significant amount of RNA, the method which was designed to differentially stain RNA and DNA, in fact, detects DNA denaturation.

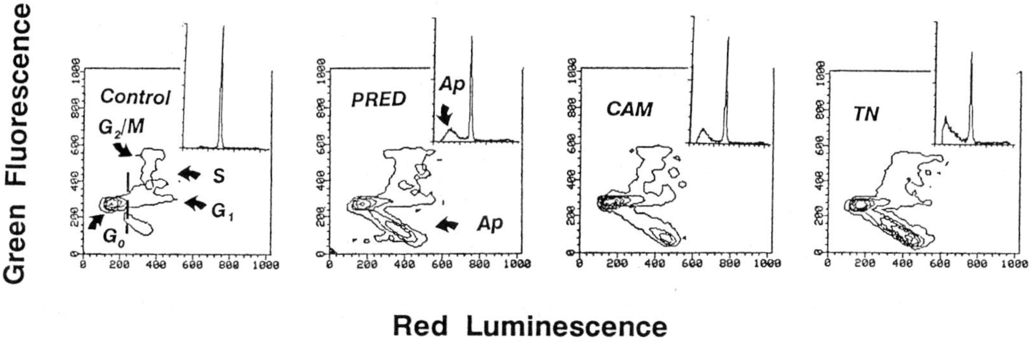

Fig. 5. Green and red luminescence of rat thymocytes treated with prednisolone (PRED), camptothecin (CAM) or teniposide (TN), following staining of their DNA and RNA with acridine orange.

Freshly isolated rat thymocytes were incubated either with 5 μg/ml of prednisolone, 0.1 μg/ml of camptothecin or 1 μg/ml of teniposide for 4 h. Their DNA and RNA was then differentially stained with acridine orange. Based on differences in RNA content it was possible to discriminate between G_0 and G_1 cells. The appearance of apoptotic cells with reduced DNA content (Ap) coincides with the selective loss of G_0 cells. This indicates that regardless of the inducer (prednisolone, camptothecin, teniposide) apoptosis of thymocytes is selective to G_0 cells. Note the increase in red, and decrease in green luminescence, of the apoptotic cells, compared with G_0 cells. Insets represent DNA frequency histograms from the respective

DISCUSSION

As mentioned in the Introduction, reduced DNA stainability served as a marker of apoptotic cells in numerous studies in which the cells were analyzed by flow cytometry. In light of the evidence that there is no change in DNA content during early stages of apoptosis (Arends *et al.*, 1990) in intact, unfixed and nonpermeabilized cells, the observed reduction in cell stainability with DNA fluorochromes has been explained as due to a change in DNA conformation and accessibility to the dyes (e.g. Telford *et al.*, 1992). Several observations, however, indicate that the reduced DNA stainability is a consequence of a partial loss of DNA from these cells due to activation of an endonuclease and subsequent diffusion of the low molecular weight DNA from the cells, following fixation or permeabilization. Thus, there is evidence that lysis of apoptotic cells by detergents releases low molecular weight DNA, associated with mono or oligonucleosomes (Arends *et al.*, 1992). A fraction of DNA in the apoptotic cell, therefore, is of low molecular weight and it is expected to dissociate from the cell after its permeabilization by detergent or fixation by ethanol and rehydration. Our present data (Fig. 2) show that prolonged rinsing after fixation results in a further reduction in DNA stainability and provides additional evidence that DNA is extracted from apoptotic cells. In all probability, therefore, the observed reduction in DNA stainability with a variety of fluorochromes is a reflection of DNA extraction rather than a specific change in nuclear chromatin involving a decrease in DNA accessibility.

Our data also indicate that the degree of DNA extraction can be manipulated by the conditions of cell rinsing. This provides an opportunity to enhance the detection of apoptotic cells based on the intercellular differences in cell stainability with DNA fluorochromes, in the cases when the distinction between G_1 and apoptotic cells is inadequate for their separation. Too extensive a rinsing, however, can reduce the stainability of apoptotic cells to such a degree that they may become undetectable especially when the measurements are triggered by a threshold based on the DNA signal, and the threshold setting is relatively high.

The metachromatic dye AO can differentially stain double stranded and single stranded nucleic acids in cells or cell nuclei (Darzynkiewicz and Kapuscinski, 1990). This dye intercalates into double stranded regions and when bound in this form, fluoresces green. AO can also condense the single stranded sections of nucleic acids and, in the condensed form, AO luminesces in red (Darzynkiewicz *et al.*, 1990). In the absence of RNA, (i.e. when permeabilized cells are preincubated with RNase prior to staining with AO), the ratio of red to total (red plus green) luminescence (α_t) represents the portion of denatured DNA in the cell (Darzynkiewicz, 1990 c). When the measurements are standardized, the sum of intensities of green and red luminescence ("total luminescence"; L_{tot}) is proportional to total DNA content (Darzynkiewicz, 1990 c). This method can be applied to evaluate the sensitivity of DNA to denaturation. As we have seen before, the DNA in condensed chromatin, such as in the case of mitotic or G_0 cells, is much more sensitive to denaturation than is DNA in interphase cells.

The present data indicate that DNA in apoptotic cells, regardless of whether apoptosis is induced by CAM or FST, is increasingly sensitive to denaturation. Apoptotic cells, therefore, can be distinguished by flow cytometry after their staining with AO, based on their increased α_t values (Figs. 3 and 4). When the data are plotted as bivariate distributions of L_{tot} *versus* α_t, it becomes also evident that the L_{tot} of apoptotic cells is diminished, consistent with the decrease in DNA content in these cells, as discussed above. Apoptotic cells, thus, are characterized by high α_t values and diminished L_{tot}. Therefore, this method can be used to both identify apoptotic cells and reveal the cell cycle phase specificity (Fig. 3), or lack thereof (Fig. 4), of apoptosis.

The increased sensitivity of DNA in apoptotic cells to denaturation creates a problem for the use of AO in the procedure to differentially stain RNA and DNA. Staining of thymocytes with AO reveals the increase in red luminescence which is paralleled by the decrease in green fluorescence of G_0 cells undergoing apoptosis. Since these cells have minimal RNA content, their red luminescence is a reflection of progressive DNA denaturation rather than an increase in RNA content (Bruno *et al.*, 1992). This method, therefore, cannot be used to assay RNA content in apoptotic cells, because the denaturation of DNA in apoptotic cells is also reflected, as is RNA, by metachromatic staining with AO. However, as is evident in the present study, the technique can be used to identify G_0 cells (Fig. 5), and to assay in this way their selective sensitivity to undergo apoptosis.

As shown by Arends *et al.* (1990), the degree of DNA denaturation in apoptotic cells is minor. Yet, the denaturation is detectable by flow cytometry perhaps because the dyes used to stain RNA or single stranded DNA (AO or pyronine Y) by virtue of their high affinity for single stranded regions, are themselves capable of potentiating DNA denaturation (Darzynkiewicz and Kapuscinski, 1990).

The mechanism by which DNA in chromatin of apoptotic cells becomes increasingly sensitive to denaturation is unknown. It is possible that the changes in chromatin during apoptosis involve proteolysis of histones or other proteins which otherwise provide counterions to DNA, stabilizing its double helical conformation (Darzynkiewicz, 1990 a). DNA depleted of these counterions is less stable, and when subjected to heat or acid, undergoes denaturation.

REFERENCES

Afanas'ev VN, Korol' BA, Mantsygin YA, Nelipovich PA, Pechtnikov VA, Umansky SR (1986) Flow cytometry and biochemical analysis of DNA degradation characteristic of two types of cell death. FEBS 194:347-350

Arends MJ, Morris RG, Wyllie H (1990) Apoptosis: The role of endonuclease. Am J Pathol 136:593-608

Bertrand R, Kerrigan D, Sarang M, Pommier Y (1991) Cell death induced by topoisomerase inhibitors. Biochem Pharmacol 42:77-85

Bruno S, Lassota P, Giaretti W, Darzynkiewicz Z (in press) Apoptosis of rat thymocytes by prednisolone, camptothecin or teniposide is selective to G_0 cells and is prevented by inhibitors of proteases. Oncology Res

Compton MM, Haskill JS, Cidlowski JA (1988) Analysis of glucocorticoid actions on rat thymocyte deoxyribonucleic acid by fluorescence-activated flow cytometry. Endocrin 122:2158-2164

Cowden RR, Curtis SK (1981) Microfluorometric investigations of chromatin structure. Histochem 72:11-23

Darzynkiewicz Z (1990 a) Probing nuclear chromatin by flow cytometry. In: Melamed MR, Lindmo T, and Mendelsohn ML (eds) Flow Cytometry and Sorting, 2nd edn. J. Wiley & Sons, Inc., New York, pp 315-340

Darzynkiewicz Z (1990 b) Differential staining of DNA and RNA in intact cells and isolated cell nuclei. Meth Cell Biol 33:285-298

Darzynkiewicz Z (1990 c) Acid-induced denaturation of DNA *in situ* as a probe of chromatin structure. Meth Cell Biol 33:337-352

Darzynkiewicz Z, Kapuscinski J (1990) Acridine orange: A versatile probe of nucleic acids and other cell constituents. In: Melamed MR, Lindmo T, and Mendelsohn ML (eds) Flow Cytometry and Sorting, 2nd edn. J. Wiley & Sons, Inc., New York, pp 291-314

Darzynkiewicz Z, Traganos F, Sharpless T, Melamed MR (1976) Lymphocyte stimulation: A rapid multiparameter analysis. Proc Natl Acad Sci USA 73:2881-2886

Del Bino G, Bruno S, Yi PN, Darzynkiewicz Z (in press) Apoptotic cell death triggered by camptothecin or teniposide. The cell cycle specificity and effects of ionizing radiation. Cell Prolif

Del Bino G, Darzynkiewicz Z (1991) Camptothecin, teniposide or 4'-(9-acridinylamino)-3-methane-sulfon-m-anisidide but not mitoxantrone or doxorubicin, induces degradation of nuclear DNA in S phase of HL-60 cells. Cancer Res 51:1165-1169

Del Bino G, Lassota P, Darzynkiewicz Z (1991) The S-phase cytotoxicity of camptothecin. Exp Cell Res 193:27-35

Dive C, Hickman JA (1991) Drug-target interactions: Only the first step in the commitment to a programmed cell death? Br J Cancer 64:192-196

Hotz MA, Traganos F, Darzynkiewicz Z (in press) Changes in nuclear chromatin related to apoptosis or necrosis induced by the DNA topoisomerase II inhibitor fostriecin in MOLT-4 and HL-60 cells are revealed by altered DNA sensitivity to denaturation. Exp Cell Res

Kaufmann SH (1989) Induction of endonucleolytic DNA cleavage in human acute myelogenous leukemia cells by etoposide, camptothecin, and other cytotoxic anticancer drugs. A cautionary note. Cancer Res 49:5870-5878

Kerr JFR, Harmon BV (1991) Definition and incidence of apoptosis: An historical perspective. In: Tomei LD, Cope FO (eds) Apoptosis: The molecular basis of cell death. Curr. Commun. Cell & Mol. Biol. Vol. 3. Cold Spring Harbor Laboratory Press, Cold Spring Harbor, NY, pp 5-29

Kerr JFR, Wyllie AH, Curie AR (1972) Apoptosis: A basic biological phenomenon with wide-ranging implications in tissue kinetics. Br J Cancer 26:239-257.

Krishan A, Frei E, III (1975) Morphological basis for the cytolytic effect of Vinblastine and Vincristine on cultured human leukemic lymphocytes. Cancer Res 35:497-501

Lennon SV, Martin SJ, Cotter TG (1991) Dose-dependent induction of apoptosis in human tumour cell lines by widely diverging stimuli. Cell Prolif 24:203-214

Lockshin RA, Zakeri Z (1991) Programmed cell death and apoptosis. In: Tomei LD, Cope FO (eds) Apoptosis: The molecular basis of cell death. Curr. Commun. Cell & Mol. Biol. Vol. 3. Cold Spring Harbor Laboratory Press, Cold Spring Harbor, NY, pp 47-60

Maniatis T, Fitsch EF, Sambrook J (1982) Molecular Cloning. (A Laboratory Manual). Cold Spring Harbor Laboratory Press, Cold Spring Harbor, NY

Ojeda F, Guarda MI, Maldonado C, Folch H (1990) Protein kinase-C involvement in thymocyte apoptosis induced by hydrocortisone. Cell Immunol 125:535-539

Schrek R, Chandra S, Molnar Z, Stefani S (1984) Two types of interphase death of lymphocytes exposed to temperatures of 37-45°C. Radiat Res 82:162-170

Swat W, Ignatowicz L, Kisielow P (1991) Detection of apoptosis of immature $CD4^{+}8^{+}$ thymocytes by flow cytometry. J Immunol Meth 137:79-87

Szende B, Zalatnai A, Schally AV (1989) Programmed cell death (apoptosis) in pancreatic cancers of hamsters after treatment with analogs of both luteinizing hormone-releasing hormone and somatostatin. Proc Natl Acad Sci USA 86:1643-1647

Telford WG, King LE, Fraker PJ (1992) Comparative evaluation of several DNA binding dyes in the detection of apoptosis-associated chromatin degradation by flow cytometry. Cytometry 13:137-143

Telford WG, King LE, Fraker PJ (1991) Evaluation of glucocorticoid-induced DNA fragmentation in mouse thymocytes by flow cytometry. Cell Prolif 24:447-459

Tomei LD, Cope FO (eds) (1991) Apoptosis: The molecular basis of cell death. Curr. Commun. Cell and Mol. Biol. Vol. 3. Cold Spring Harbor Laboratory Press, Cold Spring Harbor, NY

Traganos F, Ardelt B, Halko N, Bruno S, Darzynkiewicz Z (submitted) Effects of genistein, a tyrosine kinase inhibitor, on the growth and cell cycle progression of normal human lymphocytes and human leukemic MOLT-4 and HL-60 cells. Cancer Res

Tritton TR (1991) Cell death in cancer chemotherapy: The case of Adriamycin. In: Tomei LD, Cope FO (eds). Apoptosis: The molecular basis of cell death. Curr. Commun. Cell and Mol. Biol. Vol. 3. Cold Spring Harbor Laboratory Press, Cold Spring Harbor, NY, pp 121-137

Umansky SR, Korol' BR, Nelipovich PA (1981) In vivo DNA degradation in the thymocytes of gamma-irradiated or hydrocortisone-treated rats. Biochim Biophys Acta 655:281-290

Walker PR, Smith C, Yondale T, Leblanc J, Whitfield JF, Sikorska M (1991) Topoisomerase II-reactive chemotherapeutic drugs induce apoptosis in thymocytes. Cancer Res 52:1078-1085

Wyllie AH, Kerr JFR, Currie AR (1980) Cell death: The significance of apoptosis. In: International Review of Cytology, Vol. 68, Bourne GH, Danielli FJ, Jeon KW (eds). Academic Press, New York, pp 251-306

Wyllie AH, Morris RG, Smith AL, Dunlop D (1977) Chromatin cleavage in apoptosis: Association with condensed chromatin morphology and dependence on macromolecular synthesis. J Pathol 142:67-77

ACKNOWLEDGEMENTS

This work was supported in part by NIH NCI grants RO1 CA28704 and R37 CA23296, the Carl Inserra Fund and the "This Close" Foundation. Dr. Wojciech Gorczyca is a fellow of the Alfred Jurzykowski Foundation, on leave from the Department of Tumor Pathology, Medical Academy of Szczecin, Poland. Dr. Silvia Bruno was supported by the Istituto Nazionale per la Ricerca sul Cancro, Genova, Italy. We thank Ms. Irene Logsdon for help in the preparation of this manuscript.

GENE MAPPING AND PCR APPLICATIONS WITH FLOW-SORTED CHROMOSOMES

N.P. Carter
Department of Pathology
University of Cambridge
Tennis Court Road
Cambridge CB2 1QP
UK

Bivariate flow karyotype analysis

The bivariate analysis by flow cytometry of human chromosome suspensions stained with the two fluorochromes, Hoechst 33258 (specificity for AT-rich DNA) and Chromomycin A3 (specificity for GC-rich DNA) enables the majority of the chromosome types to be resolved and sorted (Gray *et. al.* 1979). In the flow cytometer, chromosomes pass sequentially through two spatially separated laser beams, the first operated using the UV lines (351-364 nm) to excite Hoechst fluorescence and the second operated at 457.9 nm to excite Chromomycin fluorescence. The fluorescence intensity of each chromosome is measured for both dyes independently and correlated on the bivariate flow karyotype where chromosome types are resolved by DNA content and base pair ratio (see Fig. 1).

With the exception of chromosomes 9-12 which display similar DNA content and base pair ratio, all of the chromosome types can be resolved in this way. These high resolution bivariate flow karyotypes can be produced on unmodified commercial flow sorters (Carter *et. al.* 1990). Utilising the sorting capability of such instruments, fractions highly enriched for a particular chromosome type can be produced for use in the production of DNA libraries, for gene mapping and for use in the cytogenetic analysis of metaphase chromosomes and interphase nuclei.

NATO ASI Series, Vol. H 67
Flow Cytometry
Edited by A. Jacquemin-Sablon

Inherited disease

A current goal of human molecular genetics is the location of disease-associated genes within the genome. Sequencing of the entire gene improves prenatal diagnosis, family genetic counselling and disease management and offers the potential for gene therapy. Unfortunately, much of the human genome is not expressed and structural genes are estimated to represent less than 17% of total DNA (Connor and Ferguson-Smith, 1987). Consequently, the location of disease-associated genes within the genome is a daunting task.

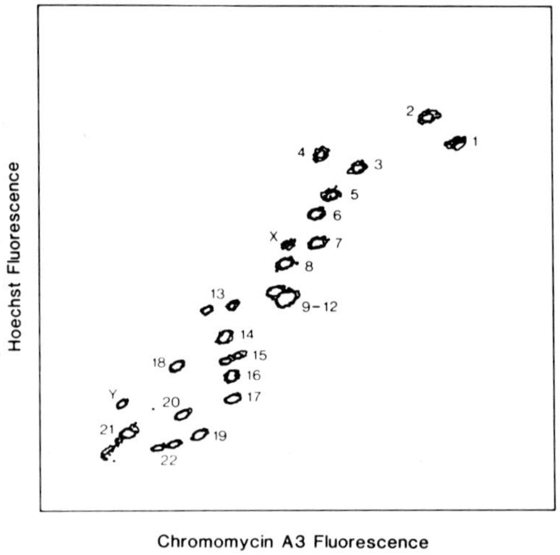

Figure 1. Bivariate flow karyotype of chromosomes from lymphoblastoid cells from a normal male (reprinted from Telenius *et. al.* 1992)

A common approach to this problem is the use of genetic linkage of DNA markers to the disease in affected families. These markers can be localised to a particular chromosome and then placed into a physical map comprising an ordered series of both expressed and anonymous DNA sequences localised along the chromosome. This physical map of the chromosome enables the markers most closely flanking the disease locus to be identified and thus defines the area of the map within which the disease-associated gene lies. Often this defined area of the physical map will contain more DNA than is practical to sequence, so it is often necessary to identify additional markers to increase the resolution of the map. As the resolution of the map improves with an increasing number of ordered sequences the smaller becomes the distance between the disease locus and the flanking markers until the disease gene itself can be localised and sequenced. It is clear that an accurate, detailed physical map of each chromosome is vital to the localisation of disease genes in this way.

Gene mapping with flow-sorted dot blots

Flow-sorted chromosomes are a valuable resource for gene mapping, both in the initial chromosomal localisation of DNA sequences and in the construction of chromosome maps. We have used flow-sorted chromosome dot-blots and the specific amplification of DNA sequences from small numbers of flow-sorted chromosomes using the polymerase chain reaction (PCR) for this purpose. Chromosomes flow-sorted directly as dots onto nitrocellulose filters can be used as targets for the direct hybridisation of cloned DNA sequences (Bernheim *et. al.*, 1983). Flow-sorted dot blots can be used in this way to provide an initial chromosomal localisation for a DNA sequence. Dot blots are produced by sorting 10,000 of each chromosome type onto nitro-cellulose filter discs, two chromosome types per disc. Spreading of the chromosomes on the disc is restricted during sorting by the application of mild aspiration from below which also acts to hold the filter in place. The dots of chromosomal DNA are denatured by alkali treatment, neutralised and then baked onto the disc. A panel of 11 discs comprises all of the different chromosome types (chromosome 9-12 being sorted as a single dot). A radio-labelled DNA probe can then be hybridised to the panel of filters, while subsequent autoradiography reveals the spot of DNA to which the probe has hybridised and thus identifies the chromosomal location of the DNA sequence (see Fig. 2).

Many sequences have been localised to specific chromosomes in this way (e.g. Lebo et.al., 1984; Collard *et. al.*, 1985; Lebo and Bruce 1987; Arai *et. al.*, 1989; Neckelmann *et. al.*, 1989; Whitcome *et. al.*, 1991). However, we have found that flow sorted dot blots work efficiently only for highly purified probes which do not contain repetitive sequences or are associated with pseudogenes, but flow-sorted dot blots have the advantage that sequence data is not required.

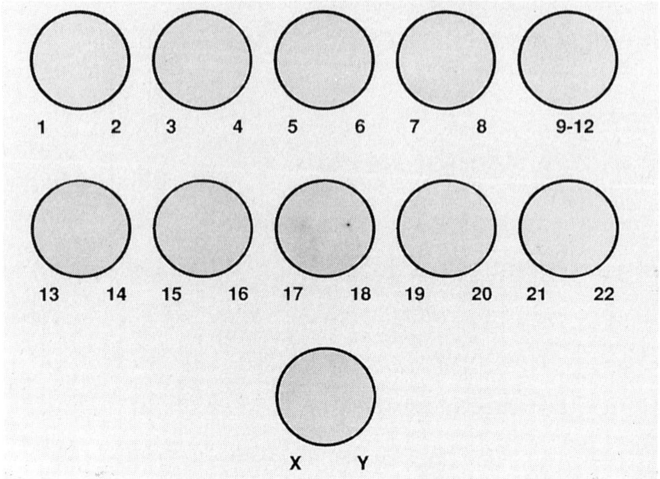

Figure 2. Localisation of the gene for human ferrochelatase to chromosome 18 by flow-sorted dot blot analysis (reprinted from Whitcome *et. al.* 1991).

Gene mapping by PCR

The alternative approach we have used for the chromosomal localisation of DNA sequences is the enzymatic amplification of the sequence from small numbers of chromosomes sorted directly into PCR tubes (Schmitz *et. al.*, 1989; Cotter *et. al.*, 1989). As few as 100 chromosomes are sufficient as target for most specific primer pairs. Chromosomes of each type are sorted directly into separate PCR tubes containing 30 µl of purified water. A panel of 21 tubes thus represents the whole genome (chromosomes 9-12 are sorted together). PCR buffer, dNTPs, specific primer pairs and Taq polymerase are added to each tube to make a 50 µl reaction which is overlayered with mineral oil and then amplified typically for 40 cycles. An aliquot of each tube (10-20 µl) is analysed by ethidium bromide stained, agarose gel electrophoresis. The detection of specific product of the correct molecular weight in one of the lanes localises the sequence to the appropriate chromosome (see Fig. 3).

We have been able to reduce the number of PCRs required for a localisation by adopting a two step analysis. In the initial analysis, pools of 200 of each of 4 chromosome types (i.e. 1-4, 5-8, 9-12, 13-16, 17-20 and 21, 22, X and Y) are used as target DNA for amplification. The PCR is performed on the chromosome pools, the reaction mixtures are run on an agarose gel and the tube in which product has been generated is identified. The same primers are then used with tubes containing 300 of the individual chromosomes within the pool and the reaction mixtures run on a second agarose gel. The tube in which product is detected maps the sequence to that particular chromosome (see Fig. 4). In this way only 10 PCR reactions are required for chromosomal localisation compared with the 21 reactions required if chromosome pools are not used.

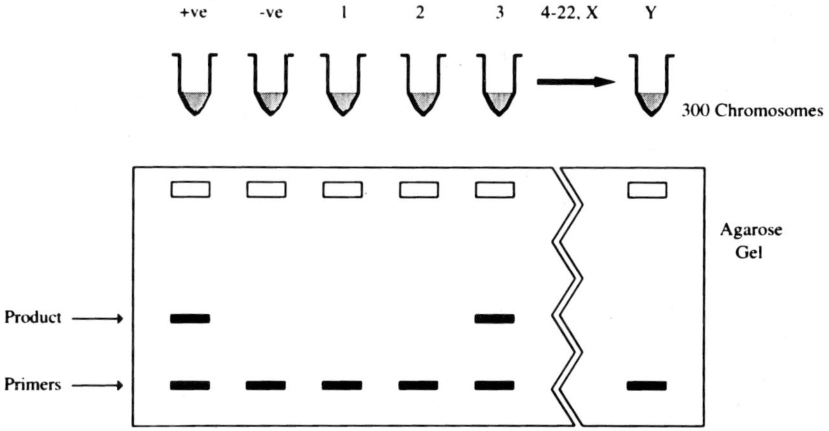

Figure 3. Schematic representation of gene mapping by PCR amplification of sorted chromosomes.

Gene mapping with translocation cell lines

To improve the resolution of gene mapping with sorted chromosomes, sequences can be localised to within sub-regions of chromosomes by the use of cell lines derived from individuals with balanced translocations. In most cases, the two abnormal translocation chromosomes are different in size from the normal chromosomes from which they are derived. Usually, these derivative

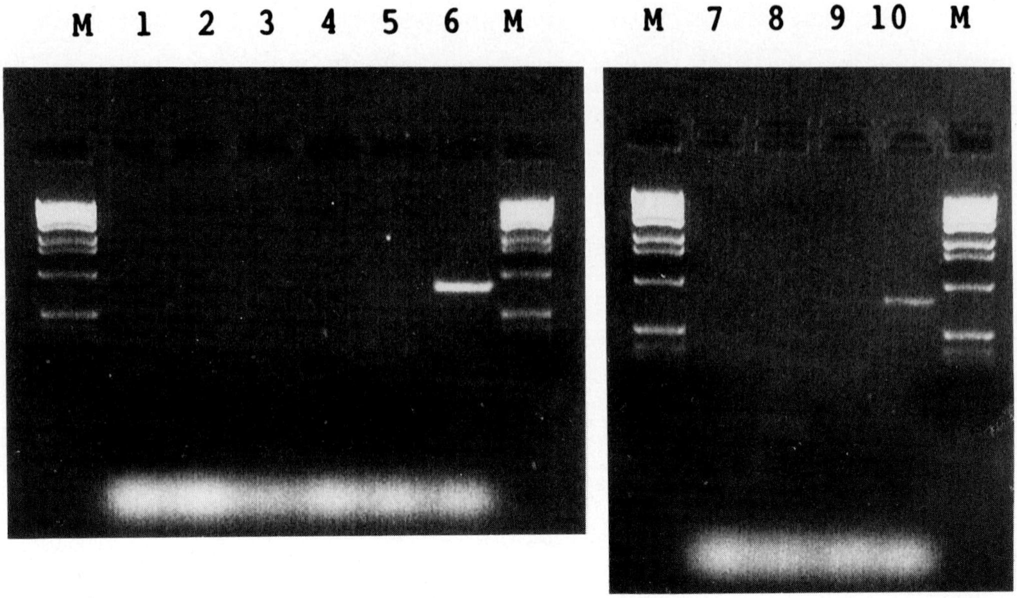

Figure 4. Agarose gel electrophoresis of the products of PCR amplification of the human SRY gene on sorted chromosomes. (M - molecular weight markers; Lane 1 - chromosomes 1-4; Lane 2 - chromosomes 5-8; Lane 3 - chromosomes 9-12; Lane 4 - chromosomes 13-16; Lane 5 - chromosomes 17-20; Lane 6 - chromosomes 21, 22, X and Y; Lane 7 - chromosome 21; Lane 8 - chromosome 22; Lane 9 - chromosome X; Lane 10 - chromosome Y).

chromosomes can be resolved on the bivariate flow karyotype (see Fig. 5) and so sorted directly onto nitrocellulose filter discs or into PCR tubes. A probe or sequence that has been mapped previously to one of the chromosomes involved in the translocation can then be used with the two derivative chromosomes to determine whether the sequence maps proximal to or distal to the translocation breakpoint (Fig. 6.).

We have used this approach extensively with both flow-sorted dot blots and with PCR amplification (Fig. 7.) to create a physical map of the long arm of chromosome 9. We have collected over 20 cell lines with different breakpoints on the long arm of chromosome 9 and thus can assign any probe or sequence into one of the 21 intervals (Fig. 8). The resolution of the physical map produced in this way is limited only by the number of breakpoints and the physical distance between them.

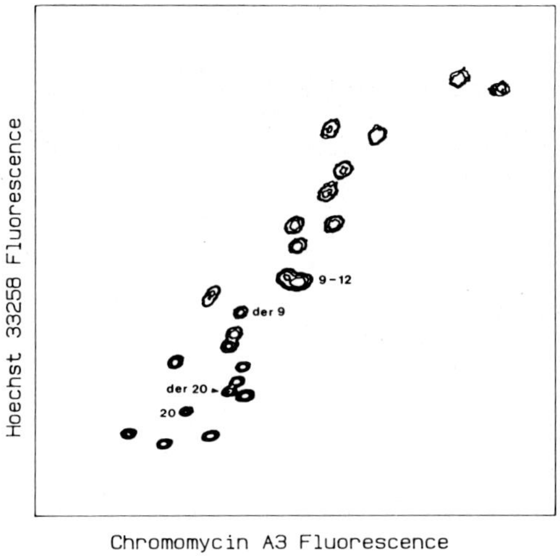

Figure 5. Flow karyotype of a lymphoblastoid cell line derived from an individual with a translocation between chromosomes 9 and 20. The normal and derivative chromosomes involved in the translocation are indicated.

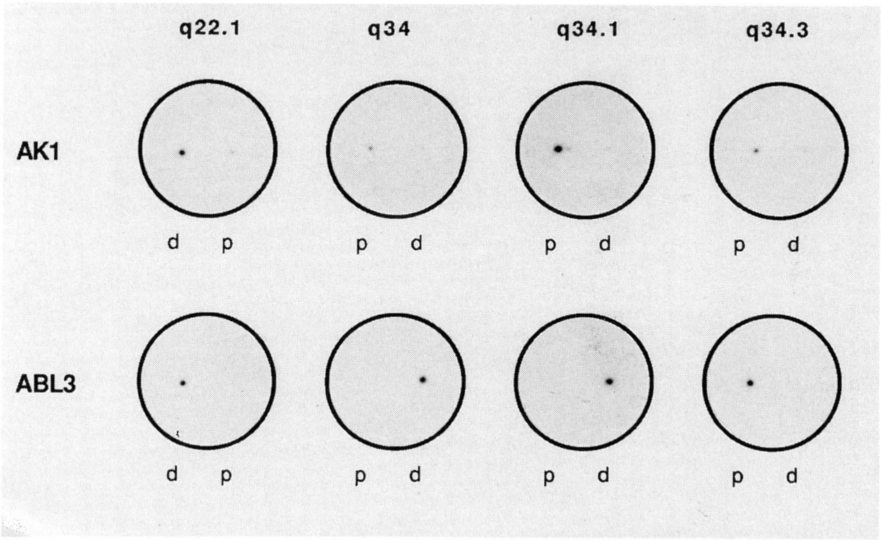

Figure 6. Hybridisation of two DNA probes to translocation chromosome flow-sorted dot blots. The chromosome dot representing the proximal part and distal part of chromosome 9 for each translocation are indicated by p and d respectively (reprinted from Carter et.al., 1990).

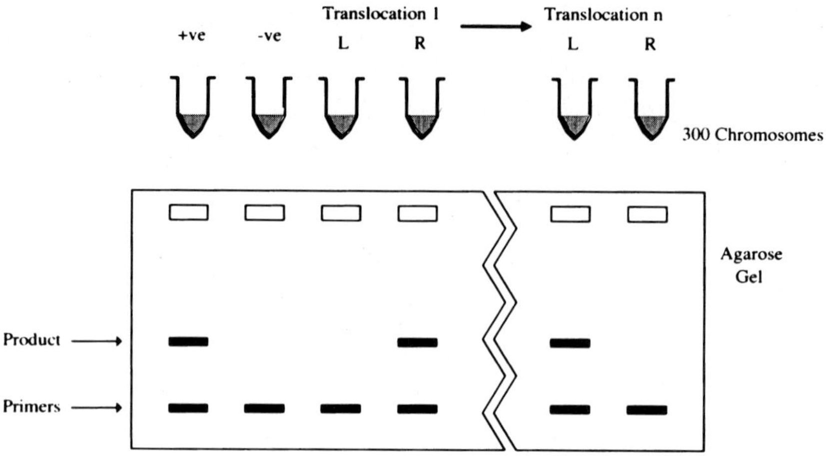

Figure 7. Schematic representation of gene mapping using PCR amplification of chromosome translocation products.

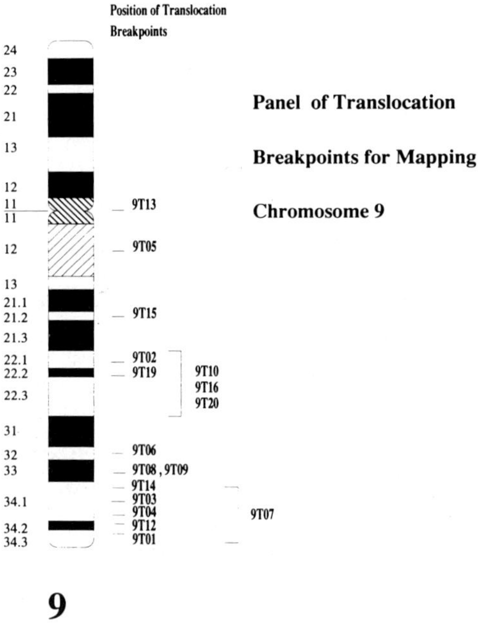

Figure 8. Panel of translocation breakpoints for mapping the long arm of chromosome 9.

Chromosome painting

Chromosome painting with flow-sorted libraries can be extremely helpful in the cytogenetic analysis of abnormal karyotypes. The availabity of libraries generated from large numbers of flow-sorted chromosomes by the Lawrence Livermore and Los Alamos Laboratories in the USA has led to their widespread use for chromosome painting of metaphase spreads using fluorescence *in situ* hybridisation (FISH). DNA from the library is labelled typically with a hapten (e.g. biotin) using nick-translation with modified dUTP. Hybridisation of the library to metaphase spreads on glass slides, using competitive *in situ* suppression of repetitive sequences with genomic or Cot 1 DNA, is visualised using a fluorescent detection system (e.g. avidin-FITC) and direct observation on the fluorescence microscope (Pinkel et.al., 1988). For analysis of metaphases from an individual with an abnormal karyotype or from a tumour, routine banding analysis identifies the appropriate chromosome-specific libraries which are then used to highlight the aberration and the chromosome breakpoints involved (Hulten et.al., 1991).

Reverse chromosome painting

While chromosome painting with libraries works well for cases where chromosome banding can identify the chromosomes involved in the rearrangement and thus indicate which chromosome libraries to use, painting with chromosome libraries cannot be used to characterise interstitial deletions or the precise origin of *de novo* unbalanced rearrangements. For such cases, we have developed a technique we have called reverse chromosome painting where the chromosome paint is generated rapidly by PCR from small numbers of the aberrant chromosome itself, flow sorted from blood cultures or cell lines from the affected individual, and the PCR products painted back onto **norma**l metaphase spreads. In this way the origin of aberrant chromosomes can be visualised directly from the hybridisation pattern on **normal** chromosomes (Carter *et. al.*, 1992).

Vital for the reverse chromosome painting technique is a PCR protocol which enables small numbers of flow-sorted chromosomes to be amplified in a general and representative way. For this purpose we have developed a novel, partially degenerate PCR primer used in a dual annealing temperature protocol. We have called this technique Degenerate Oligonucleotide-Primed PCR (DOP-PCR). The primer, 6MW (5' CCG ACT CGA GNN NNN NAT GTG G 3', where N = any base), is constructed with 6 specific bases at the 3' end and 10 specific bases at the 5' end, separated by a degenerate region of 6 bases (Telenius *et. al.*, 1992a). The target DNA (300 chromosomes) is subjected to 9 low temperature annealing cycles (94°C for 1 min, 30°C for 1 min, transition to 72°C at 0.23°C /sec, 72°C for 2 mins) followed immediately by 30 high annealing temperature cycles (94°C for 1 min, 62°C for 1 min, 72°C for 1 min). We believe that priming is restricted by the partial degeneracy to chiefly the 6 most 3' bases during the low annealing temperature cycles, resulting in priming sites on average every 2 kb (Telenius et.al., 1992b). In this way, initial product representative of the entire chromosome is generated, tailed at one end with the primer sequence and at the other with its complement. During the high annealing temperature cycles, both the 3' and the 5' bases are required for stable priming so that only product tailed during the initial cycles is amplified further. Biotinylation of the PCR product for chromosome painting is achieved simply by a second round of high annealing temperature cycles incorporating biotin-11-dUTP in the reaction mix.

We have been applying this reverse chromosome painting technique for the analysis of routine cases submitted for diagnosis to the East Anglian Regional Cytogenetics Laboratory. Fig. 9 shows the flow karyotype obtained from a mother referred for genetic analysis after two fetal losses due to neural tube defects. Chromosome banding analysis indicated an insertion from the long arm of chromosome 13 into the short arm of chromosome 1 and while the deleted chromosome 13 is resolved clearly on the flow karyotype the chromosome 1 with the insertion is not resolved from the normal chromosome 1. Reverse painting DOP-PCR products from the combined chromosome 1 peak onto normal chromosomes is shown in Figure 10a. In addition to the two normal chromosomes 1, signal was detected from distal 13q31 to distal 13q33, clearly demonstrating the precise origin of the insertion into the aberrant chromosome 1. The involvement of the same region also could be demonstrated from the reverse painting of the deleted chromosome 13 onto normal metaphase spreads as shown in Fig. 10b. In this case the origin of the insertion is demonstrated by the segment of the normal chromosome 13 not painted. While the origin of the insertion is demonstrated clearly, the position of the insertion into chromosome 1 cannot be visualised by the reverse painting technique. However, conventional painting of the patient's chromosomes with a normal chromosome 13 paint (DOP-PCR product) allows us to visualise the insertion and determine the chromosome breakpoints (Fig. 10c).

We have found the reverse chromosome painting technique particularly powerful for the analysis of *de novo* chromosomal aberrations. Fig. 11 shows the flow karyotype obtained from a 1 year old boy with hypotonia and delayed motor development whose GTL banded karyotype showed additional material on the short arm of one chromosome 21.

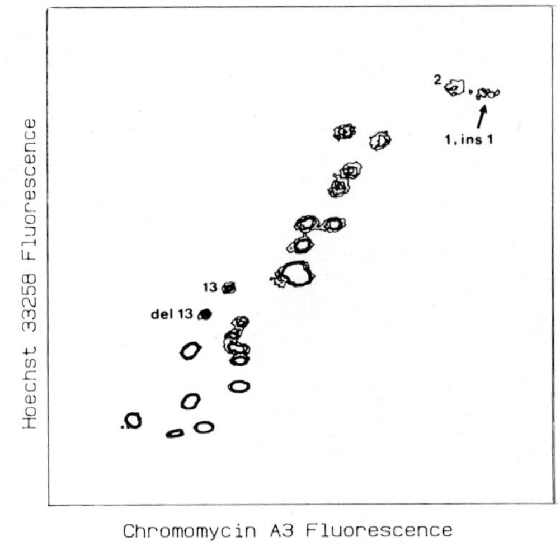

Figure 9. Flow karyotype of a patient with an insertional translocation between chromosomes 13 and 1. The normal and derivative chromosomes involved in the translocation are indicated (reprinted from Carter *et. al.*, 1992).

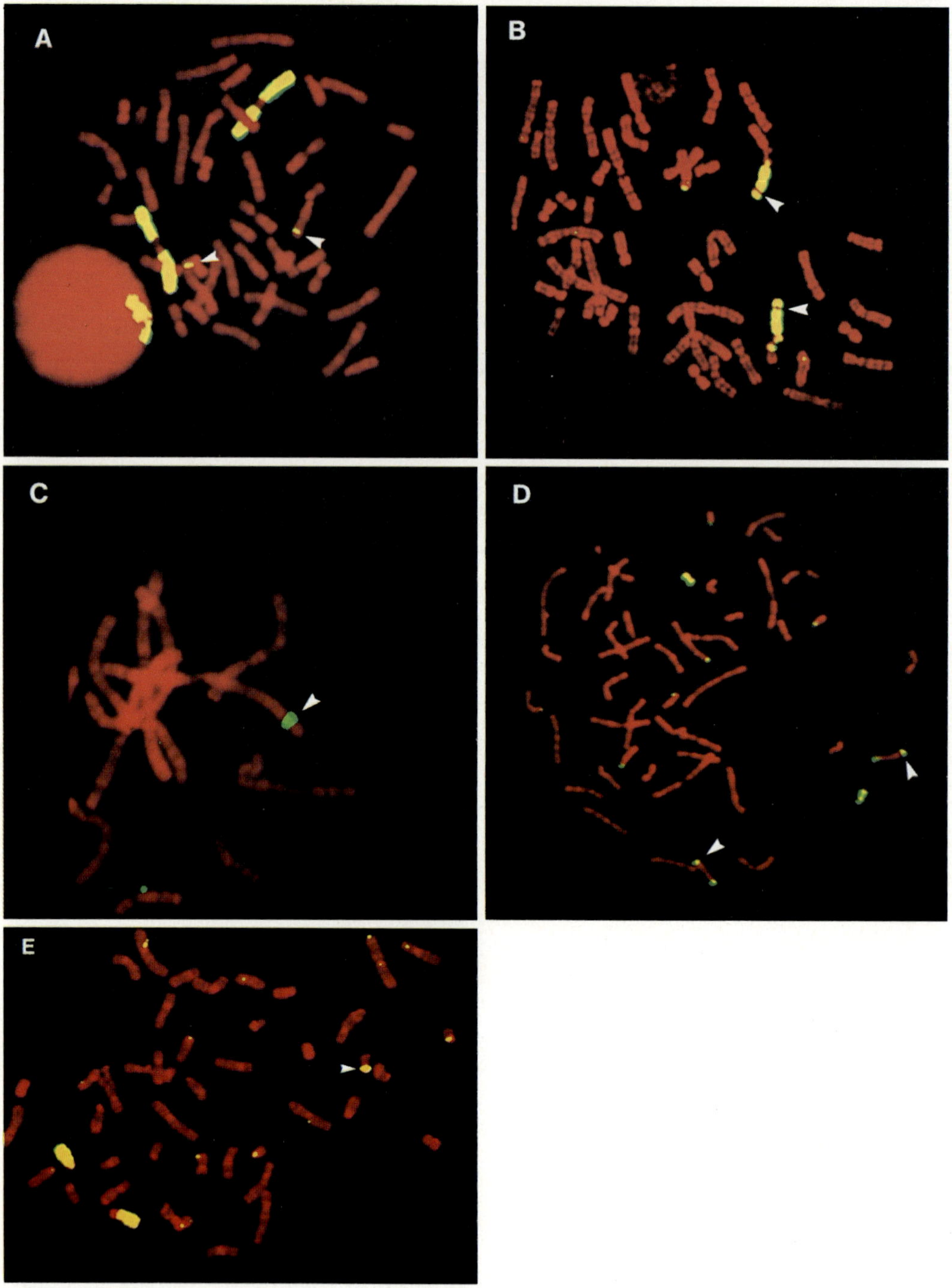

Figure 10. Analysis of abnormal chromosomes using reverse and normal chromosome painting. See text for details of each panel (reprinted from Carter *et. al.*, 1992)

This 21p+ chromosome is resolved from the normal chromosome 21 in the flow karyotype and thus both chromosomes could be sorted separately for reverse chromosome painting. The DOP-PCR product from the normal chromosome 21 from the patient painted onto normal chromosomes showed hybridisation only to chromosomes 21. However, the derivative chromosome 21 paint showed hybridisation on normal metaphase spreads to chromosomes 21, acrocentric chromosome short arms and to the region q32.1->qter of chromosome 14 (Fig. 10d). Based on this analysis, the child has a duplication and partial trisomy of chromosome 14 involving region 14q32.1->qter. The diagnosis of chromosome 14 as the source of the duplication was confirmed by applying a normal chromosome 14 DOP-PCR paint onto the patient's chromosomes. In addition to the two chromosomes 14, signal was detected on the short arm of one chromosome 21 (Fig. 10e). While this conventional painting analysis was able to confirm the chromosome 14 origin of the duplication, only reverse chromosome painting was able to demonstrate the region of chromosome 14 involved.

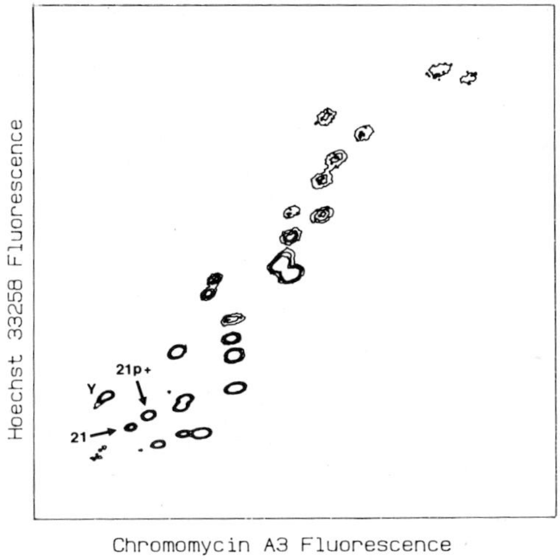

Figure 11. Flow karyotype of a child with additional material on one chromosome 21. The normal and abnormal chromosomes 21 are indicated (reprinted from Carter *et. al.*, 1992).

Conclusion

In this chapter I have attempted to demonstrate the usefulness of flow sorted chromosomes in gene mapping and molecular cytogenetics. For the next few years, the construction of increasingly refined physical maps of each chromosome will be an imperative for the various Human Genome Projects in progress around the world. Flow sorting of naturally occurring translocation products and the mapping of cloned sequences using these chromosomes will continue to have an useful role in the construction of physical maps. Such mapping is facilitated greatly by the ability to design primer pairs from published sequences for subsequent PCR amplification with small numbers of flow sorted chromosomes.

I have also shown how DOP-PCR of small numbers of flow sorted chromosomes can be used for the rapid analysis of abnormal chromosomes. This reverse painting technique has particular application for the analysis of *de novo* chromosomal abnormalities and for the analysis of tumour cell lines where multiple rearrangements make conventional banding analysis difficult and inaccurate. The ability to amplify sorted chromosomes in a general and representative way opens up possibilities for direct cloning as well as coincidence, deletion and breakpoint cloning. Using these techniques it should be possible to saturate rapidly small chromosomal regions with new markers to facilitate the localisation of disease associated genes.

Acknowledgements

The work described here was carried out with support from the Medical Research Council of the UK, the UK Human Genome Project and the Cancer Research Campaign. I would particularly like to acknowledge Professor M.A. Ferguson-Smith; H. Telenius who designed the DOP-PCR primers and protocol; A.H. Pelmear, M.T. Perryman, L.R. Willatt and M.A. Leversha for FISH analyses; the staff of the East Anglian Regional Genetics Service for referral and cytogenetic analysis of patients. I am grateful to G. van den Engh for the program ANALIST used to produce flow karyotype contour plots.

References

Arai N, Nomura D, Villaret D, DeWaal Malefijt R, Seiki M, Yoshida M, Minoshima S, Fukuyama R, Maekawa M, Kudoh J, Shimizu N, Yokota K, Abe E, Yokota T, Takebe Y, Arai K (1989) Complete nucleotide sequence of the chromosomal gene for human IL-4 and its expression J Immunol 142:274-282

Bernheim A, Metezeau P, Guellaen G, Fellous M, Goldberg ME, Berger R (1983) Direct hybridization of sorted human chromosomes: Localization of the Y chromosome on the flow karyotype Proc Natl Acad Sci USA 80:7571-7575

Carter NP, Albertson DG, Smith SJ, Rhodes DA, Cox TM (1991) Assignment of the human ferrochelatase gene (FECH) and a locus for protoporphyria to chromosomes 18q22 Genomics 11:1152-1154

Carter NP, Ferguson-Smith MA, Perryman MT, Telenius H, Pelmear AH, Leversha MA, Glancy MT, Wood SL, Cook K, Dyson HM, Ferguson-Smith ME, Willatt LR (1992) Reverse chromosome painting: a method for the rapid analysis of aberrant chromosomes in clinical cytogenetics J Med Genet (in press)

Carter NP, Hampson RM, Harris RM, Yates JRW, Ferguson-Smith MA (1990) Gene mapping using flow[a]sorted chromosomes Proc Royal Microscopical Soc 1:511-514

Carter NP, Ferguson-Smith ME, Affara NA, Briggs H, Ferguson-Smith MA (1990) Study of chromosome abnormality in XX males using bivariate flow karyotype analysis and flow sorted dot blots Cytometry 11:202-207

Collard JG, de Boer PAJ, Janssen JWG, Schijven JF, de Jong B (1985) Gene mapping by chromosome spot hybridization Cytometry 6:179-185

Connor JM, Ferguson-Smith MA (1987) Essential Medical Genetics Blackwell Scientific Publications

Cotter F, Nasipuri S, Lam G, Young BD (1989) Gene mapping by enzymatic amplification from flow-sorted chromosomes Genomics 5:470-474

Gray JW, Langlois RG, Carrano AV, Burkhart-Schultz K, Van Dilla MA (1979) High resolution chromosome analysis: one and two parameter flow cytometry Chromosoma 73:9-27

Hulten MA, Gould CP, Goldman ASH, Waters JJ (1991) Chromosome *in situ* suppression hybridisation in clinical cytogenetics J Med Genet 28:577-582

Lebo RV, Bruce BD (1987) Gene mapping with sorted chromosomes Methods Enzymol 151:292-313

Lebo RV, Gorin F, Fletterick RJ, Kao F-K, Cheung M-C, Bruce BD, Kan YW (1984) High-resolution chromosome sorting, DNA spot-blot analysis assign McArdle's syndrome to Chromosome 11 Science 225:57-59

Neckelmann N, Warner CK, Chung A, Kudoh J, Minoshima S, Fukuyama R, Maekawa M, Shimizu Y, Shimizu N, Liu JD, Wallace DC (1989) The human ATP synthase beta subunit gene: Sequence analysis, chromosomal assignment and differential expression Genomics 5:829-843

Pinkel D, Landegent J, Collins C, Fuscoe J, Segraves R, Lucas J, Gray J (1988) Fluorescence *in situ* hybridisation with human chromosome-specific libraries: Detection of trisomy 21 and translocations of chromosome 4 Proc Natl Acad Sci USA 85:9138-9142

Schmitz A, Olschwang S, Chaput B, Thomas G, Frelat G (1989) Oncogene detection by enzymatic amplification on flow sorted chromosomes Nucleic Acids Res 17:816

Telenius H, Pelmear AH, Tunnacliffe A, Carter NP, Behmel A, Ferguson-Smith MA, Nordenskjold M, Pfragner R, Ponder BAJ (1992a) Cytogenetic analysis by chromosome painting using degenerate oligonucleotide-primed-polymerase chain reaction amplified flow-sorted chromosomes Genes, Chromosomes Cancer (in press)

Telenius H, Carter NP, Bebb CE, Nordenskjold M, Ponder BAJ, Tunnacliffe A (1992b) Degenerate oligonucleotide-primed PCR (DOP-PCR): general amplification of target DNA by a single degenerate primer Genomics (in press)

CYTOMETRY, ANTITUMOUR DRUGS AND DNA TOPOISOMERASES

P.J. Smith
Medical Research Council Clinical Oncology and Radiotherapeutics Unit
MRC Centre,
Hills Road,
Cambridge, UK, CB2 2QH

SUMMARY

Flow cytometry has a significant role to play in the of study cellular factors that determine the cytotoxic potential of anticancer drugs. This brief overview provides specific examples of techniques for the analysis of: drug uptake, nuclear localization of drug molecules and drug-induced cell death. Emphasis has been placed on the study of anticancer drugs recognised as potent DNA topoisomerase II poisons together with examples of how target enzyme expression can be analysed in single cells with respect to cell cycle age. An important advantage of the single cell analytical techniques described is the ability to distinguish population heterogeneity, often a confounding factor in the analysis of human tumour specimens.

INTRODUCTION

The clinical problem of acquired resistance to initially useful cytotoxic agents remains of major significance. The ability to identify and monitor cellular features that determine or reflect the responses of human tumour cells to cytotoxic agents is a fundamental step in the development of rational approaches to anticancer chemotherapy. Flow cytometric based techniques for the evaluation of drug sensitivity in human tumour biopsies could offer a means of identifying the cellular responsiveness of specified subpopulations.

Cell lines selected for resistance to a single cytotoxic agent are often found to display cross-resistance to a range of structurally unrelated agents, a phenomenon termed multidrug resistance (MDR; Endicott and Ling, 1989). Cells displaying MDR are resistant to cytotoxic agents such as anthracyclines, vinca alkaloids, epipodophyllotoxins and dactinomycin. There is growing evidence that the MDR phenotype is associated with over-expression of a membrane glycoprotein, termed P-glycoprotein, which appears to play a role in drug efflux

NATO ASI Series, Vol. H 67
Flow Cytometry
Edited by A. Jacquemin-Sablon

(Endicott and Ling, 1989). On the other hand cells which are selected for resistance to one topoisomerase II inhibitor (eg an epipodophyllotoxin) sometimes express cross-resistance to other classes of drugs that interact with topoisomerase II, but not to the other agents involved in classical MDR. This "atypical MDR" appears to be mediated through changes in topoisomerase II (Danks *et al.* 1988).

Topoisomerase enzymes control conformational changes in DNA and aid the orderly progression of DNA replication, gene transcription and the separation of daughter chromosomes at cell division. Type II topoisomerases can effect unknotting, decatenation or relaxation of supercoiled DNA molecules by a process of introducing transient double strand breaks through which the strands of an intact helix can pass (Wang, 1988). DNA topoisomerase II is the major type II enzyme found in eukaryotes and is a major target for several classes of apparently unrelated anticancer drugs such as the epipodophyllotoxins (eg VP16-213) and anthracyclines (eg Adriamycin). Such drugs act to poison topoisomerase II by trapping enzyme molecules on DNA leading to the generation of potentially lethal lesions, probably unrepaired double strand breaks. Understanding how drugs generate such complexes and why they are toxic to actively growing cancer cells is a major challenge for the development of modern approaches to chemotherapy (Smith, 1990). Furthermore there is evidence that a reduction in topoisomerase II activity may result from either quantitative or qualitative changes in topoisomerase II. Indeed several cell lines showing atypical-MDR demonstrate a reduced activity of topoisomerase II associated with a decrease in the nuclear level of the enzyme (Pommier *et al.*, 1986; Danks *et al.* 1988).

Importantly, the intrinsic sensitivity of actively proliferating tumour cells to topoisomerase II poisons appears to be a function of the availability of the target enzyme (Liu, 1989; Smith and Makinson, 1989) and the delivery of bioactive drug to the nuclear compartment (Endicott and Ling 1989). Three related areas, all pertinent to the evaluation of the chemoresponsiveness of tumour cells, are considered in this paper, namely: i) the measurement of the intracellular delivery of an anticancer drug and is access to sensitive nuclear targets, ii) the availability of nuclear DNA topoisomerase II and its relationship with the cell cycle, and iii) the measurement of inactivation of human cells by a topoisomerase II poison.

MATERIALS AND METHODS

Cell culture drug treatments: The established small cell lung cancer (SCLC) cell line, NCI-H69/P (designated H69; kindly provided by Dr P R Twentyman, MRC Centre, Cambridge) was grown under standard suspension culture conditions in RPMI medium supplemented with 10% foetal calf serum, glutamine and antibiotics. SCLC cells grew as small aggregates of less than 50 cells and large spheroid formation was prevented by repetitive pipetting of the culture medium every 2-3 days. The brain tumour lines G1 and G19 were derived from grade III and grade IV gliomas respectively and cultured in Ham's F10 medium supplemented with 10% foetal calf serum. The drug etoposide (VP16-213; Vepesid; Bristol Myers Pharmaceuticals, Syracuse, NY) was provided as a 34 mM stock solution. Adriamycin (Farmitalia Carlo Erba, St.Albans, UK) was stored at -20°C at 500mg ml^{-1} in distilled water. Hoechst dye No. 33342 was obtained from CP Laboratories (Bishop's Stortford, UK) and prepared freshly as a 2mM aqueous stock solution.

Analysis of drug uptake by flow cytometry: Following drug treatments H69 cells were resuspended in phosphate buffered saline (PBS) and analysed immediately using a flow cytometer incorporating an Innova 70 argon laser (Coherent) tuned to 488 nm. Signals for forward and 90° light scatter and fluorescence above 630 nm were collected for 1x10^{4} cells using the 90° light scatter parameter as the master signal. Median fluorescence intensities for populations of single cells were derived.

Fluorescence imaging of intracellular drug accumulation by confocal laser scanning microscopy: The confocal imaging system used was a BioRad MRC-600 (BioRad Microscience Ltd, Hemel Hempstead) scanning assembly incorporating a 25 mW argon ion laser and coupled to a Nikon Optiphot II fluorescence microscope with excitation at 514 nm wavelength. Following drug treatments H69 cells were resuspended in PBS and Cytospin preparations mounted in DABCO. Slides were immediately examined under the confocal laser scanning microscope without neutral density filtration, using confocal apertures of 2 to 7 and a x60 oil immersion objective lens. Images of the cells were stored after Kalman filtration on optical cartridge and photographs taken subsequently using Ilford FP4 film.

Flow cytometric analysis of mitoxantrone-DNA interaction by Hoechst 33342-DNA. After the treatment of cultures with mitoxantrone, washed cells were permeabilised by freeze-thawing using the method described previously (Smith, Nakeff and Watson 1985). Cells were resuspended in buffer (10 mM tris-HC1, pH8; 100 mM NaCl; 10 mM EDTA; 1 mg/ml bovine serum albumin) and diluted to 2.5 x 10^{5} cells/ml containing 5 mM Hoechst 33342 prior to flow cytometric analyses. Samples were analysed using a flow cytometer as described previously (Smith *et al.* 1990).

Single cell analysis of nuclear DNA topoisomerase II content: Samples of approximately 1 x 10^{6} cells were washed in buffer and gently permeablised using the technique described previously (Minford *et al.* 1986). Permeablised cells were fixed in 50% methanol (v/v) and agitated for 30 min at 4°C. Fixed cells were washed once in PBS and resuspended in 20 μl anti-topoisomerase II (p170 form) affinity purified antibody (kindly supplied by Cambridge Research Biochemicals, ICI, Alderley Park, UK;1:4 dilution) and held for 1h at room temperature. Antibody treated samples were washed once in PBS and resuspended in 20 μl FITC-conjugated sheep anti-rabbit IgG (1:100 dilution; Sigma Chemicals, whole molecule) and held at room temperature for 30 min. Finally samples were pelleted and resuspended in PBS containing ribonuclease A and propidium iodide (5 µg/ml) to stain nuclear DNA. Controls samples were processed as above but without the first anti-topoisomerase II antibody treatment. The analysis of samples by flow cytometry has been described previously (Smith and Makinson 1989) providing dual fluorescence analysis of cell populations gated for the elimination of debris and cell clumps. The right-angle fluorescence (RF) parameters monitor DNA content (630 nm RF) and second antibody binding (530 nm RF).

Hoechst 33342 uptake viability assay: Exponentially growing cultures, at 5-8 x 10^5 cells/ml, were exposed to VP16 (VP16-213; Vepesid; etoposide; Bristol-Myers Pharmaceuticals, Syracuse, NY, USA) for 24 h prior to resuspension at 2 x 10^5 cells/ml and subsequent treatment with 10 µM Hoechst 33342 at 37°C. Samples were then analysed by flow cytometry as described described previously (Smith *et al.* 1991; Morgan *et al.* 1989; Smith Nakeff and Watson 1985). The emission spectrum of Hoechst 33342 stained DNA extends over a range of 400 nm to beyond 600 nm and the cytometer system used analysed 90° and forward scatter pulse widths, and pulse areas for fluorescence at 410 nm or >630 nm wavelengths.

DRUG UPTAKE, SUBCELLULAR DISTRIBUTION AND DNA BINDING

The natural fluorescence of agents such the anthracycline Adriamycin has enabled uptake studies, using fluorescence microscopy (Fig 1) or flow cytometry (Fig 2), to be performed on intact, viable cells. This approach allows the distribution of drug molecules to be monitored both within intracellular compartments and throughout cell populations. It is possible to use direct imaging and flow cytometric methods to detect intracellular deposits of drugs (eg mitoxantrone) with very low intrinsic levels fluorescence, although it should be appreciated that excitation of such drug molecules may involve charge transfer reactions due to interactions with various cellular ligands.

Mitoxantrone is an anticancer anthraquinone drug and is used clinically in the treatment of non-Hodgkin's lymphomas, acute myeloid leukaemias and advanced breast cancer (Shenkenberg and Von Hoff 1986). Compared with other anticancer DNA intercalating agents mitoxantrone is unusually potent (Epstein and Smith 1988). It is probable that various properties of the drug contribute to its high cytotoxicity, including: the unusual mode of binding to DNA (Kapusinski *et al.* 1981; Lown et al, 1985, Bowden *et al.*, 1985), the induction of topoisomerase related DNA damage (Tewey *et al.* 1984, Smith *et al.* 1990), the long term inhibition of DNA synthesis (Fox and Smith, 1990) and the persistence of the drug within cells (Wallace, Lindh and Durr 1987; Roberts *et al.* 1989). Thus methods for evaluating the nuclear delivery of mitoxantrone and its sequestration are pertinent to the biological action of this anthraquinone and to the detection of MDR.

Unlike Adriamycin, mitoxantrone shows only low levels of intrinsic fluorescence. However using high intensity excitation conditions coupled with sensitive detection devices, it is possible to image anthraquinone deposits within intact cells. Fig 1 compares the intracellular fluorescence of cells treated with either Adriamycin or mitoxantrone. It is clear that both agents distribute within cells, predominantly within the cytoplasm but also labelling some intranuclear structures and nuclear membranes. The confocal photomicrographs indicate that at least a 20-fold greater concentration of mitoxantrone is required to achieve levels of fluorescence equivalent to those in Adriamycin treated H69 cells. This difference in fluorescence intensity can be quantified using flow cytometry (Fig 2).

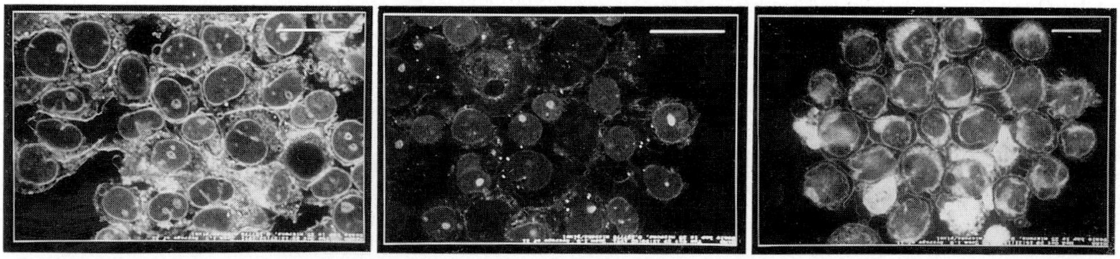

Fig 1 Intracellular fluorescence, detected by laser scanning confocal microscopy, within groups of H69 SCLC cells following a 1 h drug exposure. Left-hand panel: 1 μM Adriamycin. Middle panel: 1 μM mitoxantrone. Right-hand panel: 20 μM mitoxantrone. Scale bars represent 25 μm.

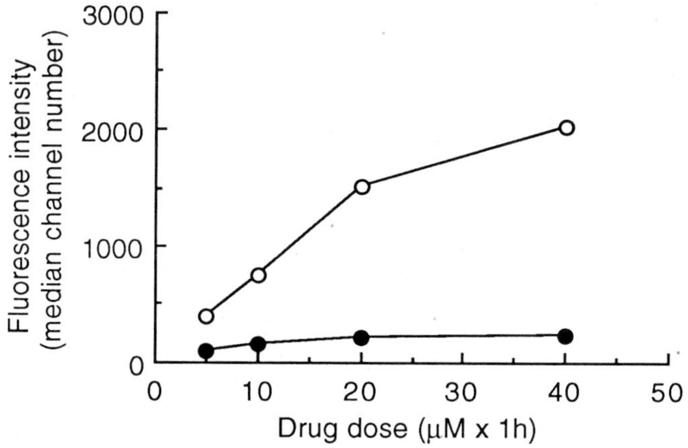

Fig 2 Drug-associated whole cell fluorescence (488nm excitation, 630nm emission), in H69 SCLC populations, detected by flow cytometry. Data are median population values from a single representative experiment. Symbols: ○ , Adriamycin ; ● , mitoxantrone.

To monitor the mode of DNA-mitoxantrone interaction with cellular DNA it is possible to use the ability of intercalating agents to compete with the UV-excitable DNA-specific fluorochrome Hoechst 33342 for binding sites on nuclear DNA (Preisler 1978). A recent adaptation of this general approach is the application of spectral analysis (Smith *et al.* 1990) of Hoechst 33342 fluorescence in single cells to provide preliminary information on the nuclear binding characteristics of mitoxantrone. Fig 3 shows fluorescence quenching determined for violet and red regions of the Hoechst 33342 emission spectrum. The kinetics of staining and rate of spectral shift were consistent with the nuclei being freely accessible to and in equilibrium with the Hoechst 33342 dye. The use of permeabilised cells helps to reduce any potential interference of the nuclear fluorescence signals by cytoplasmic pools of ligand. The results show that the degree of quenching observed in H69 nuclei was dependent on the mitoxantrone dose and the wavelength monitored, with a saturation of quenching potential occurring for the longer wavelength emissions.

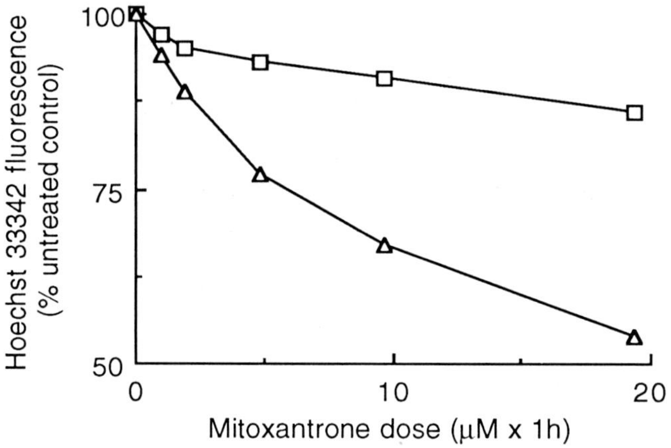

Fig 3 Quenching of Hoechst 33342-DNA fluorescence as a function of mitoxantrone treatment of H69 SCLC cells. Data (derived from Smith *et al.* 1990) are median fluorescence intensities for permeabilised cells, derived from mitoxantrone treated cultures, stained with 5μM Hoechst 33342 and fluorescence emissions monitored at 490 nm (Δ) and 630 nm (□) wavelengths.

The violet- or red-biased fluorescence of Hoechst 33342-DNA have been related to the excitation of ligand molecules at sites where the ligand is bound either within, or more external to the minor groove of the helix respectively (Bontemps, Houssier and Fredericq 1975). The results can be interpreted as reflecting the presence of high affinity sites for mitoxantrone on nuclear DNA, these sites being saturated at low drug doses and representing sites at which the competing Hoechst 33342 molecules are attempting to bind more external to the minor groove.

We have observed that the differential quenching of Hoechst 33342-DNA fluorescence according to the region of the emission spectrum (data not shown) can be exploited to identify cell populations with reduced levels of drug-DNA binding due to the expression of classical multidrug resistance (Endicott and Ling 1989).

INTRANUCLEAR TARGET AVAILABILITY

Consistent with a role for DNA topoisomerase II in chromosome segregation, the level of enzyme is regulated throughout the cell cycle (Heck *et al.* 1988). Whole population assays for topoisomerase II content, such as Western blotting, do not take into account the cell cycle dependency of the enzyme and potential differences between tumour populations due to differences in the distribution of cells throughout the cell cycle.

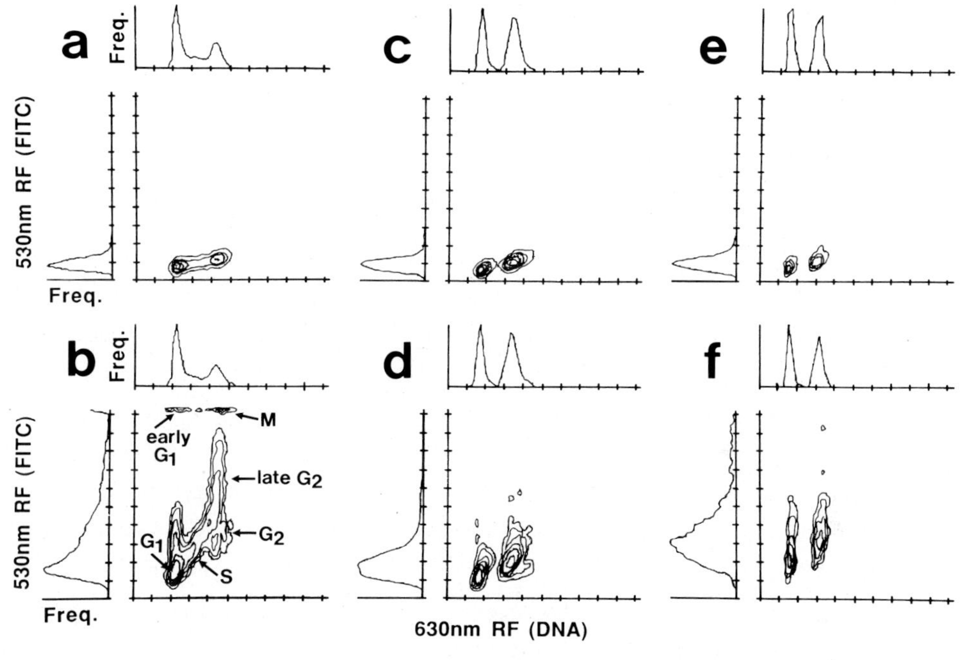

Fig 4 a-f Contour plots of dual fluorescence analyses vs frequency for SCLC cells (H69; panel a and b) and two human brain tumour lines (GL19, panels c and d; GL1, panels e and f). The right-angle fluorescence (RF) parameters monitor DNA content (630 nm RF) and binding of an FITC labelled second antibody (530 nm RF) directed at a bound anti-topoisomerase II antibody. In all cases cell poulations were gated to exclude cell debris and clumps. Opposite each axis is the projected monodimensional frequency distribution for the axis parameter scaled to the maximum channel within the data set. Panels: a, c and e, no first antibody treatment; b, d and f, with first antibody treatment.

Fig 4 a-f show the simultaneous analysis of cellular DNA content and nuclear DNA topoisomerase II by flow cytometry. Fig 4 a and b show typical results for asynchronous H69 SCLC cells. The plots indicate that 530nm right-angle fluorescence is dependent upon the presence of the first antibody and reveal the expected distribution of the p170 form of DNA topoisomerase II (Drake *et al.* 1989) throughout the cell cycle: low levels in G1, levels increasing through S phase, higher levels in G2 with a subset in G2 with very high levels as the cells enter mitosis. Upon cell division, cells in early G1 retain high levels of topoisomerase II but these fall rapidly as cells mature in G1. Fig 4 c-f also shows corresponding contour plots for two glioma cell lines displaying differences in the overall levels of protein whilst continuing to demonstrate the cell cycle dependency of the enzyme. Clearly the differences between tumour cells with respect to cell age and type will have a significant impact of the potential drug sensitivity of particular subpopulations. It is important to note the the profiles shown in Fig 4 represent "steady-state" levels of DNA topoisomerase II and that cells in a "G_0-like" state can modulate topoisomerase II levels dramatically when they undergo growth factor stimulation (Epstein *et al.* 1989).

MONITORING CELL INACTIVATION

In any analysis of human tumour populations it is important to distinguish intact, potentially viable cells from those which are "dead" or at least metabolically disrupted/inert. Dye exclusion assays for cell viability have been used extensively in various areas of cell biology (Dive, Watson and Workman 1990; Ellwart and Dormer 1990) and essentially depend upon the reduced ability of a reporter molecule to penetrate intact membranes or for an agent to be actively excluded from viable cells.

A recently described technique (Smith *et al.* 1991) makes use of the phenomenon that a red shift occurs in the fluorescence emission spectrum of a DNA specific dye (Hoechst 33342) as an increasing number of dye molecules bind to nuclear DNA. Intact, viable cells undergo a time-dependent spectral shift that can be distinguished from the rapid shift observed in cells with damaged membranes by the use of multiparametric flow cytometry. Spectral shift analysis provides a rapid method for assessing the responses of tumour cells to cytotoxic agents. A secondary advantage of the method is that it provides an estimate of the protection of cellular DNA from a model DNA binding agent (namely Hoechst 33342) that participates in the multidrug resistance phenotype (Endicott and Ling, 1989, Morgan *et al.* 1989).

Fig 5 shows a typical analysis for an untreated H69 culture, containing a small burden of dead cells, in which cell-sized particles were analysed for the presence of slowly staining populations (SSPs) or rapidly staining populations (RSPs), the latter comprising non-viable cells which undergo a rapid red shift in the Hoechst 33342-DNA emission spectrum.

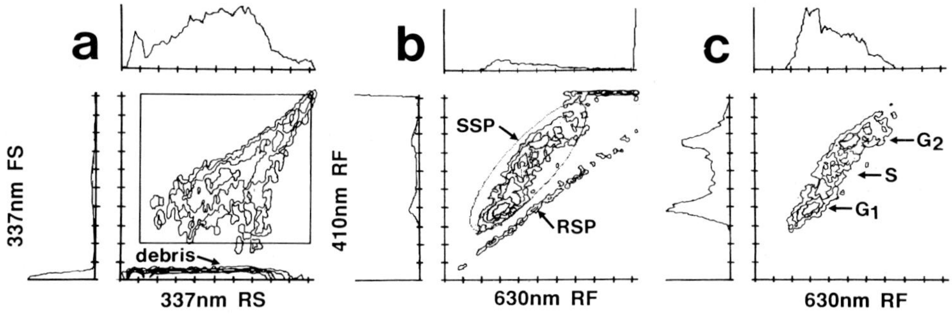

Fig 5 Contour plots showing the differential fluorescence of control H69 cell subpopulations following Hoechst 33342 staining for 15 min. Panel (a) shows the initial gating of cell sized particles on the basis of light scatter characteristics. The fluorescence characteristics of cells within the rectangular gate are displayed in panel (b) indicating the presence of a slowly staining population [SSP] with a high (410 nm RF/630 nm RF) ratio and a rapidly staining population [RSP] demonstrating no significant change in ratio during a 2-60 min incubation period. Panel (c) shows the abstracted SSP within the elliptical gate while the projected monodimensional histogram for 410 nm fluorescence represents the cell cycle distribution of the SSP.

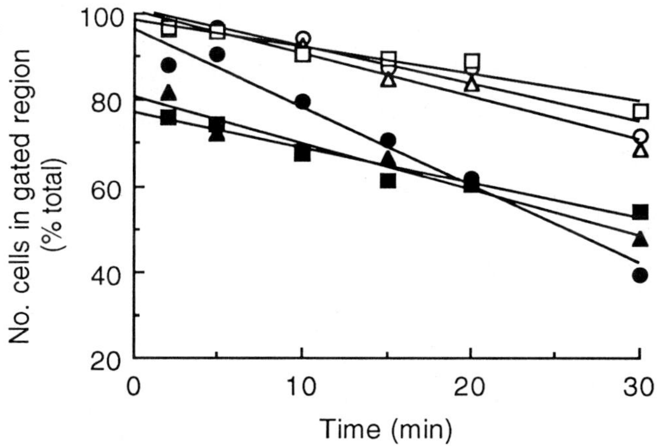

Fig 6 The effects of a 24 h exposure to VP16-213 on the percentage of H69 cells in the viable SSP gated region shown in Fig 5 , monitored during a 30 min uptake period in 10 μM Hoechst 33342. Symbols: ○ , control; △ □ ● ▲ ■ , represent, 0.0625, 0.125, 0.25, 0.5, and 2 μM VP16-213 x 24 h respectively.

The relative frequency of cells in the two populations can be used as an estimate of the loss of viable cells from a treated population. An example of this approach is shown in Fig 6 for the effects of the epipodophyllotoxin topoisomerase II poison VP16-213 (etoposide) on SCLC cells. Extrapolation of the curves yields an estimate of the proportion of inactive cells (eg a loss of 24% of viable cells after an exposure to 2 μM VP16-213 for 24 h). Furthermore, comparison of the results for the cell cycle distribution for viable cells (eg Fig 5c, 410 nm RF projection) with the distribution obtained by conventional ethidium bromide staining of nuclei of all cells within a sample provides an indication of the cell cycle-dependency of cell kill (data not shown).

CONCLUSIONS

An important goal of any rational approach to modern chemotherapy must be the deployment of methods to evaluate the potential responsiveness of tumour cells prior to treatment, together with methods for obtaining early indications of the likely effectiveness of a selected schedule. The examples discussed above show how flow cytometric techniques can contribute to such an approach particularly with respect to the anticancer topoisomerase II poisons. The use of complementary imaging methods, such as laser scanning confocal microscopy, to analyse biologically relevant levels of drugs in intact cells also provides a non-invasive approach which further reveals the relationship between drug accumulation and cellular architecture.

ACKNOWLEDGEMENTS

I thank Professor N M Bleehen for his support, Dr J V Watson for making the flow cytometry facility available, Dr LH Patterson, Department of Pharmacy, Leicester Polytechnic, for his invaluable discussions on anthraquinone pharmacology, and Ms HR Sykes for technical assistance. The samples of glioma cells were kindly supplied by Professor NM Bleehen, Dr J Ramsay and Mrs R Ward.

REFERENCES

Bontemps J, Houssier C, Fredericq E (1975) Physico-chemical study of the complexes of '33258 Hoechst' with DNA and nucleohistone. Nuc Acids Res 2:971-984

Bowden GT, Roberts R, Alberts DS, Peng Y-M, Garcia D (1985) Comparative molecular pharmacology in leukaemic L1210 cells of the anthracene anticancer drugs mitoxantrone and bisantrene. Cancer Res 45:4915-4920

Danks MK, Schmidt CA, Cirtain MC, Suttle DP, Beck WT (1988) Altered catalytic activity of and DNA cleavage by DNA topoisomerase II from human leukaemic cells selected for resistance to VM-26. Biochemistry 27:8861-8869

Dive C, Watson JV, Workman P (1990) Multiparametric analysis of cell membrane permeability by two colour flow cytometry with complementary fluorescent probes. Cytometry 11:244

Drake FH, Hofmann GA, Bartus HF, Mattern MR, Crooke ST, Mirabelli CK (1989) Biochemical and pharmacological properties of p170 and p180 forms of topoisomerase II. Biochemistry 28:8154-8160

Ellwart JW, Dormer P (1990) Vitality measurement using spectrum shift in Hoechst 33342 stained cells. Cytometry 11:239-243

Endicott JA, Ling V (1989) The biochemistry of P-glycoprotein-mediated multidrug resistance. Annu Rev Biochem 58:137-171

Epstein RJ, Smith PJ (1988) Estrogen-induced potentiation of DNA damage and cytotoxicity in human breast cancer cells treated with topoisomerase II interactive antitumour drugs. Cancer Res 48:297-303

Epstein RJ, Smith PJ, Watson JV, Waters C, Bleehen NM (1989) Oestrogen potentiates topoisomerase-II-mediated cytotoxicity in an activated subpopulation of human breast cancer cells: implications for cytotoxic drug resistant in solid tumours. Int J Cancer 44:501-505

Fox ME, Smith PJ (1990) Long-term inhibition of DNA synthesis and the persistence of trapped topo II complexes in determining the toxicity of the antitumour DNA intercalators mAMSA and mitoxantrone. Cancer Res 50:5813-5818

Heck MM, Hittelman WN, Earnshaw WC (1988) Differential expression of DNA topoisomerases I and II during the eukaryotic cell cycle. Proc Natl Acad Sci USA 85:1086-1090

Kapusinski J, Darzynkiewicz Z, Traganos F, Melamed MR (1981) Interaction of a new antitumour agent 1-4-dihydroxy-5,8-bis[[2-[2-hydroxyethyl]amino]-ethyl]amino]-9,10-anthracenedione, with nucleic acids. Biochem Pharmac 30:231-240

Liu LF (1989) DNA topoisomerase poisons as anti-tumour drugs. Ann Rev Biochem 58:351-375

Lown JW, Hanstock CC (1985) High field ^{1}H-NMR analysis of the 1:1 intercalation complex of the antitumor agent mitoxantrone and the DNA duplex $[d(CpGpCpGp)]_2$. J Biomol Struct Dynam 2:1097-1106

Minford J, Pommier Y, Filipski J, Kohn KW, Kerrigan D, Mattern M, Michaels S, Schwartz R, Zwelling LA (1986) Isolation of intercalator-dependent protein-linked DNA strand cleavage activity from cell nuclei and identification as topoisomerase II. Biochemistry 25:9-16

Morgan SA, Watson JV, Twentyman PR, Smith PJ (1989) Flow cytometric analysis of Hoechst 33342 uptake as an indicator of multi-drug resistance in human lung cancer. Br J Cancer 60:282-287

Pommier Y, Kerrigan D, Schwartz RE, Swach JA, McCurdy A (1986) Altered DNA topoisomerase II activity in Chinese hamster cells resistant to topoisomerase II inhibitors. Cancer Res 46:3075-3081

Preisler HD (1978) Alteration of binding of the suprvital dye Hoechst 33342 to human leukemic cells by adriamycin. Cancer Treat Rep 62:1393-1396

Roberts RA, Cress AE, Dalton WS (1989) Persistent intracellular binding of mitoxantrone in a human colon carcinoma cell line. Biochem Pharmac 38:4283-4290

Shenkenberg TD, Von Hoff DD (1986) Mitoxantrone: a new anti- cancer drug with significant clinical activity. Ann Intern Med 105:67-81

Smith PJ (1990) DNA topoisomerase dysfunction: a new goal for antitumor chemotherapy. BioEssays 12:167-172

Smith PJ, Makinson TA (1989) Cellular consequences of overproduction of DNA topoisomerase II in an ataxia-telangiectasia cell line. Cancer Res 49:1118-1124

Smith PJ, Morgan SA, Fox ME, Watson JV (1990) Mitoxantrone-DNA binding and the induction of topoisomerase II associated DNA damage in multi-drug resistant small cell lung cancer cells. Biochem Pharmac 40:2069-2078

Smith PJ, Morgan SA and Watson JV (1991) Detection of multidrug resistance and quantification of responses of human tumour cells to cytotoxic agents using flow cytometric spectral shift analysis of Hoechst 33342- DNA fluorescence. Cancer Chemother Pharmac 27:445-450

Smith PJ, Nakeff A, Watson JV (1985) Flow cytometric detection of changes in the fluorescent emission spectrum of a vital DNA-specific dye in human tumour cells. Exp Cell Res 159:37-46

Tewey KM, Chen GL, Nelson EM, Liu LF (1984) Intercalative antitumour drugs interfere with the breakage-reunion of mammalian DNA topoisomerase II. J Biol Chem 259:9182-9187

Wallace RE, Lindh D, Durr FE (1987) Development of resistance and characteristics of a human colon carcinoma subline resistant to mitoxantrone in vitro. Cancer Invest 5:417-428

Wang JC (1985) DNA topoisomerases. Annu Rev Biochem 54:665-697

Detection of Apoptosis using Fluorescent In Situ Nick Translation

R.R. Jonker, J.G.J. Bauman and J.W.M. Visser
TNO Institute of Applied Radiobiology and Immunology
PO Box 5815
2280 HV Rijswijk
The Netherlands

Introduction

Apoptosis is the most common mechanism of cell death (Kerr et al., 1972). It is also known as Programmed Cell Death and it is dependent on specific enzymatic activity. Apoptosis takes place in embryogenesis, tumor regression, tissue atrophy and cell death induced by cytotoxic T-lymphocytes (Duvall and Wyllie, 1986). It can be induced by a number of factors, including glucocorticosteroids, tumor necrosis factor, lymphotoxin and radiation. The tumor suppressor gene bcl-2 blocks apoptosis as has been reviewed by Williams (1991). Wild-type p53 also functions as a tumor suppressor gene and induces apoptosis (Yonish-Rouach et al. 1991). Sellins and Cohen (1987) proved that, in contrast to necrosis, apoptosis is an active process by showing that induction of gene expression takes place after γ-irradiation. This gene expression is required for DNA fragmentation by endonucleases, an early event in apoptosis. Growth factors promote cell survival which means that they suppress apoptosis in haemopoietic cells (Williams et al., 1990, Nieto et al., 1989). Resting lymphocytes are very sensitive for irradiation and die via apoptosis. This process is known as Interphase Death. Sellins and Cohen (1987) suggested that this mechanism has a role in the elimination of cells that could be potentially harmful to the organism, because they might possibly turn into tumor cells by errors in the DNA repair process. Therefore they called this mechanism a 'better dead than wrong' mechanism that defends the organism against these potentially dangerous irradiated clonogenic lymphocytes.
The main characteristics of apoptosis are degradation of the chromatin in nucleosome-sized fragments (160 base pairs), segmentation of the nucleus, condensation of the cytoplasm and membrane blebbing. Several methods have been described to detect apoptosis, based on the various characteristics of apoptosis. DNA fragmentation can be detected by analyzing DNA from apoptotic cells with agarose gel electrophoresis (Duke et al., 1983). Cells in apoptosis show a specific ladder-pattern of oligonucleosome sized fragments. Dependent on the fraction of cells that are in apoptosis, this method routinely requires more than 10^6 cells and is hard to quantify. Many procedures have been developed based on the differential permeability of living and dead cells to fluorescent dyes. Cells in apoptosis but also cells in necrosis will be scored positive using these procedures. Flow cytometry allows for rapid analysis of this difference. Propidium iodide and ethidium bromide (Swat et al., 1991) are the regular dyes for staining dead cells. 7-Amino-Actinomycin D can also be used (Schmid et al., 1992). Due to nuclear disorganization and shrinkage of the cytoplasm, also the light scatter properties of cells in apoptosis change (Swat et al., 1991) and can be measured flow cytometrically. DNA fragmentation also leads to decreased DNA contents, as assessed by propidium iodide staining (Nicoletti et al., 1991).

NATO ASI Series, Vol. H 67
Flow Cytometry
Edited by A. Jacquemin-Sablon

We developed a method for detection of apoptosis that combines single cell detection with specificity for DNA fragmentation. It facilitates the detection of apoptosis in cell-suspensions as is also applicable for paraffin embedded material. DNA fragmentation in single cells was detected with *in situ* nick translation (ISNT) (Kerem et al., 1983) With ISNT, the fragmentation of DNA is translated into a fluorescent signal. Endogenous endonucleases. These enzymes cut in the linker-DNA between nucleosomes resulting in DNA breaks and digestion of the chromatin into oligonucleosome-sized fragments. We added E. coli DNA polymerase to incorporate nucleotides starting at these DNA breaks. E. coli DNA polymerase consists of two subunits that have exonuclease and polymerase activity respectively. The exonuclease activity enables the enzyme to remove one strand from double stranded DNA starting at single strand breaks to produce single stranded DNA. The polymerase activity leads to incorporation of nucleotides. The enzyme is able to fill in recessed 3' ends of double strand breaks. Biotin labeled dUTP was added to be incorporated and this could be detected with avidin-FITC and measured flow cytometrically. The resulting fluorescence intensity of the cells had a linear relationship with the number of DNA strand breaks. The fluorescence intensity of cells in apoptosis was up to a 100 fold stronger than that of normal cells.

Materials and Methods

Cells:

Single cell suspensions of thymocytes from B/CBA mice were made by pressing the thymus through nylon gauze. Bone marrow cells were obtained by flushing the femurs and tibias with Hanks' Balanced Salt Solution buffered at pH 6.9 with 10 mM HEPES buffer (HH). Cell viability was measured by measuring propidium iodide exclusion, which was routinely higher than 99%. Cells were irradiated with doses of 0-5 Gy gamma rays. During incubations following irradiation cells were cultured in RPMI with 10% fetal calf serum.

DA-1 is a Moloney Leukemia Virus induced murine lymphoid leukemic cell line. It is an IL-3 dependent cell line that is cultured in 10% WEHI-conditioned RPMI medium with 10% fetal calf serum. WEHI is a murine IL-3 producing cell line.

Fixation:

Cells were fixed in 1% formaldehyde in HH for 15 minutes at 0 ^{0}C. This was followed by fixation in 70% ethanol, 0 ^{0}C for one hour. Storage of cells in 70% ethanol was at -20 ^{0}C.

In Situ Nick Translation:

The In Situ Nick Translation procedure was performed in nick translation buffer. This buffer contained 5 mM $MgCl_2$, 10 mM ß-mercapto-ethanol, 50 mM Tris pH 7.8 and 10 μg/ml bovine serum albumin. This buffer also contained 1 unit E. coli polymerase (GIBCO BRL, Paisley, UK), 0.2 nmoles biotin-11-dUTP and 0.2 nmoles of respectively. unlabeled dATP, dCTP and dGTP. Routinely, 200,000 cells were suspended in 12.5 μl buffer and this mix was incubated for 90 minutes at 15 ^{o}C. Afterwards, cells were washed with HH and resuspended in 40 μl staining buffer. This staining buffer consisted of 2.5 μg/ml avidin-FITC (Vector, DCS grade) in 4*SSC, 0.1% Triton-X-100 and 5% nonfat dry milk. Staining was performed for 30 minutes at room temperature. After this, cells were washed again with HH containing 0.1 % Triton-X-100. DNA was counterstained with the fluorescent dye Diamino Phenol Indole (DAPI, 1 μg/ml, 30 minutes at 4°C in HH). The samples were stored on ice and measured on the flow cytometer. In case of simultaneous detection of cell surface markers, digoxigenin-11-dUTP was incorporated instead of biotin-11-dUTP. Detection was performed with anti-digoxigenin (Boehringer Mannheim, Germany).

Flow cytometry

Flow cytometry was performed on the home-built RELACS flow cytometer, equipped with two lasers. The first laser was tuned to 488 nm and was used for the detection of the light scatter properties of the cells and for the measurement of the FITC signal of the cells. This FITC signal was measured with a 530 nm band pass filter (Corion , Hollistone, MA). The second laser was tuned at UV light (350-365 nm) and was used to measure the DAPI fluorescence of the cells. The DAPI signal was measured using two KV408 nm long pass filters (Schott Glaswerke, Mainz, FRG). The data of 6000 cells measured on the flow-cytometer were stored in list mode on a computer disk. During data analysis afterwards, cells with the normal G_0/G_1 DNA content were selected on basis of their DAPI fluorescence. Of these cells, the FITC fluorescence was analyzed. The percentage of cells with a high fluorescence was computed.

Results and Discussion

Radiation induced apoptosis in murine thymocytes and bone marrow cells was used to analyze the applicability of the ISNT procedure in the detection of apoptosis. Figure 1 shows dot plots of irradiated cells and control bone marrow cells. In the irradiated sample, many cells show an increased FITC fluorescence, due to incorporation of biotin-11-dUTP. Similar results were obtained with thymocytes (data not shown).

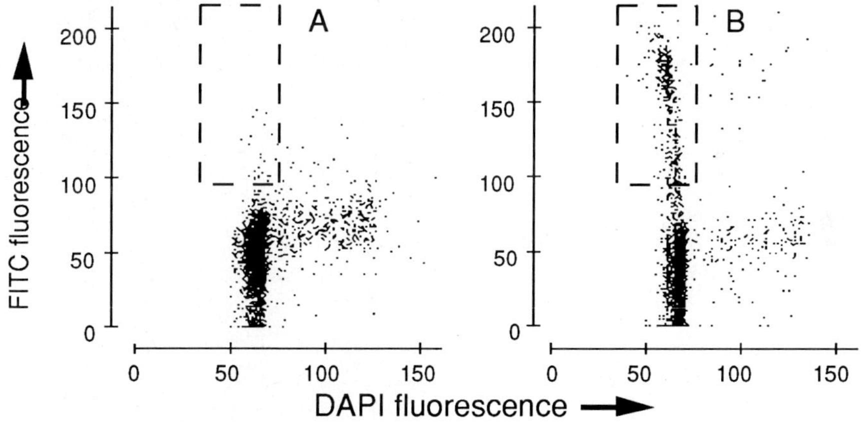

Figure 1: Detection of apoptosis in mouse bone marrow cells.
Murine bone marrow cells were irradiated and incubated in medium for five hours. After incubation the cells were fixed, the ISNT procedure was performed and the cells were measured. Bivariate displays of DAPI fluorescence (DNA contents) versus FITC fluorescence (DNA fragmentation) are displayed. A: Unirradiated control. B: 5 Gy irradiated bone marrow cells, cultured for six hours. The box indicates cells in apoptosis.

The results of ISNT were correlated with gel electrophoresis of DNA from irradiated cells (figure 2). Gel electrophoresis is the standard procedure for detection of apoptosis. DNA from irradiated thymocytes was isolated and analyzed as described by Sellins and Cohen (1987). A clear ladder-pattern of oligonucleosome-sized fragments was obtained from 5 Gy irradiated thymocytes. The smallest band is about 160 base pairs long, which is the size of a nucleosome. The longer bands have a length of multiple nucleosome fragments. Unirradiated, 5 hr-incubated thymocytes showed a weak signal with this pattern. This is due to apoptosis induced by the isolation and culturing procedure and by naturally occurring apoptosis. In the unirradiated control thymocytes, no ladder-pattern was visible. These results are in accordance with the results obtained with the *In Situ* Nick Translation procedure.

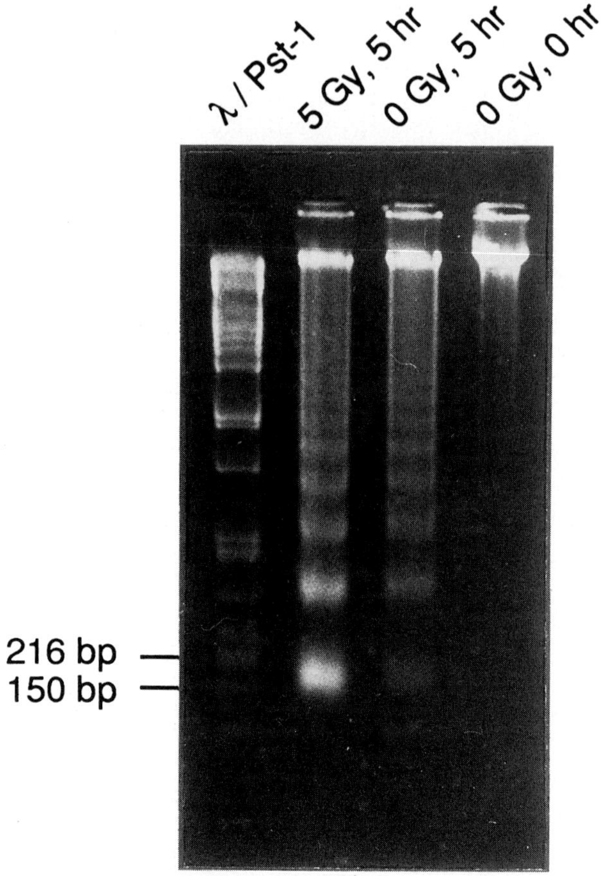

Figure 2: Gel analysis of DNA from irradiated murine thymocytes. Apoptosis in mouse thymocytes was detected by analysis of cellular DNA with agarose gel electrophoresis. DNA was isolated from 1.10^6 thymocytes. Unirradiated thymocytes were fixed shortly after isolation. 5 Gy irradiated and unirradiated thymocytes were fixed after 5 hour incubation at 37 ^{0}C.

Figure 1 also shows that ISNT combines cell cycle analysis and apoptosis detection. It was concluded that a strongly increased FITC signal, corresponding to extensive DNA degradation, together with a slightly decreased DNA content are characteristics of cells in apoptosis. The percentage of cells in apoptosis was computed. In figure 3, this percentage is depicted in relation to the radiation dose. Thymocytes show a stronger response to irradiation than bone marrow cells. They are known to be very radiosensitive. The bone marrow cells contain radiosensitive lymphocytes but also other less sensitive cell types.The high background at 0 Gy is due to the 6 hour incubation procedure. In unirradiated control thymocytes, only 0.7 % of the cells were in apoptosis.

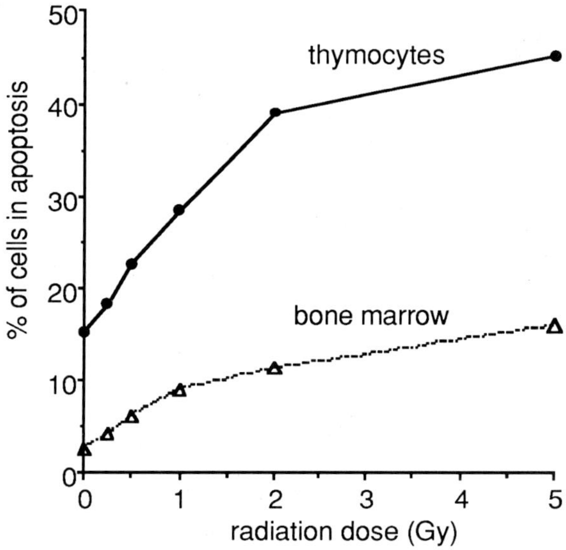

Figure 3: **Detection of apoptosis in mouse thymocytes and bone marrow.** Murine thymocytes and bone marrow cells were irradiated and incubated in medium for six hours. After incubation the cells were fixed, the ISNT procedure was performed and the cells were measured. The percentage of apoptotic murine bone marrow cells and thymocytes as determined by the ISNT procedure is plotted as a function of the radiation dose.

Radiation-induced apoptosis could also be detected in cell lines as can be seen in figure 4, where histograms are depicted of the FITC fluorescence of DA-1 cells. DA-1 is an IL-3 dependent cell line. DA-1 cells were irradiated and afterwards cultured in absence of IL-3. After 0.5 and 5 Gy, a clear positive population of brightly fluorescent cells could be seen. The fluorescence distribution of unirradiated control cells is comparable to the distributions of negative cells in the irradiated samples. These are normal living cells without extensive DNA fragmentation. From these results, it was concluded that ISNT results in a clear discrimination of normal living cells and cells in apoptosis. DNA fragmentation of cells in apoptosis is extensive and fast. Not many cells show an intermediate fluorescence. Therefore, ISNT offers an objective criterion for detection of apoptosis. In figure 5 the percentage of DA-1 cells in apoptosis is shown as a function of the radiation dose. It is clear from this figure that In Situ Nick Translation is a sensitive technique. Down to 0.25 Gy, an increased number of cells in apoptosis in DA-1 cells was measured, when cells are incubated without IL-3. There is a clear relationship between radiation dose and percentage of cells in apoptosis as measured with ISNT. Without IL3, DA-1 cells are very sensitive to radiation induced apoptosis. At 5 Gy, 62% of the DA-1 cells die via apoptosis after 5 hours. However, when IL-3 is added to the medium, apoptosis was almost completely blocked. Only 2.9% of the cells is in apoptosis after a dose of 5 Gy in the presence of IL-3. These data show that radiation induced apoptosis can be strongly inhibited with growth factors.

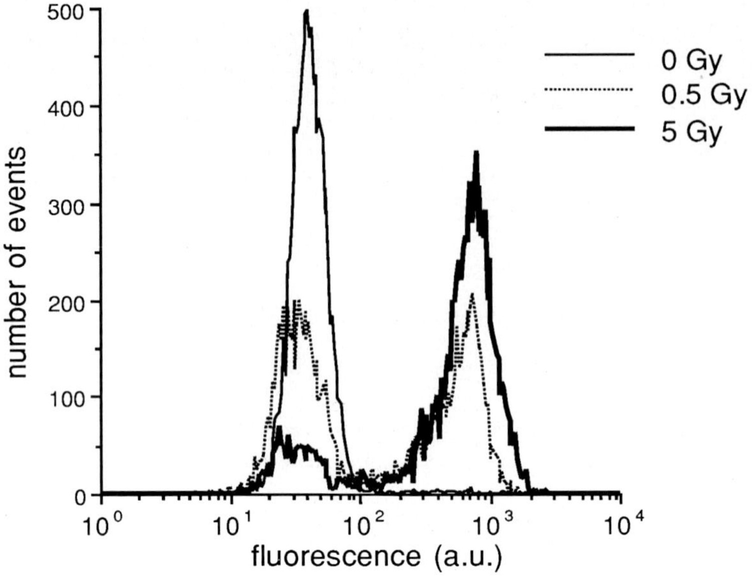

Figure 4: **Detection of apoptosis in DA-1 cells.**
DA-1 cells were irradiated and incubated for 5 hours without IL-3. The FITC fluorescence histograms are depicted of 0.5 Gy and 5 Gy irradiated DA-1 cells and of control cells.

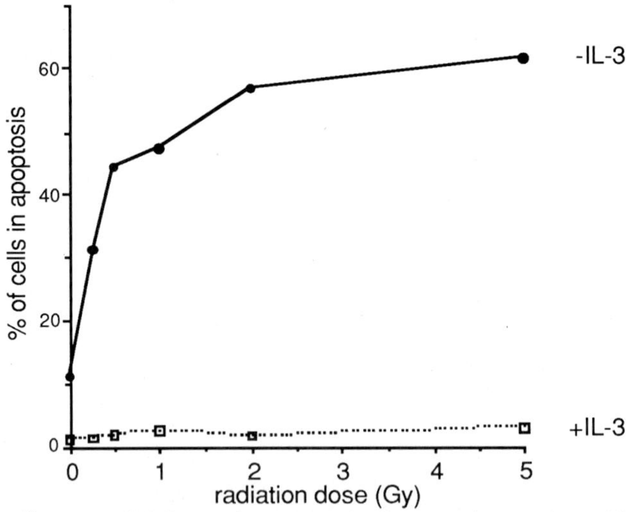

Figure 5: Inhibition of apoptosis in DA-1 cells by IL-3.
DA-1 cells were irradiated and incubated for 5 hours with or without IL-3. The percentage of apoptotic DA-1 cells as determined by the ISNT procedure is depicted in relation to the radiation dose and the presence or absence of IL-3.

Simultaneous flow cytometric detection of apoptosis and cell surface markers is important to determine which cell types die via apoptosis. We used the method of Bayer et al (1990) for labeling of cell surface markers in such a way that this labeling survives fixation and DNA labeling. The *In Situ* Nick Translation method for the detection of apoptosis was used. Murine bone marrow was irradiated, incubated in culture medium for 5 hours and labeled with the monoclonal antibody 15-1.1. MCA 15-1.1 detects an unknown antigen that is expressed brightly on monocytes, dull on granulocytes and is absent on lymphocytes.
Biotinylated Goat anti-Mouse antibody was used to detect MCA 15-1.1. Instead of biotin-labeled dUTP, digoxigenin labeled dUTP was incorporated. Staining was with streptavidin-phycoerythrin (PE) for MCA 15-1.1 and anti-digoxigenin-FITC for digoxigenin. Figure 5 shows the FITC and PE fluorescence of murine bone marrow. The histogram of the PE distribution clearly shows a positive (dull) peak, some highly fluorescent cells and a negative population the lymphocytes. The FITC histogram shows a large negative peak and a positive peak with 32% of the cells. Of the MCA 15-1.1 negative cells, 49% was positive for apoptosis. Of the MCA 15-1.1 positive cells, only 14% was positive for apoptosis. So, the MCA 15-1.1 negative cells, mainly lymphocytes, showed to be more sensitive for radiation induced apoptosis than MCA 15-1.1 positive cells, mainly granulocytes and monocytes. This is in agreement with the radiosensitivity of lymphocytes that die of apoptosis.

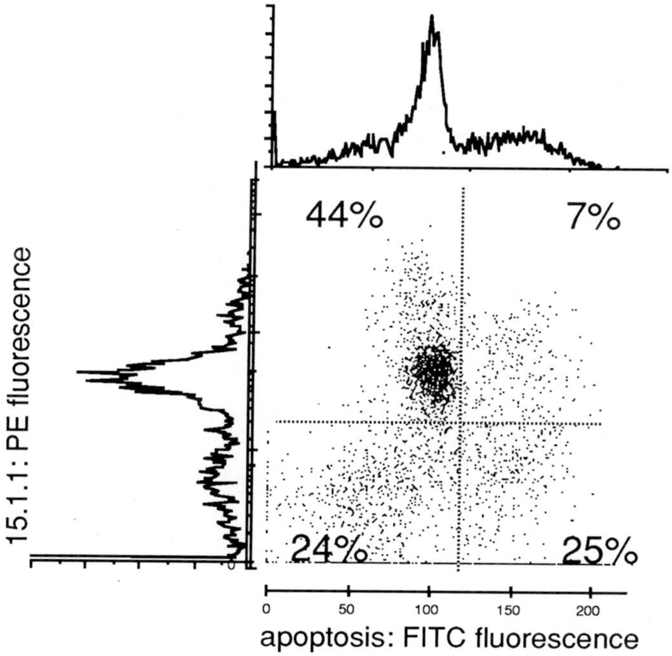

Figure 6: **Simultaneous detection of apoptosis and cell surface marker.** Apoptosis was detected with the ISNT procedure and cells in apoptosis were labeled with FITC. The antibody 15.1.1 was labeled with phycoerythrin.

Summary

In this study we demonstrated the use of *in situ nick translation* for flow cytometric detection of DNA fragmentation in irradiated cells. This fragmentation is an early characteristic of apoptosis and Programmed Cell Death. We showed that this procedure specifically labels cells with extensive DNA fragmentation. This fragmentation precedes and is associated with DNA loss, which was also measured. The resulting fluorescent signal is clearly distinguishable from that of normal cells. We showed that the results of the ISNT procedure correlate well with existing procedures. The ISNT results correlated well with DNA ladders obtained with gel electrophoresis. As is demonstrated in this study, In Situ Nick Translation offers a quantitative and objective method to detect apoptosis. It is a very sensitive and specific method and allows for the detection of low percentages of cells in apoptosis. This procedure allowed for correlation of cell cycle data with data on apoptosis, since a good DNA-contents histogram could be obtained simultaneously. The method also allows for simultaneous detection of cell surface markers and apoptosis. The *in situ nick translation* method for detection of cell death in individual cells is very useful in studies to determine the response of cells to induction and inhibition of cell death.

References

Bayer JA, De Vries P, Herweijer H and Bauman JGJ (1990) The use of E. coli exonuclease III to generate single stranded DNA in BrdUrd cell-cycle analysis permits simultaneous detection of cell surface antigens. J. Immunol. Methods 132: 13-24.

Duke RC, Chervenak R and Cohen JJ (1983) Endogenous endonuclease-induced DNA fragmentation: An early event in cell mediated cytolysis. Proc. Natl. Acad. Sci. USA 80: 6361-6365.

Duvall E and Wyllie AH (1986) Death and the cell. Imm. Today 7: 115-119.

Kerem BS, Goitein R, Richler C, Marcus M, Cedar H (1983) In situ nick translation distinguishes between active and inactive X chromosomes. Nature 304, 88-90.

Kerr JFR, Wyllie AH, Currie AR (1972) Apoptosis: a basic biological phenomenon with wide ranging implications in tissue kinetics. Br. J. Cancer 26: 239.

Nicoletti I, Migliorati G, Pagliacci MC, Grignani F, Riccardi C (1991) A rapid and simple method for measuring thymocytes apoptosis by propidium iodide staining and flow cytometry. J. Imm. Methods 139: 271-279.

Nieto MA, Lopez Rivas A (1989) IL-2 protects T lymphocytes from glucocorticoid-induced DNA fragmentation and cells death. J. of Immunology 143: 4166-4170.

Schmid I, Krull WJ, Uittenbogaart CG, Braun J, Giorgi JV 1992. Dead cell discrimination with 7-Amino-Actinomycin D in combination with dual color immunofluorescence in single laser flow cytometry. Cytometry 13: 204-208.

Sellins KS, Cohen JJ (1987) Gene induction by γ-irradiation leads to DNA fragmentation in lymphocytes. J. of Immunology 139: 3199-3206.

Swat W, Ignatowicz L, Kisielow P (1991) Detection of apoptosis of immature $CD4^+8^+$ thymocytes by flow cytometry. J. Imm. Methods 137: 79-87.

Williams GT, Smith CA, Spooncer E, Dexter TM, Taylor DR (1990) Haemopoietic colony stimulating factors promote cell survival by suppressing apoptosis. Nature 343: 76-79.

Williams GT (1991) Programmed cell death: Apoptosis and oncogenesis. Cell 65: 1097-1098.

Yonish-Rouach E, Resnitzky D, Lotem J, Sachs L, Kimchi A, Oren, M (1991) Wild-type p53 induces apoptosis of myeloid leukaemic cells that is inhibited by interleukin-6. Nature 352, 345-347.

PART IV

DATA MANAGEMENT

EXPERT SYSTEMS AND

CELL SORTING TECHNIQUES

FLOW CYTOMETRY DATA ANALYSIS. BASIC CONCEPTS and ANALYSIS of MONO-DIMENSIONAL IMMUNOFLUORESCENCE HISTOGRAMS

James V. Watson
MRC Clinical Oncology Unit,
The Medical School,
Hills Road,
CAMBRIDGE, CB2 2QH,
England.

Flow cytometers produce very large numbers of numbers. The majority of investigators using these instruments are biologists. Unfortunately, handling and interpreting numbers is not generally a "strong point" for biologists. And yet, this is the most important single aspect of flow cytometry (FCM) as a series of numbers (ie. data) is meaningless until it is interpreted within the context of the biological experiment. This is the first problem which applies particularly where there are large numbers of numbers that have to be represented by some "average" value as a summary. Moreover, there will always be "dispersion" about the average in biological systems and this too must be represented by some form of measurement of that dispersion.

The most commonly asked question in FCM data analysis is whether there is more than one population in a sample. That, after all, is why FCM is used as the single cell measurement attributes enable "clusters" to be identified. This applies particularly to the two most commonly used types of assays namely, immunofluorescence and DNA histogram analyses. The deceptively simple question "how many clusters ?" is very easily answered if no two populations overlap. This is the second major problem because many populations overlap as evidenced by the G1/S and S/G2+M interfaces in a DNA histogram. This problem is also encountered in many immunofluorescence analyses and the immunologists, very worthily, have sought to improve their reagents to "pull apart" more effectively populations which exhibit overlap. Addition of extra parameters (multi-parameter analysis) can help with this problem. However, there may

NATO ASI Series, Vol. H 67
Flow Cytometry
Edited by A. Jacquemin-Sablon

be overlap throughout the biology in which case no improvement in reagents or addition of extra parameters will enable an absolute discrimination to be made. This scenario can only be resolved in probabalistic terms where we would need to know the means and variances of each putative population so that statistical analysis can be applied. This, to date, has received relatively little attention considering its importance.

The third major problem is a whole class of problems. These concern the means by which the large numbers of numbers in multi-dimensional data sets are handled, how sub-sets (clusters) are defined and how correlations and cross-correlations are preserved within multi-dimensional data matrices so that subsequent statistical analysis, if required, can be applied to those clusters. In this chapter the first two classs of problems will be considered and multi-variate plus cluster analysis and expert systems will be covered in subsequent chapters.

BASIC CONCEPTS

Large numbers of numbers must be handled by some sort of summary system because just supplying the raw data, say 10000 fluorescence and/or light scatter recordings from a flow cytometric analysis run, would be unintelligible. This section was included to re-familiarise the reader with some very basic number handling concepts and to set the scene for converting numbers into information.

1) **Central tendancy; mean, median and mode** The most commonly used measurement of central tendancy is the arithmetic mean of the population. This is obtained by summing all the measurments in a particular class of measurement, x, and dividing by the number of observations, N, where $\bar{x}$, the arithmetic mean, is given by the following formula,

$$\bar{x} = \sum_{k=1}^{k=N} (x_k)/N$$

The symbol Σ represents the summation of all the k x values from 1 to N.

There are two further methods of expressing central tendancy, the mode and the median of the distribution. The mode is the point in the distribution where the frequency is at a maximum and the median is the point where half the area lies to each side. In symmetrical distributions the mean, mode and median are the same point but, this is not true for "skewed" distributions and the relationship between these parameters is shown in figure 1.

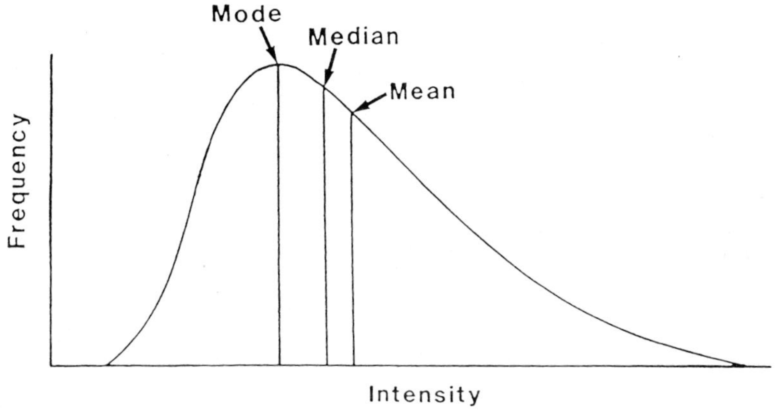

Figure 1. Relationships between the mean, median and mode in a skewed distribution.

It is particularly important to make these distinctions in flow cytometry where some cells may exceed the dynamic range of the analogue-to-digital conversion step and be scored "off-scale" and appear in the last channel of the histogram. The problem is that we do not know how far "off-the-end" they are thus, the true mean cannot be calculated and the median **must** be used as the measure of central tendancy. Figure 2 shows a number of flow cytometric distributions for various cell surface determinant assays in which all cells should have been labelled and where the mean, median and mode have been calculated and displayed on each individual panel. The potential problems are seen in panels A, C and D. In panel A some cells were scored with either zero or very little fluorescence and the mode was calculated as 16. In panels C and D there

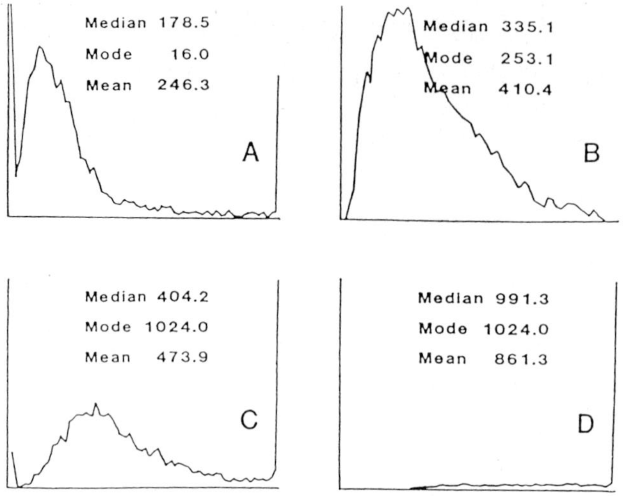

Figure 2. Examples of flow cytometric distributions for various cell surface determinant assays where the mean, median and mode are recorded on each individual panel.

are very bright cells which are off scale at the right and in each distribution the mode is 1024 which is meaningless. Also, the means of these distributions are underestimated as it is not possible to know how far off-scale the off-scale cells would have been recorded if this had been possible. The only measure of central tendancy for the data shown in panels A, C and D is the median. As soon as more than 50% of the population is "off-scale" there is no measurement of central tendancy which can be used.

2) **Absolute distance between points** Measuring the absolute distance between two points is not as immediately straight forward as it might seem even though we are dealing with Euclidean as opposed to relativistic distances. Consider two points marked on a sheet of paper. We can place a ruler between these points so that the "zero" mark is adjacent to one point and read off the number on the scale adjacent to the second point. If we now turn the sheet of paper through 180^0 and repeat the excercise we will get a very similar, but not an identical, answer. This begs two further questions. Why might the results be very slightly different and where is ZERO ? The answer to the first question is that there will always be an error associated with your placement of the "zero" mark

against one point and your reading of the scale adjacent to the second point. The answer to the second question is that ZERO is exactly where you want it to be. There is no such reality as ZERO, it is a convention just as the whole of any measuring system is a convention.

At first sight the introduction of a discussion about where ZERO might be may seem a little esoteric, unecessary or even futile but is very important in flow cytometry. We do not directly measure whatever commodity we are measuring, we measure light. This strikes the photocathode in the photomultiplier which then emits electrons in direct proportion to the number of photons striking the cathode. This is the point at which the light flux is transduced (changed) into the electronic signal. The electron flux emitted from the photocathode is then amplified through a dynode chain within the photomultiplier to give a current thence to produce a voltage which is amplified. Modern electronics are capable of amplifying a signal by many orders of magnitude depending on the various settings you happen to choose and instruments can have a total dynamic range, in terms of number of molecules measurable, of from 10^2 - 10^{16} per cell. This range of 14 orders of magnitude must be viewed through a "window" which, even with log-amplifiers, is still only 10^4 wide. Hence, ZERO on the particular viewing window you are using is an arbitrary point dictated by the assay you are performing which is somewhere within the 10^2 to 10^{16} total possible dynamic range available to you.

How then can we measure the absolute distance between two points within this arbitrary measurement system particularly as we do not know where ZERO happens to be located ? The square of a positive number is positive and the square of a negative number is also positive. Thus, we obtain the absolute distance between two points by subtracting one from the other (it doesn't matter which from which), squaring the result then taking the square-root. This gets round the problem of where ZERO might happen to find itself, it could be on the left, or the right or up or down, or in or out, or anywhere, it is no longer of any consequence. This relationship, which is extremely simple and of fundamental importance, is as follows

$$D_a = \sqrt{(x_1 - x_2)^2}$$

where D_a is the absolute distance between points and where x_1 and x_2 are the two measurements in a mono-dimensional data space.

Let us now suppose we are working in a two dimensional data space of (x-y) coordinates. The absolute distance between any two points with coordinates (x_1, y_1) and (x_2, y_2) respectively is given by,-

$$D_a = \sqrt{(x_1 - x_2)^2 + (y_1 - y_2)^2}$$

This should have a familiar ring to it, Pythagoras of Samos did it first in BC 600 or there abouts and it should now be intuitively obvious that we can generalize this relationship for any number of sets of different measurements. Moving up to the 3-dimensional coordinates of the (x-y-z) data space we get,-

$$D_a = \sqrt{(x_1 - x_2)^2 + (y_1 - y_2)^2 + (z_1 - z_2)^2}$$

When we go beyond 3-dimensional coordinates we enter Euclidean hyperspace which is just a fancy term (but it sounds impressive) to describe sets of 4 or more coordinates. It is not something which is easily visualised as we live in, and are only familiar with, 3-dimensional space however, extension of the Pythagorean relationship to hyperspace is perfectly valid and is used in some cluster analysis techniques.

3) **Dispersion and its representation** Just as we must have some way of summarizing the average, say the arithmetic mean of a set of numbers, in order to make those numbers "handlable" so we must also have some way of

measuring and representing dispersion in that set of numbers. Clearly, we cannot give each individual dispertion and we have to represent this by some sort of "average" dispertion value. Such a measurement of "spread" is required whenever data are distributed as there has to be some way of expressing how far the overall characteristics of the population differ from the average.

The simplest measure of dispersion is given by the mean deviation, d, where,-

$$d = \Sigma|x - \bar{x}|/N$$

The expression $|x - \bar{x}|$ represents the subtraction of the mean, $\bar{x}$, from a given point, x, to give the deviation of this point from the mean and the symbol Σ represents a summation of the deviations of all the points from the mean. The two vertical lines, $||$, in the expression means ignore the sign after the subtraction and always regard the answer as positive. I'm sure you will recognise that, in this context, $|x - \bar{x}|$ is equivalent to $\sqrt{[(x - \bar{x})^2]}$, the absolute distance between the mean and the point x. The mean deviation is obtained by dividing the summation by N, the number of observations.

The mean deviation gives equal relative weight to all deviations from the mean whatever their magnitudes. A further measure of spread, which gives greater weight to the larger deviations, is given by the mean squared deviation which is the VARIANCE, σ^2, the formula for which follows.

$$\sigma^2 = \Sigma(x - \bar{x})^2/N$$

Thus, the variance is defined as the overall average squared distance (deviation) of all the points from the mean of all the points.

There are a number of ways in which variance is defined depending on the type of problem and distribution being considered. This particular definition of variance relates to observations or measurments where deviations from the mean are likely to be equal either side of the mean.

The standard deviation, s, is obtained from the variance by taking the square-root hence, it is the root mean squared deviation.

$$s = \sqrt{\Sigma(x - \bar{x})^2 / N}$$

4) **Student's t** This statistical test was the invention of W.S. Gosset, a research chemist and not a mathematician, who very modestly published under the pseudonym of "Student" (1908). It was later consolidated by Fisher (1925a,b) and enables an assessment of the significance of differences between two distributions with different sample sizes to be made. It is defined formally as the absolute difference between means divided by the standard error of difference which is given by the following formula.

$$\text{Student's } t = \frac{|\bar{x}_1 - \bar{x}_2|\sqrt{N}}{\sigma}$$

where the expression $|\bar{x}_1 - \bar{x}_2|$ represents the absolute distance between the means (irrespective of sign) which is divided by the standard error of difference of the mean of the population, $\sigma/\sqrt{N}$, which places $\sqrt{N}$ on the top of the fraction.

IMMUNOFLUORESCENCE DATA ANALYSIS

The time-honoured method for analysis of mono-dimensional immunofluorescence data has been to run a control sample, place a delimiter "by eye" at the upper boundary of the control and subsequently to score all cells to the right of this delimiter in test-samples as "positive". This approach is based on asking the following questions.

1) Is there more than 1 population ?

2) If there is more than 1, how many cells are in each ?

Results are then presented as "this sample contained 15.3% positive cells". If the two distributions being compared are completely separated there is obviously no problem with this approach. However, if there is overlap between two populations, which is more frequent than seems to be generally appreciated by manufactures, this method is not only hopelessly inadequate but also potentially dangerous particularly in the clinical context where results may influence therapeutic decisions. We will see later that this proceedure can produce errors of up to one order of magnitude under some conditions. Moreover, when results are presented in this manner we do not know either how far apart the "negative" and "positive" might be or the dispersion in either population. Hence, the informed reader has absolutely no information which would enable an objective assessment to be made as to the validity of the assertion that 15.3% of the population is positive. I submit we should be presenting results in the following type of format. "The analysis was compatible with two populations whose means were separated by a factor of 1.75 where the predicted positive fraction constituted 15.3% of the total at the 95% confidence interval". We can now get a "feel" for how different the populations actually were and, obviously, we would be more confident about interpreting the result of 15.3% positive if this had been obtained at the 99% confidence interval.

However, in order to effect this approach we must restructure our thinking to include the following.

1) Is there more than 1 population ?

2) If there is more than 1, how far "separated" are they ?

3) What is the significance of that separation ?

4) If the populations are significantly separated what are the estimates of the relative proportions of cells in each ?

5) What significance can we attach to the estimated proportions ?

In order to present results in this manner we need to know the mean and variance of each population and then perform significance testing and the analytical methods described in this section were developed to estimate the proportions of cells in two populations embedded within an overlapping histogram of the type generated by flow cytometric immunofluorescence analysis where the labelling is "weak". Both methods allow subsequent objective statistical assessment of the validity of the extimates to be made using Kolmogorov-Smirnov statistics, Student's t and $\Sigma\chi^2$. The first proceedure was designed for data sets obtained with linear amplifiers (constant CV analysis) and the second was designed for data with constant variance as obtained with log amplifiers.

As a prelude to both analytical proceedures the K-S statistic, which assesses the statistical difference between two cumulative frequency curves, is applied to test if the control and labelled samples are different. The two data sets used for illustrative purposes were chosen for a number of reasons. Firstly, in both it was considered impossible to place a vertical delimeter by eye with any degree of reliability. Secondly, the linear amplifier generated data have some local historical interest. They were the first produced in Cambridge with a flow cytometer (in 1976) using Milstein's IgM monoclonal antibody (then designated Ig5.C3(H6/31)) which was specifically directed towards "IgD-like" molecules (Pearson, Galfre, Ziegler and Milstein, 1977). Thirdly, the constant variance data obtained with log amplifiers were chosen, for reasons which will become apparent later, as the control histogram conformed very closely to a normal distribution with skew and kurtosis of -0.18 and 2.86 respectively.

These data are shown together in figure 3 with the linear and log amplifier data in panels A and B respectively and where the control data sets were obtained with an irrelevant antibody prior to staining with the

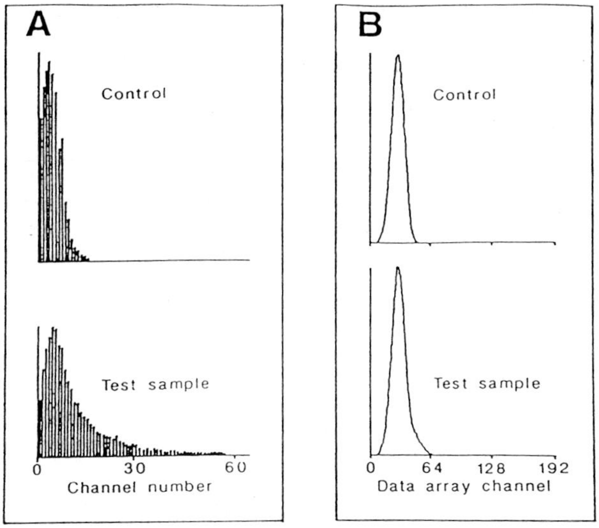

Figure 3) Panels A and B respectively show immunofluorescence data obtained with linear and log amplifiers. In both panels the controls and test sample data are displayed as the top and bottom histograms respectively. The scale units on the abscissa of panel B are styled "data array channel" where 64, 128, 192 and 256 correspond to 10^1, 10^2, 10^3 and 10^4 respectively.

fluorescenated second antibody. The control and test samples are marked on each panel for clarity. The scale units on the abscissa of panel B are styled "data array channel" where 64, 128, 192 and 256 correspond to 10^1, 10^2, 10^3 and 10^4 respectively.

The respective cumulative frequency distributions of these histograms are shown in panels A and B of figure 4 where the maximum difference between control and test sample, both normalised to unity, was 0.3813 at

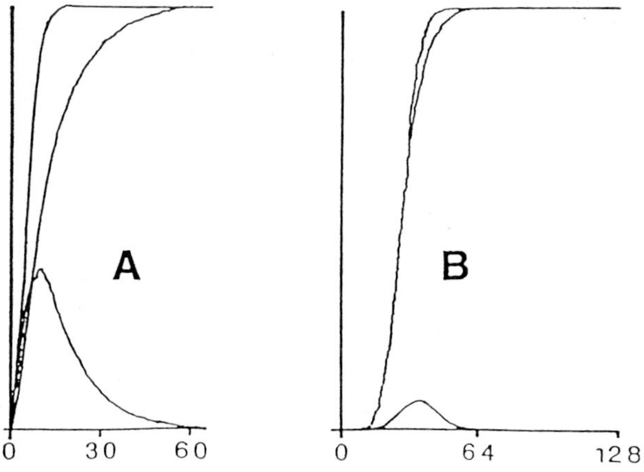

Figure 4) Cumulative frequency distributions of the data shown in figure 3. The histogram in each panel shows the difference between the control and test sample data.

channel 8 in panel A. There were 9070 and 10229 cells in the control and test sample respectively and the difference was associated with $p < 0.001$. A comparably analysis for the panel B data gave a maximum difference of 0.0655 at channel 37 with 8976 and 8570 cells in the control and test sample respectively and this difference was associated with $p < 0.05$. However, although both test distributions would be regarded as significantly different from their controls the analysis does not give us a quantitation of the proportions which constitute the labelled fractions which is what we want to know.

1) **Constant CV analysis** In this section we consider a method of analysis for deconvoluting histograms with two overlapping distributions obtained with a linear amplifier to estimate the proportions of cells in the two populations. The coefficient of variation (CV) of such distributions is constant and as a consequence the analogue to digital conversion (ADC) will produce a positive skew in the data. If the data were in fact normally distributed they will end up looking like a log-normal distribution due to the positive skewing effect. What we see is the data output from the instrument and we have to ask the question "What is the real biological distribution from which the output data was obtained ?". This question must be addressed before we can start fitting a distribution to the data. Schuette et al (1983) also considered this problen in resolving closely spaced DNA histogram peaks with low CVs (< 0.05) in delineating aneuploid tumour components close to a diploid peak. Their method used an iterative computational procedure but, for CV values of 0.04 they required up to 1600 passes through the data to obtain satisfactory resolution. With CVs in the region of 0.2, which is typical for immunofluorescence, the number of passes through the data would increase astronomically rendering a purely iterative technique totally impracticable. The approach in our laboratory for the higher CVs seen with surface marker analysis has been to use a semi-analytic approach where we are testing the hypothesis of Gaussian distribution in the data

(Watson and Walport, 1985). The various steps in the proceedure are as follows.

Step 1) Assume a Gaussian distribution

Step 2) Calculate the distribution for given mean and SD

Step 3) Transform this to a distribution with constant CV as would be generated by the flow cytometer. This requires that each channel of the true distribution be redistributed according to the CV of that particular channel. The new distribution (defined as skewed-normal) is then obtained by a channel-by-channel summation of the redistributed data. In practice we have limited the spreading to ±4SD of a given channel although for some very extreme examples we have encountered some computational rounding down problems as predicted by Schuette et al. (1983).

Step 4) Compute the mean of the skewed-normal distribution which should be identical to the original as an equal number of cells are redistributed either side of the mean although the mode and the median will differ from the original. This estimated mean is satisfactory as long as only a small proportion of the total is redistributed to the first and last channels of the finite range into which the data must be confined in order to simulate the instrument's finite range. In extreme examples a systematic error was noted for which corrections could be applied.

Step 5) Calculate the variance and hence CV of the skewed-normal distribution.

Step 6) Apply systematic corrections for the CV found at step 5 and for the mean (if necessary) found at step 4 and generate the reconstructed normal distribution from these values.

Step 7) Take the reconstructed distribution and transform it according to the procedure at step 3 and compare it with the initial skewed-normal distribution generated at step 3.

The control data of figure 3A were analysed using the above technique which predicted a mean and standard deviation of 5.27 and 2.1 respectively. $\Sigma\chi^2$ was 10.5 for the observed differences between the data and the skewed-normal distribution with 14 degrees of freedom. Entering the χ^2 table with this number of degrees of freedom we find a value of 26 at the 0.05 probability level and 32 at the 0.01 probability level. Hence, these results are compatible with the hypothesis that the data were derived from a normal distribution.

We now assume that any unlabelled cells in the test sample of figure 3A are also normally distributed and that these can be described with $\bar{x} = 5.27$ and $s = 2.1$. A first, very approximate, value for the proportion of unlabelled cells in the test sample was obtained by doubling the area to the left of this mean value in the test sample and comparing this with the whole area. The left side of the mean was used as this is the area of the test sample which is least likely to have overlap of labelled cells. A normal distribution was then generated with $\bar{x} = 5.27$ and $s = 2.1$ containing this proportion and the distribution was "skewed" as in step 3. The resulting skewed-normal distribution was then subtracted from the test data set on a channel-by-channel basis. There were no negative numbers in the channels of the test sample from which the subtractions were carried out hence, this initial proceedure had underestimated the unlabelled cell fraction (UF) which was taken as the minimum of the iterative range. The maximun in the range was obtained by incrementing successively the initial estimate of the UF by 10% until negative values were obtained. If the initial UF estimate had produced negative values after the subtraction this would have been taken as the maximum and the minimun would have been found by successive subtraction of 10%.

The analysis continued by using initial increments of 0.05 within the estimated range of 20% for the unlabelled population thus, 5 values were considered. A skewed-normal distribution of the unlabelled fraction was generated for each of these values which was then subtracted, on a channel-by-channel basis, from the test distribution. The remaining distribution was then analysed using the skewed-normal technique to give a mean and standard deviation of the predicted labelled fraction. A theoretical normal distribution for the latter was generated containing (1.0 - UF) x N cells, where N is the number of cells in the test distribution. This was now skewed then added, again on a channel-by-channel basis, to the skewed-normal distribution of the unlabelled fraction. This theoretical distribution was now compared with the experimental test sample data and $\Sigma\chi^2$ for observed (experimental) versus expected (theoretical) was calculated. The value for the unlabelled fraction associated with the minimum $\Sigma\chi^2$ was now used as the mid-point of the final range which was taken to be ± 0.05 of this mid-point. The increment was reduced to 0.01 and the analysis was repeated which gave an estimate of the unlabelled fraction in the test sample with a resolution of $\pm$ 1%.

The final result of the analysis is shown in figure 5 where the symbols represent the experimental data and where the dashed curves represent the labelled and unlabelled fractions. The continuous curve which passes through the experimental data points is the theoretical reconstruction of the experimental data which gave a total $\Sigma\chi^2$ of 42.7 with 60 degrees of freedom. This is associated with $p > 0.95$ which means that the fit was good. The labelled fraction was predicted to contain

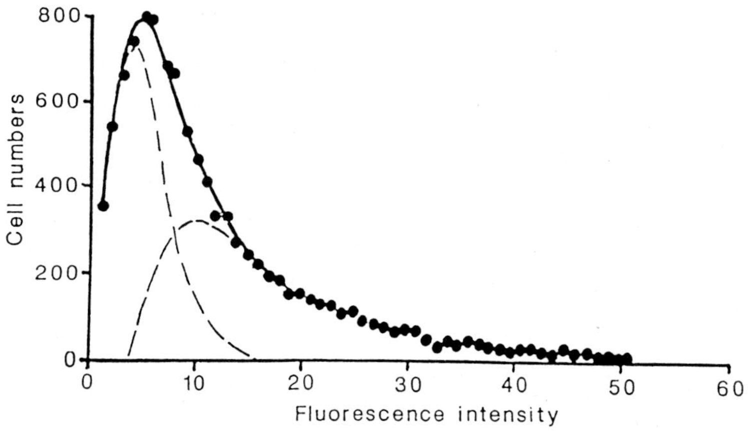

Figure 5) Result of the analysis of the data in figure 3A.

5287 cells (52%) with a mean in channel 17.2 and standard deviation of 6.92. We are now in a position to assign a significance level to this result using Students's t where $n_1 = 4942$, $\bar{x}_1 = 5.27$ and $s_1 = 2.1$ are the parameters for the unlabelled fraction and where $n_2 = 5287$, $\bar{x}_2 = 17.2$ and $s_2 = 6.92$ are the parameters for the labelled fraction. Thus, the standard error of difference is given by,-

$$SE = \sqrt{\left(\frac{4942 \times 2.10^2 + 5287 \times 6.92^2}{4942 + 5287 - 2}\right) \times \left(\frac{1}{4942} + \frac{1}{5287}\right)} = 0.1026$$

which gives $t = (17.2 - 5.27)/0.1026 = 116$ with 10227 degrees of freedom which is a very highly significant result, $P < 0.0001$. Thus, with these numbers of cells involved, we can be more than 99.99% certain that there were two populations in the test sample with 52% being labelled.

2) **Constant variance analysis** A different approach is used with constant variance data which does not have the added complication of ADC skewing as the data were obtained using log amplification. Hence, the mean and standard deviation of a distribution can be obtained directly without having to use the skew analysis technique. The mean of the control in panel B of figure 3, $\bar{x}_1$, was 29.1 with a standard deviation of 6.69. The mean of the test sample in figure 3B was 30.5, a shift of only 1.4 linear "data array channel" units but the standard deviation cannot meaningfully be calculated as this is not a single distribution as was revealed by K-S analysis in figure 4B. Linear units were used on the abscissae of figures 3B and 4B as the ensuing calculations must be performed in these linear units.

Ratio analysis of means (RAM) This aspect of the analysis is derived from simple mechanics where moments are taken about a point (Watson, 1992). Consider the weightless beam depicted in figure 6 where two masses, W_1 and W_2 respectively, are suspended from the beam at points $\bar{x}_1$ and $\bar{x}_2$. If the origin of the scale (zero) is set at the extreme left of

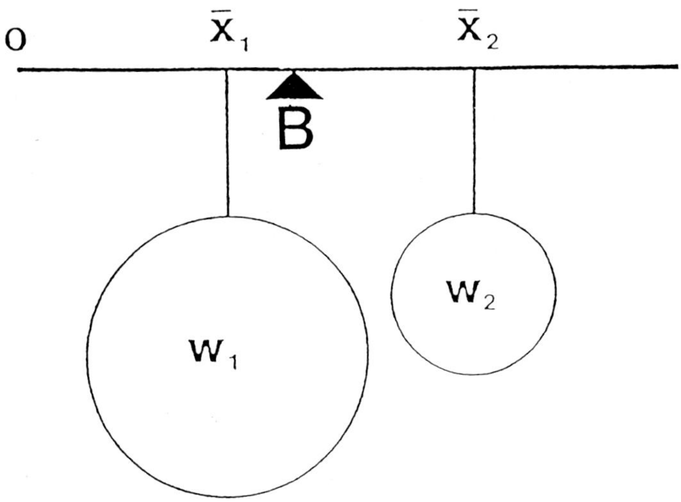

Figure 6) Depiction of a weightless beam where two weights, W_1 and W_2, are suspended from points $\bar{x}_1$ and $\bar{x}_2$ respectively. B is the balance point when these weights have a mass ratio of 0.75:0.25.

the beam then the balance point, B, can be obtained by taking moments as in the following relationship.

$$W_1(B-\bar{x}_1) = W_2(\bar{x}_2-B) \qquad 1$$

On rearrangement we get,-

$$B = (W_1\bar{x}_1 + W_2\bar{x}_2)/(W_1 + W_2) \qquad 2$$

As a simple worked example let us suppose that $\bar{x}_1$, $\bar{x}_2$, W_1 and W_2 have values of 100, 200, 75 and 25 respectively. We can now calculate B from equation 2 to be 125. Thus, the balance point (as depicted in figure 6) is 25 distance units from $\bar{x}_1$ and 75 distance units from $\bar{x}_2$.

It should be noted that neither the shapes nor the densities of the masses depicted as hanging from the beam in figure 6 nor, indeed, their orientations are of any consequence as long as the forces through the centers of gravity of the masses pass through the points $\bar{x}_1$ and $\bar{x}_2$. This

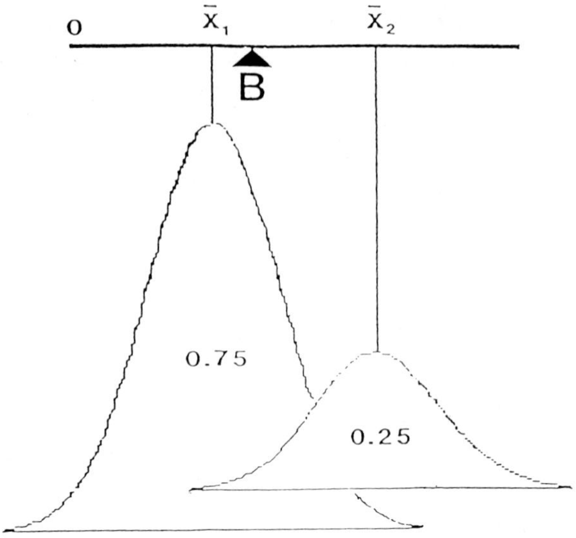

Figure 7) The same beam as shown in figure 6 where the weights have been replaced with masses in the shape of normal distributions cut from a uniform material. The beam still balances at the same point B inspite of the "overlap" between the distributions.

is illustrated in figure 7 where the masses are depicted as normal distributions with the same variance cut from a material of uniform thickness. Although these distributions are overlapping the beam will still balance at the same point as their areas have a 75:25 ratio and the material is uniform. If you are not convinced by this argument turn (in your minds eye) the two distributions, which are flat, through 90^0 so that they are no longer overlapping.

Let us now suppose that the distances $\bar{x}_1$, $\bar{x}_2$ and B are known for a normalized total mass of unity where $W_1 + W_2 = 1.0$. We can now calculate the relative proportion of W_2 from equation 1 by substituting W_1 with $(1.0 - W_2)$. This gives,-

$$(1.0 - W_2)(B - \bar{x}_1) = W_2(\bar{x}_2 - B)$$

which on simplification reduces to,-

$$W_2 = (B - \bar{x}_1)/(\bar{x}_2 - \bar{x}_1) \qquad 3$$

The masses in figure 7 were depicted as normal distributions in order to illustrate the connection between the concept of taking moments about a point in mechanics and immunofluorescence histogram analysis. We can see from equation 3 that the proportion, W_2, which will now be referred to as labelled cells in a mixed population of labelled plus unlabelled, is defined by three "distances" namely, $\bar{x}_1$, $\bar{x}_2$ and B. $\bar{x}_1$ is the mean of the control unlabelled fraction. B is the mean of the test sample which contains both labelled and unlabelled cells and $\bar{x}_2$ is the mean of the labelled cell fraction. Values for $\bar{x}_1$ and B can be obtained directly from the experimental data but we also require a value for $\bar{x}_2$ which is calculated as described in the next section.

Labelled fraction mean calculation This component of the analysis arose from a desire to be absolutely sure that the K-S statistic for analysing differences between distributions was being correctly implemented within our software package. Consider the top panel of figure 8 which shows two normal distributions where both have standard deviations of 16 and means in channels 64 ($\bar{x}_1$) and 192 ($\bar{x}_2$) respectively. In order to test that the correct K-S difference was being computed cummulative frequencies of the left histogram and of both

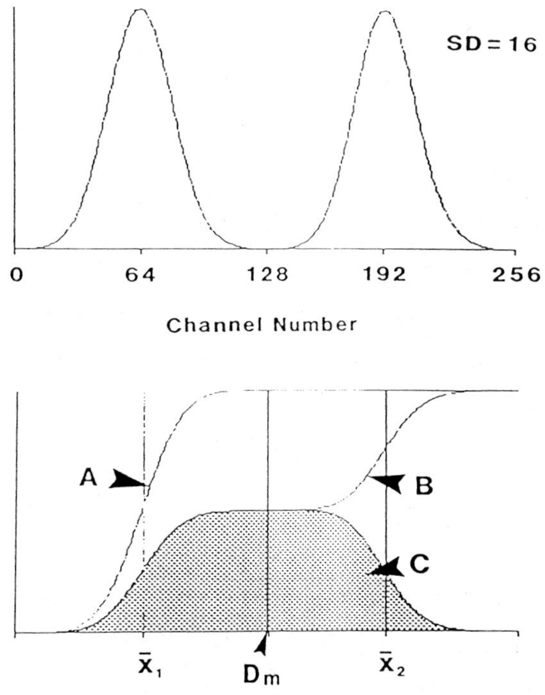

Figure 8

histograms combined were displayed. These are shown in the lower panel as curves A and B respectively where both are normalized to unity. A difference histogram was then generated by subtracting curve B from curve A which is shown as the stippled area, C. It was now noted that the mean of this difference histogram, D_m, (middle vertical line) appeared to be very close to the mid-point of the means (flanking vertical lines) of the histograms. In fact, as can just be seen in this figure, D_m is just to the left of the mid-point between the histogram means.

This observation was potentially important as, if confirmed, it would enable the exact position of the mean of the second distribution, $\bar{x}_2$, to be calculated as,

$$\bar{x}_2 = (2.0 \times D_m) - \bar{x}_1 \qquad 4$$

Hence, an approximate value for W_2 would be obtainable via equation 3.

A large series of simulations was carried out for varying standard deviations, ratios of means and relative proportions in control and test samples where cumulative frequency subtraction of the two distribtuions was carried out. In each case D_m was calculated to be exactly 0.5 channels to the left of the mid point between $\bar{x}_1$ and $\bar{x}_2$. This relates to the fact that the distributions being considered are not strictly continuous as they have a class interval (abscissa scale units) of unity. However, in the limiting case, where the distributions are absolutely continuous the class interval is infinitely small and equation 4 should hold exactly. However, the latter may be corrected for class interval (CI) as shown below in equation 5.

$$\bar{x}_2 = (2.0 \times (D_m + (CI/2.0))) - \bar{x}_1 \qquad 5$$

Figure 9 shows four simulations where the class interval correction has been included. The standard deviation was 16 in each distribution and the unlabelled (a, mean, $\bar{x}_1$, in channel 64) and labelled (b) fractions each constitute 50% of the total. The histograms are shown in the left

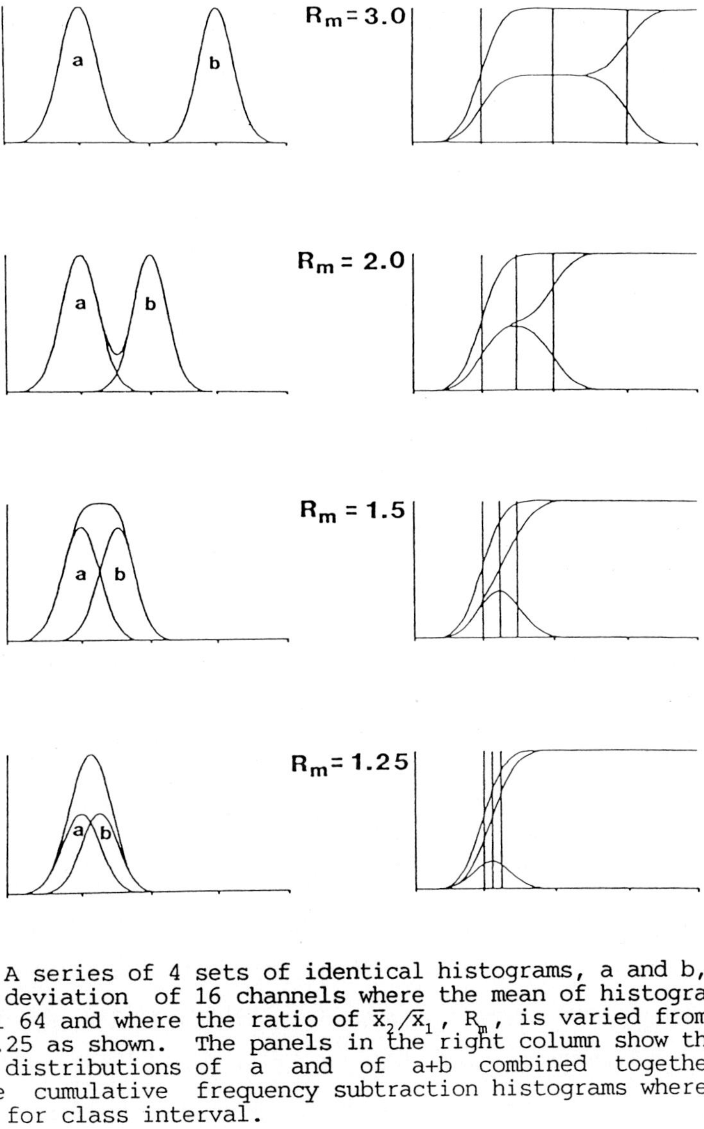

Figure 9) A series of 4 sets of identical histograms, a and b, both with standard deviation of 16 channels where the mean of histogram a is fixed in channel 64 and where the ratio of $\bar{x}_2/\bar{x}_1$, R_m, is varied from 3.0, 2.0, 1.5 and 1.25 as shown. The panels in the right column show the cumulative frequency distributions of a and of a+b combined together with the respective cumulative frequency subtraction histograms where D_m has been corrected for class interval.

column with the corresponding cummulative frequency distributions in the right column. Ratios of the means ($\bar{x}_2/\bar{x}_1$, R_m) are shown between each pair of displays and these were set at 3.0, 2.0, 1.5 and 1.25 respectively from top to bottom. Note that as the labelled fraction, b, becomes closer to the unlabelled, a, so the maximum frequency of the cumulative frequency subtraction distribution decreases however, this only occurs when the distributions a and b overlap. Nevertheless, the relationship between $\bar{x}_1$, D_m and $\bar{x}_2$ is maintained.

The solution for $\bar{x}_2$ shown in equation 5 can now be substituted into equation 3 which on simplification for FCM data with a class interval of unity gives,

$$W_2 = (B - \bar{x}_1)/(2.0\times(D_m + 0.5 - \bar{x}_1)) \qquad 6$$

This equation gives the proportion of labelled cells, W_2, in a mixed population of labelled plus unlabelled in terms of $\bar{x}_1$, the mean of the control sample; B, the mean of the test sample containing labelled plus unlabelled cells and D_m, the mean of the subtracted cumulative frequency distribution. It is important to note that equation 6 should only be applied if the variances of the control and labelled cell distributions are equal. This condition is most likely to be met in FCM when log amplifiers are used as the standard deviation is independent of intensity. Equation 6 cannot be applied directly for data obtained with linear amplifiers where the coefficient of variation (CV) is constant.

Figure 10 shows the results of the cumulative frequency subtraction analysis for the control and test sample distributions of figure 3B where $\bar{x}_1$, D_m and $\bar{x}_2$ and marked by the vertical lines on the figure together with their values which were 29.1, 37.9 and 46.7 respectively. The calculated value of 46.7 for $\bar{x}_2$ was now used with the value of 29.1 for the control mean, $\bar{x}_1$, and the mean of 30.5 for the test sample, B, to calculate the labelled fraction as 8.0%. The predicted unlabelled and labelled populations are shown in figure 11 together with the experimental data.

In the first log-amplification decade (1-64) we can convert to fluorescence intensity units, FIU, with the relationship $FIU = 10^{(x/64)}$ where x is the data array channel number. Thus 29.1, 30.5 and 46.7 translate to 2.849, 2.996 and 5.366 fluorescence intensity units respectively. Hence, in FIU the test sample mean, 2.996, is only 5.2%

greater in intensity than the control, 2.849 FIU. Nevertheless, in spite of this very small shift the analysis predicts 8.0% of the total population to be labelled.

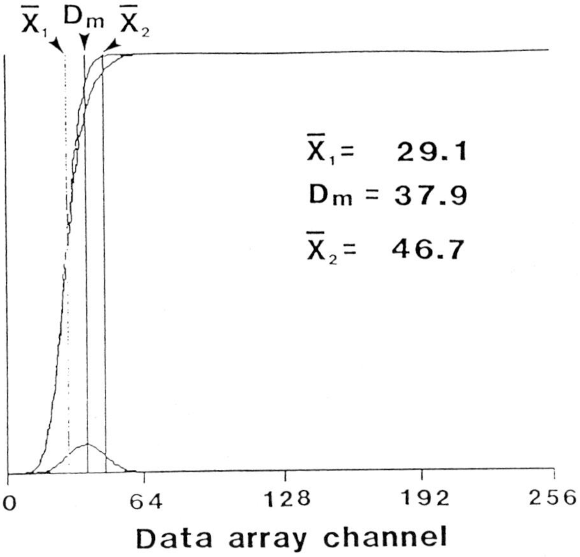

Figure 10) Cumulative frequency distributions of the control and test sample shown in figure 5 with the cumulative frequency subtraction histogram displayed close to the X-axis. The latter was analysed to give D_m and hence $\bar{x}_2$ which are shown, together with $\bar{x}_1$, as the vertical lines as indicated.

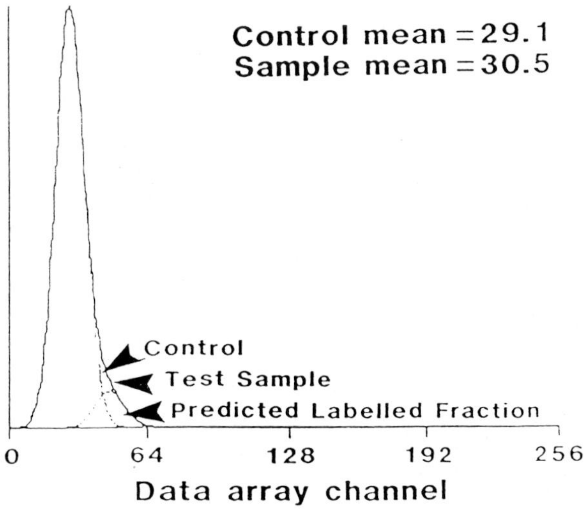

Figure 11) A control and test sample data set as indicated together with the predicted labelled fraction histogram. The means of the control and test sample were 29.1 and 30.5 respectively. Note that the abcissa scale units are expressed as "data array channels" where 64, 128, 192 and 256 correspond to 10^1, 10^2, 10^3 and 10^4 fluorescence intensity units (FIU) respectively.

Statistical verification We now have to ask if this is reasonable and what significance we can place on the result. This is particularly important when analysis of data predicts a minority labelled subset embedded within the test sample. There are three ways in which the predictions can be verified namely K-S analysis, Student's t and simulation of the experimental data using $\Sigma\chi^2$ as the criterion of fit.

Kolmogorov-Smirnov The cumulative frequency distributions of the control and test sample were reanalysed over a range of ± 3 standard deviations about the mean of the predicted labelled distribution, $\bar{x}_2$. These data are shown in figure 12 together with the difference histogram close to the abscissa. K-S analysis, with the number of cells involved, showed that the two cumulative frequency distributions over this limited range had a probability of being different at the 99% confidence limit.

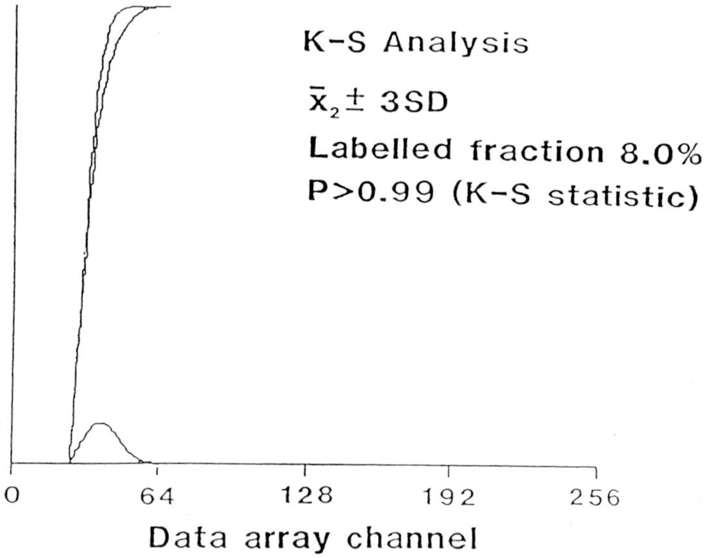

Figure 12) Kolmogorov-Smirnov analysis of the control and test sample distributions over the interval $\bar{x}_2$ $\pm$3SD. The difference histogram is shown close to the X-axis and the maximum height of this is the maximum difference between the two cumulative frequency distributions which was significant at $p > 0.95$.

Student's t The predicted results were also submitted to Student's t analysis assuming the variance was equal in both distributions. The standard error of difference, SE, was calculated as detailed in section 4.1.2 and this is shown below.

$$SE = \sqrt{\left(\frac{9171 \times 6.69^2 + 797 \times 6.69^2}{9171 + 797 - 2}\right) \times \left(\frac{1}{9171} + \frac{1}{797}\right)} = 0.2463$$

The difference between means was 17.6 channels which when divided by 0.2463, the standard error of difference, gives $t = 71.46$ with 9966 degrees of freedom. This is significant at $p < 0.0001$.

Simulation with $\Sigma\chi^2$ A final verification of the analysis proceedure was carried out by generating two normal distributions with $\bar{x}_1$ and $\bar{x}_2$ of 29.1 and 46.7 respectively both with a standard deviation of 6.69 where the proportions in the two distributions were 0.92 and 0.08. These distributions were summed on a channel-by-channel basis and the synthetic data set was normalized to a total area of 9968 to correspond to the number of cells in the test sample data set. A $\Sigma\chi^2$ calculation was then used as the criterion for the difference between the data and the histogram synthesised from the results of the data analysis. This gave a probability of $p < 0.0001$ that the difference between the simulated and experimental data could have arisen by chance.

Errors with conventional methods A further analysis of 95 synthetic histograms was carried out to tabulate the errors incurred with conventional analysis where the labelled fraction is calculated using a vertical delimiter set at 2.5 standard deviations above the mean of the control distribution. The mean of the latter was set in channel 32 and both the control and "labelled" fraction distributions were synthesised with a standard deviation of 8 channels (CV = 0.25 in channel 32). The labelled fraction was then varied from 5% to 95% in increments of 5% for ratios of labelled to unlabelled fraction means of 2.5, 2.0, 1.75, 1.5 and

1.25. These results are shown in figure 13 where the labelled percentages calculated by vertical delimiter are plotted on the ordinate versus labelled percentages calculated by this method (ratio analysis of means) plotted on the abscissa. As long as the data are normally distributed and the variance is constant the error calculated by RAM is < 0.5%. We can see from figure 13 that when the means ratio was 2.5 the error incurred by vertical delimiter was zero as the channel separation between the two means was 43 (75-32). This is greater than 5 standard deviations (5x8), 2.5 above the lower mean and 2.5 below the upper mean. However, as the separation between the means decreased the errors increased dramatically and with a separation of about 1.3 the error to be expected reach approximately one order of magnitude.

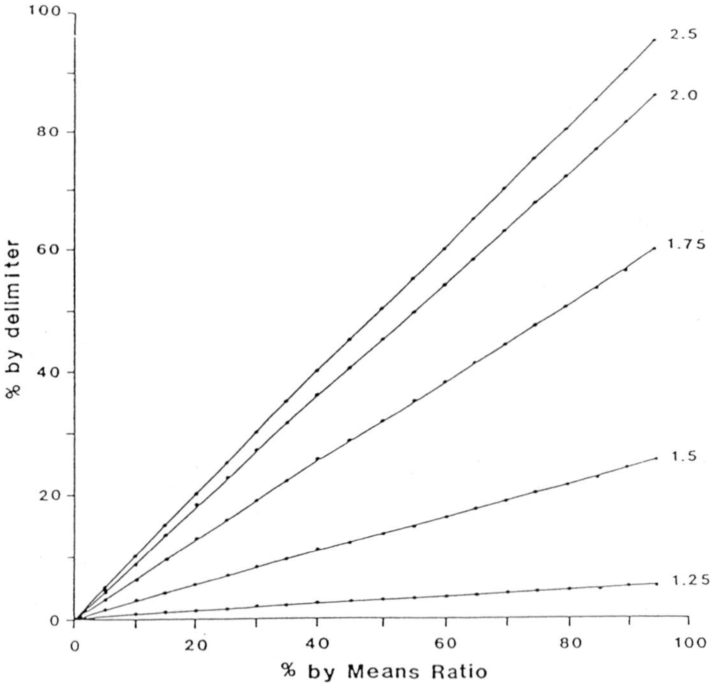

Figure 13) Labelled percentages calculated by vertical delimiter plotted on the ordinate versus labelled percentages calculated by RAM on the abscissa for means ratios of 2.5, 2.0, 1.75, 1.5 and 1.25.

Concluding remarks The methods described in this chapter were undertaken to help in the analysis of immunofluorescence data by presenting results in the format elluded to earlier. For the first data set we can say that the analysis was compatible with two populations where the respective means were in channels 5.27 and 17.2 and the labelled fraction comprised 52% of the total. Moreover, the separation between the means, with the standard deviations involved, 2.1 and 6.92 respectively, could have occurred in less than 1 trial in 10,000 on a random basis. In the second data set the analysis was also compatible with two populations whose means were separated by 1.6 data array channels where the predicted positive fraction constituted 8.0% of the total at, or above the 99.9% confidence interval. If we had just "eye-balled" the data in figure 3 and placed a vertical delimiter we could not possibly have come to the conclusion that 52% and 8.0% of the respective populations were labelled. And yet, this objective statistical analysis tells us there is a probability of > 0.999 that there were two populations in each sample. One of the reasons why we can have this degree of certainty is that a large number of cells was analysed which is one of the frequently voiced advantages of the technology. However, the large number "good statistics" attribute of the technology is very infrequently used to its full advantage. The objective statistical assessments described in this chapter covers a comparison of the experimentally derived control and test sample data using Kolmogorov–Smirnov statistics, Student's t assessment of the differences between the predicted labelled and unlabelled distributions and a $\Sigma\chi^2$ to test for differences between the predicted and experimental data. In the constant variance analysis this represents a 3-way comparison namely experimental versus experimental, theoretical versus theoretical and experimental versus theoretical.

At present the second method of analysis can only be used for data sets with constant variance and the analysis was illustrated with distributions where both skew and kurtosis conformed very well with a

Gaussian profile and, indeed, these data were chosen for this property. However, the technique can also be applied to data which depart from normality as long as the skew is close to zero and the variance is constant. For example if, at the extreme, the distributions being analysed have rectangular profiles the cumulative frequency curves will be oblique straight lines. This will result in a rhomboid or triangular cumulative freqency subtraction histogram but with constant variance the shape will be symetrical and the slopes of the lateral two oblique sides will be equal and opposite. Hence, the mean of the subtraction histogram will still be mid-way between the means of the component distributions subject to class interval correction.

In general, the distribution of biological behaviour tends to be skewed to the right and of log-normal type and any analogue to digital conversion step in combination with linear amplification introduces a further positive skew (Schuette et al, 1983, Watson and Walport, 1985). With log-amplification the positive skewing tends to be eliminated and log-normal distributions present themselves in Gaussian form which is highly convenient for the constant variance analysis described here. Furthermore, inspection of many published immunofluorescence data sets suggests that the variance is not infrequently constant as the shapes of well separated control and test sample distributions tend to be very similar. However, with overlapping data for which the constant variance method is intended, it is not possible directly to test if the variance of the unlabelled and labelled fractions is constant. It is likely that this will be so with log-amplification, at least to within reasonable limits, but this is an assumption not an axiom. The final phase of this analysis attempts to address this point by reconstructing the test sample distribution from the results of the analysis where an objective method of assessing the difference between experimental and reconstructed data using $\Sigma\chi^2$ is used. This is important in order to identify discrepancies which are most likely to be due to non-constant variance, a problem which, from initial studies, appears to be associated with sytematic errors for which corrections should be able to be made.

REFERENCES

Pearson, T., Galfre, G., Ziegler, A. and Milstein, C. (1977) "A myeloma hybrid producing antibody specfic for an allotypic determinant on "IgD-like" molecules of the mouse" EUROP. J. IMMUNOL. 7; 684-690

Fisher, R.A. (1925a) "Applications of "Student's" distribution" METRON 5; 90-104

Fisher, R.A. (1925b) In "Statistical Methods for Research Workers" 1st edition, Chapter V, Oliver and Boyd, Edinburgh.

"Student" (1908) "The probable error of a mean" BIOMETRICA 6; 1-25

Schuette,W.H., Shackney, S.E., MacCollum, M.A. and Smith, C.A. (1983) "High resolution method for the analysis of DNA histograms that is suitable for the detection of multiple aneuploid G1 peaks in clinical samples" CYTOMETRY 3; 376-386

Watson, J.V. and Walport, M.J. (1985) "How does flow cytometry interpret Gaussian distributed information" J. IMMUNOL. METH. 77; 321-330

Watson, J.V. (in press 1992) "Flow cytometry Data Analysis" CAMBRIDGE UNIV. PRESS, Chapter 7 (Projected publication date December)

Watson, J.V. (1992) "Proof without prejudice revisited; Ratio analysis of means plus Kolmogorov-Smirnov statistics for immunofluorescence histogram analysis" CYTOMETRY (In the press)

Analysis of Multiparameter Flow Cytometric Data

James H. Jett and Robert C. Habbersett
Life Sciences Division
Los Alamos National Laboratory
Los Alamos, New Mexico 87545 U.S.A.

Introduction

Today, most flow cytometric (FCM) measurements are multiparameter measurements of cells or cellular components. Analysis of these complex data requires that the experimenter become intimately familiar with the data in order to extract meaningful information. Here we will discuss some aspects of the analysis of multivariate flow cytometric data. However, drawing conclusions through the statistical analysis of measurements made on numerous samples will not be discussed. The statistics of intra-sample comparisons are discussed in numerous nonparametric statistics texts. As will become clear, the approach presented here will be primarily graphical versus statistical.

The first problem encountered in the analysis of multivariate flow cytometric data is that there is typically a large amount of data. Data files can easily contain measurements of 10,000 cells. Many of the traditional statistical approaches to multivariate data are constructed around the premise of small numbers of measurements. For example, some multivariate data visualization techniques make figures such as faces for each data point in N-dimensional space in which the characteristics of the figures vary according to the values of the parameters measured. Clearly, techniques such as this can not be applied to visualize measurements of 10,000 cells.

In flow cytometry, the number of measurements that can be made on each cell can range from a single parameter to numbers approaching a thousand. For whole cell measurements, the multi-laser system described by Watson(1988) can measure 8 or more parameters per cell. Steinkamp(1982) has described a system that can record 8 parameters selected from 12 primary possibilities. A multi-angle light scatter cytometer was capable of quantifying light scattered by cells at 32 angles (Salzman 1975).

Flow systems that record one or two dimensional morphological information from cells or chromosomes produce prodigious amounts of data. The slit scan system described by Robinson, et al. (1990) records three orthogonal one dimensional cellular profiles which results in more than 750 bytes of information per cell. The profile data is analyzed to reduce the number of parameters per cell by extracting morphological information such as cell diameter and nuclear diameter from the profiles.

Slit scanning techniques have also been used in attempts to differentiate

NATO ASI Series, Vol. H 67
Flow Cytometry
Edited by A. Jacquemin-Sablon

human choromsomes. Lucas, et al. (1991)and Johnson, et al. (1985) have described systems in which fluorescence and light scatter profiles were recorded from chromosomes oriented in the flow stream. Analysis of the profiles yielded information on the centromeric index which was used to separate chromosomes with overlapping total fluorescence intensities.

Analysis of multi-parameter flow cytometric data is a complex process. There are several reasons why this type of data is not easy to analyze. It is hard to "get hold" of such data. For more than 3 parameters, it is not possible to completely visualize the data for the large numbers of events that are typically recorded in an experiment. Even for 3 parameter data, a single 3 dimensional display does not necessarily reveal all of the intricate structure of the data.

Techniques for the analysis of these complex data are tedious. Most techniques require manual intervention. Thus, a knowledgeable person must be involved in this tedious activity of analyzing the data. Some automated techniques are beginning to emerge. Clustering on multiparameter data to determine the number of subpopulations present is an active field of research (Salzman 1992). The application of artificial intelligence/expert system methods to flow data is just beginning.

Many statistical analyses of multiparameter data produce more numbers to be digested than might be desirable. A clear understanding of the applicability of the statistical procedures to FCM data is needed. This is particularly true when a standard statistical analysis package is used.

Another reason that multiparameter data is hard to comprehend is that there is no theoretical structure or model that describes the data. This can be contrasted with the situation that exists for single parameter DNA distributions. The cell cycle serves as a foundation upon which a model for the observed data can be built. Such model building is not possible for multivariate data. For example, the relationship, if it exists, between cellular DNA, RNA and protein content is not known. Similar statements can be made about the relationships among cell surface antigens and cell size or DNA content. Without a theoretical framework, analysis of multivariate data must ultimately resort to phenomenological and statistical analysis.

Finally, one must have in mind a clear question or set of questions that the data are expected to answer. For example, the question could be how many subpopulations are represented in this data set? What are the regions in N-space that define each subpopulation? What fraction of the cells are CD4 positive and CD8 negative? Have the limits in N-space that define subpopulation X changed from sample to sample? Whatever question is posed, it must ultimately be posed in such a manner that the analysis is relevant and manageable.

What is multivariate FCM data ?

Multivariate data is simply the data generated by the measurement of multiple properties of a number of objects. In flow cytometry, that translates to multiple measurements of each cell or chromosome. There are numerous examples from a variety of areas of cell biology of the utility of multivariate measurements. Perhaps the most popular multivariate flow cytometric measurement is immunophenotyping in which the levels of expression of several cell surface antigens are determined along with light scatter measurements and other parameters (Ault 1990). Crissman, et al. (1985) have measured cellular DNA/RNA/protein content to determine the relationships between those entities in cells that are in exponential growth and in cells that have gone into a state of unbalanced growth due to confluence or some other outside influence.

Multivariate FCM data are stored primarily as a simple list of the measurements, hence the name list mode data. Essentially all multivariate data of more than 2 parameters is stored in list mode. It is possible to store bivariate data in histogram form without loosing any information except the order in which the measurements were made. No correlation information is lost. For low resolution histograms such as 64 by 64 channel histograms, the most efficient way to store the bivariate data is as histograms. For higher resolution data, the trade off point depends on the number of events recorded.

A list mode file contains information in something similar to the format shown below. The header contains information about the structure of the data that follows, the sample analyzed and the machine making the measurements. There is a standard header format, the FCS format, that has been accepted by the flow community (Dean, et al. 1990). A list mode file for 4 parameter data will have a structure something like the following.

List mode data format:

Header information
Parameter #1 for cell #1
Parameter #2 for cell #1
Parameter #3 for cell #1
Parameter #4 for cell #1
Parameter #1 for cell #2
Parameter #2 for cell #2
Parameter #3 for cell #2
Parameter #4 for cell #2
Parameter #1 for cell #3
Parameter #2 for cell #3
......
Parameter #2 for cell #N
Parameter #3 for cell #N
Parameter #4 for cell #N

The list mode format preserves all of the information that is measured. For example, the correlations between different parameters is retained. The sequence order of the data is retained. This can be useful in kinetics experiments of for diagnostic purposes when one parameter is a measure of instrument performance such as laser power.

When is FCM data not multivariate ?

There are instances when FCM data appears to be multivariate data when in fact it is not. An example is the measurement of 2 parameters such as light scatter and fluorescence intensity that are recorded as independent histograms. Data recorded in this manner is the equivalent, with one exception, of performing two sequential measurements on the same population of cells in which only one of the parameters is recorded. The exception is that in the sequential measurements, the data are recorded for two samples drawn from the same population of cells whereas when the two measurements are made simultaneously, they represent a single sample of the population. However, allowing for sampling variations, the measurements are equivalent.

In the example above, these measurements do not produce complete multivariate data. The missing element is the correlation between the measured parameters for each cell. If there is no correlation between the two parameters, the independent histograms contain all of the information possible. However, if the measured parameters are correlated, it is necessary to record the data as pairs of numbers.

Finally, the dimensionality of the data is not magic. Another experiment could record the same parameters along with additional parameters, thus increasing the dimensionality of the data. In the other direction, the dimensionality can be reduced. If parameters are highly correlated, the dimensionality of the data set can be lowered by removing the dependent parameters. Figure 1 demonstrates both positive and negative correlation between a pair of hypothetical parameters. Since the value of parameter 2 can be determined by the value of parameter 1, parameter 2 can be removed from the data set with no loss of information.

Another approach to reducing the dimensionality of a data set is to perform a principal components analysis of the data set. The principal component analysis determines a new set of parameters which are linear combinations of the original measurements. A characteristic of the principal components is that the first principal component accounts for as much of the variation of the data set as is possible under the conditions of the transformation. The next component, which is orthogonal to the first, accounts for as much of the remaining variation as possible. Figure 2 demonstrates the principle. This approach has been applied to flow cytometric data by Mann (1985) to reduce the dimensionality of a data set. Judicious choice of the new axes by principal components analysis or by manually selecting the angle of rotation can resolve more subpopulations in a single histogram. However, there are several reasons not to perform a conventional principal components analysis of FCM data. A FCM data set often represents several subpopulations, each of which has its own variance/covariance matrix. Thus, using the variance/covariance matrix for the complete data set is not appropriate to calculate the principal components is not appropriate.

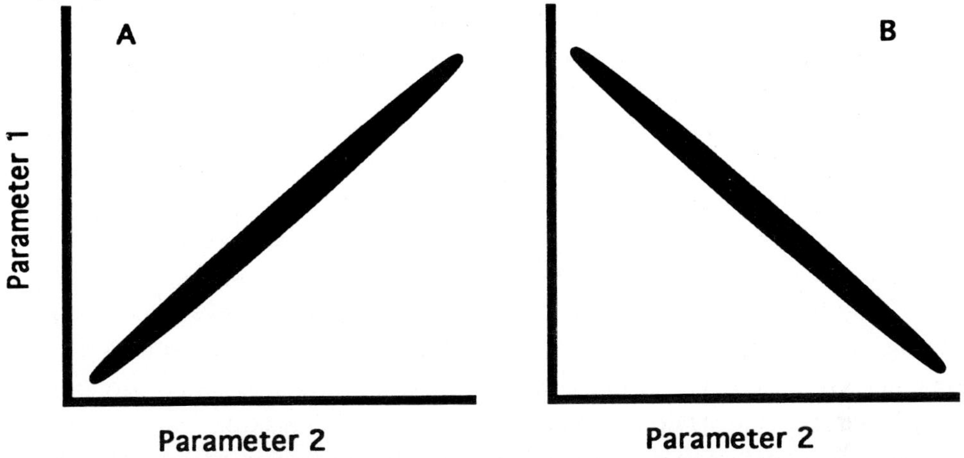

Figure 1: Correlation between measured parameters. Panel A demonstrates a high degree of positive correlation between the two parameters. Measurement of one parameter determines the other parameter with a high degree of accuracy. Panel B is a hypothetical plot for two negatively correlated parameters.

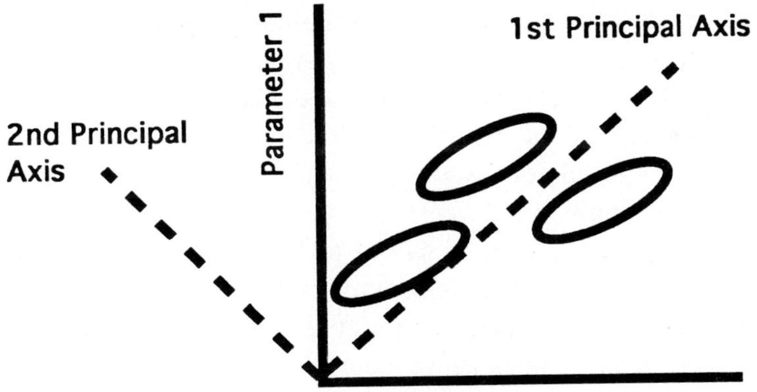

Figure 2: Principal axes transformation. The dashed lines denote the first and second principal axes. The first principal axis accounts for the maximum amount of variation for the data set. Projection of the data to this axis would reveal 3 distinct populations.

How can multivariate FCM data be viewed ?

To date, no completely automated technique for analyzing any type of multivariate flow cytometric data has been developed. Salzman (1992) discusses clustering and expert system approaches in this volume. Since no general automated method of analysis exists, the first step in the analysis of multivariate data is visualization of the data. In fact, graphical presentation of data is an important field in statistics. Chambers, et al. (1983) state that "There is no single statistical tool that is as powerful as a well-chosen graph. Our eye-brain system is the most sophisticated information processor ever developed, and through graphical displays we can put this system to good use to obtain deep insight into the structure of data."

The ongoing revolution in computer hardware and software has brought forth a whole new world of graphical tools. The platforms range from personal computers such as the Macintosh and the IBM PC and its clones to more powerful workstations and onto supercomputers. The graphs presented later were generated on a SUN workstation using the Integrated Data Language (IDL, Research Systems, Inc. Boulder CO) data analysis package. IDL in just one example of the type of powerful data analysis packages that are available today. In addition to being a data presentation package, IDL has its own data analysis language which contains many functions that facilitate the analysis of list mode FCM data.

Since visualization of data is an important step in gaining an understanding of the data and its structure, a number of visualization techniques will be presented. They will start with 2 parameter data and progress to higher dimensions. Data plots can be categorized as either plots of individual data points or as summary plots. As will be seen, plots of individual data points get saturated quickly as the number of data points plotted increases. Some types of displays can be overwhelmed with as few as 50 to 100 events - a small sample of a typical 5,000 event data file. Other types of plots do not reach saturation or the saturation is less objectionable for numbers of events in the hundreds. Summary plots almost always imply histograms presented in some form.

Presentation of bivariate FCM data can be as individual measurements or as summaries of a complete set of measurements. The dot plot shown in Figure 3 is an example of plotting individual events. A dot or other plotting symbol is placed at the appropriate coordinates for each event recorded up to some maximum number of events. The data used to generate this type of plot is the 2 parameter list mode data or its equivalent. When such a plot gets dense in any region, visual information concerning relative numbers of events in that region is lost. This is due to saturation of the plot in the region. The data in Figure 3 are not preselected from the data set. They represent a complete sampling of the population. If the data set has been subdivided by some means such as manual partitioning or automated clustering, the subpopulations can be represented by different symbols (Figure 4) or by different colors. These are referred to as brush plots (Chambers, et al. 1985).

Figures in the chapter by Salzman (1992) demonstrate this type of plotting for multiple bivariate dot plots.

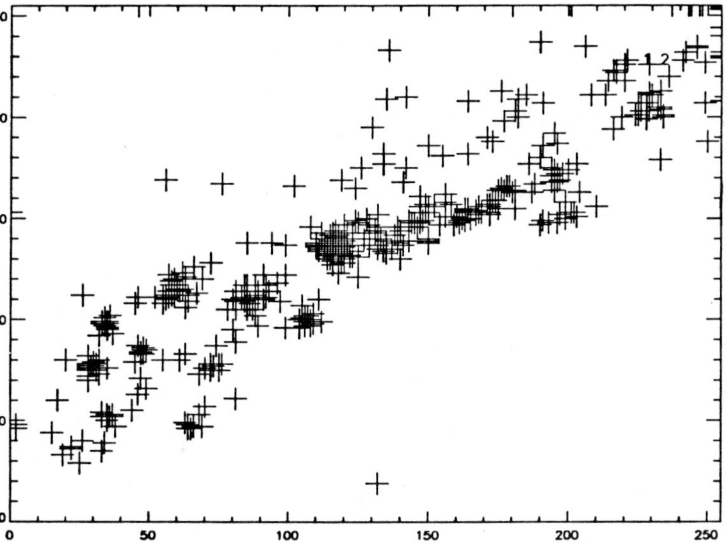

Figure 3: Bivariate dot-plot of Hoechst versus chromomycin fluorescence for 500 chromosomes.

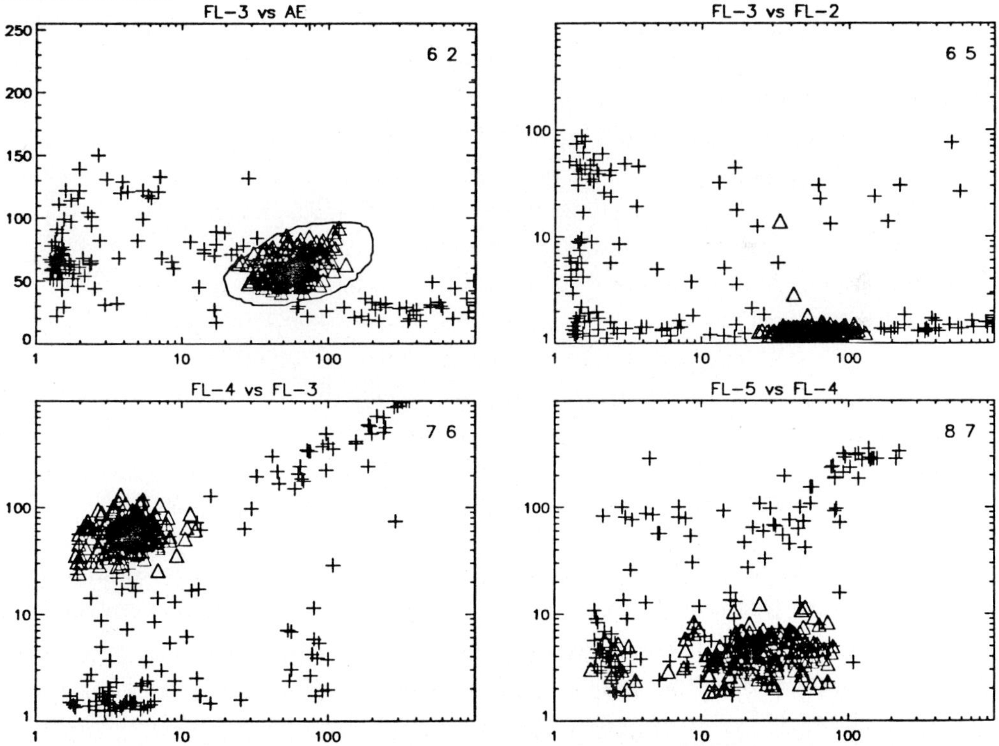

Figure 4: Dot-plots of 4 combinations of parameters from an 8 parameter file.

Histograms of any dimension are used to present in summary form complete data sets. For digitized univariate data, the histogram contains all of the measured information except the order in which the data were recorded. The same is true for bivariate histograms since the histogram retains the correlation between the two measured parameters. Up to three parameters can be visualized in a single histogram. For univariate and bivariate histograms, one axis is number of events per channel, cell or bin.

Bivariate histogram data can be visualized in at least 3 ways. The simplest to grasp is the pseudo 3-dimensional isometric plot of the surface defined by the number of events per bin as a function of the 2 measured parameters. Figure 5 displays the results of a bivariate chromosome karyotype measurement plotted in this manner. For this particular data set, no peaks in the surface are hidden from the point of view used to generated the figure. In general, that is not true and the data should be viewed from several different angles. Figure 6 shows several views of the chromosome data that combines an isometric view, a contour plot at the top and univariate histograms on the sides of the isometric plot.

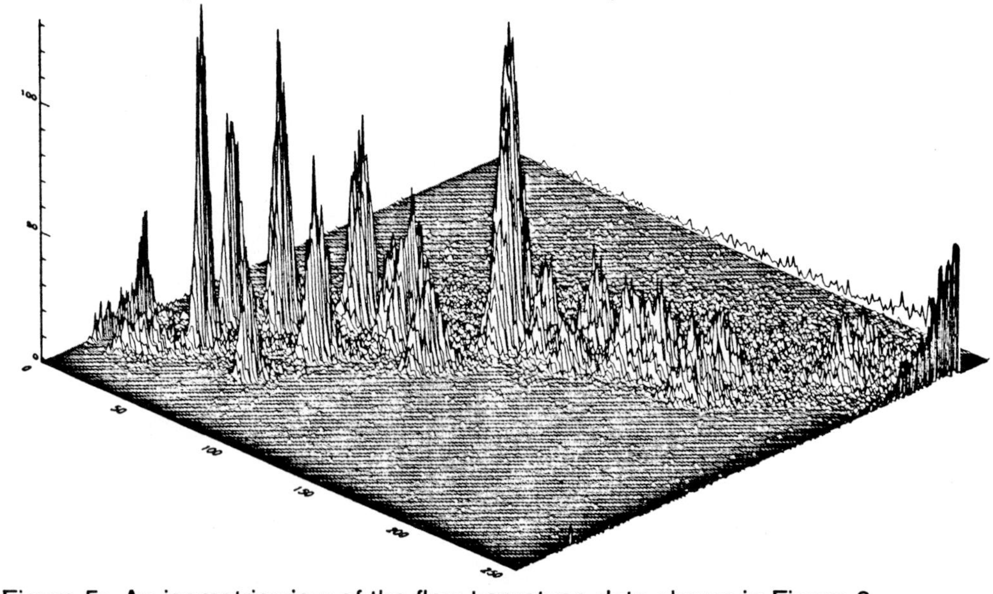

Figure 5: An isometric view of the flow karyotype data shown in Figure 3.

The problem of populations hidden from view is solved by displaying the data as a contour plot. Contour plots can be thought of as being formed by viewing an isometric plot from a point directly above it and drawing lines - contours - around the peaks at constant elevations. Thus, a contour presentation of the data depicted in Figure 5 is shown in Figure 7. The contour levels were selected to emphasize the peaks at the expense of showing the events in the plateaus surrounding the peaks.

Figure 8 replots the same data set with additional lower level contours that demonstrate that there are events on the wings of the peaks. Unfortunately, setting the lowest contour level above the debris level can be used to give an impression of the cleanliness of the data that is false. Contour levels can be set according to a variety of algorithms. They can be at drawn given percentages of the number of events occurring in the channel that contains the maximum number of events. Contour levels can be logarithmically spaced or set to include equal volumes (Moore and Kautz 1986).

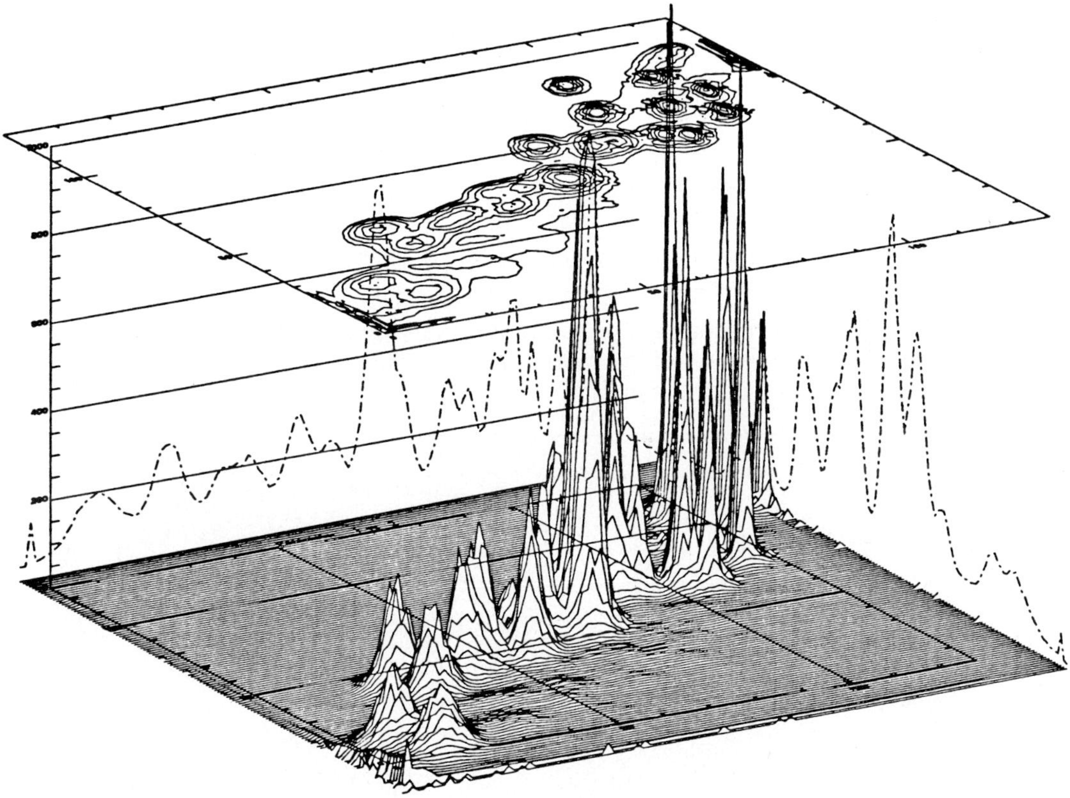

Figure 6: An isometric view of the flow karyotype data shown in Figure 3 combined with a bivariate contour plot at the top and univariate histograms on the sides.

Finally, bivariate histograms can be presented as “image” plots. In this plotting mode, which at first has the appearance of a dot-plot, a histogram is formed at the full resolution of the measurements and a color coded dot is plotted in the bivariate space for each channel. The color of the dot is determined by the number of the events in the channel and a scale that is predetermined in a manner similar to the determination of the contour levels.

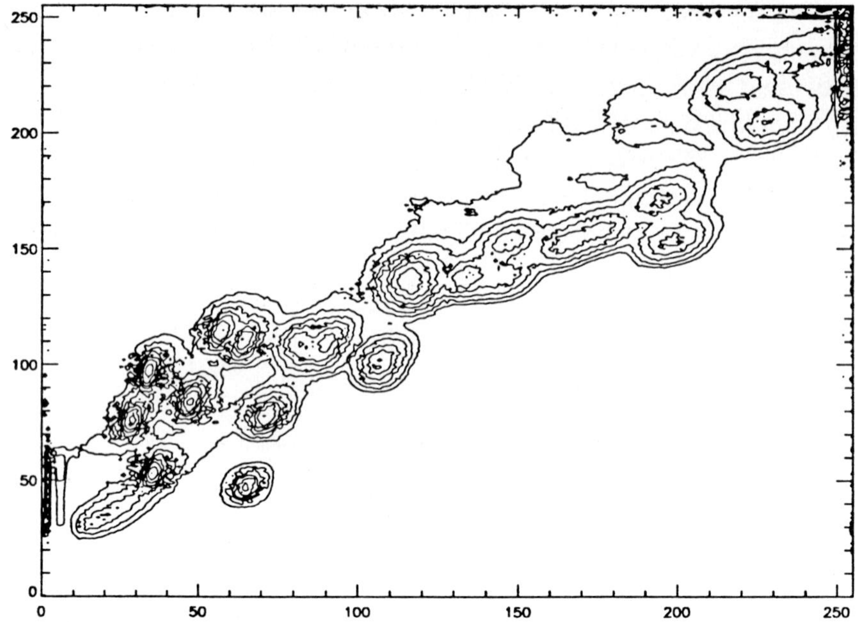

Figure 7: A contour representation of the flow karyotype data shown in Figure 3 plotted at the full resolution of the measurement which was 256 channels. The lowest contour is at 3 counts per bin.

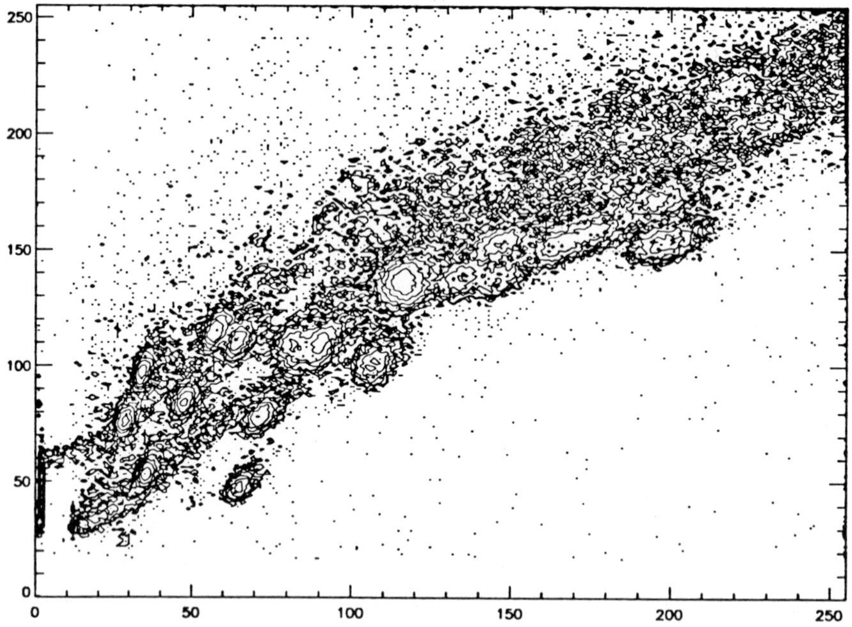

Figure 8: A contour representation of the flow karyotype data shown in Figure 3. The lowest contour is at 1 count per bin. With the lower contour level. the background that lies under the ridge of peaks can bee seen.

Trivariate data can be plotted in modes that are similar to the bivariate data. Three dimensional dot plots can be constructed but they loose much of their impact if they are static. That is, if they are viewed from only one point of view, the perception of depth that is necessary to see the third dimension is often lost. Figure 9 demonstrates this fact. There is no depth cue in the size of the plotting points or by any other means. If the figure is set in motion. the distant points become readily apparent as the figure rotates. A text devoted to the subject of dynamic graphics (Cleveland and McGill, 1988) discusses approaches to display if data sets including microcomputer programs such as MACSPIN (Donoho, et al. 1988) for the Macintosh.

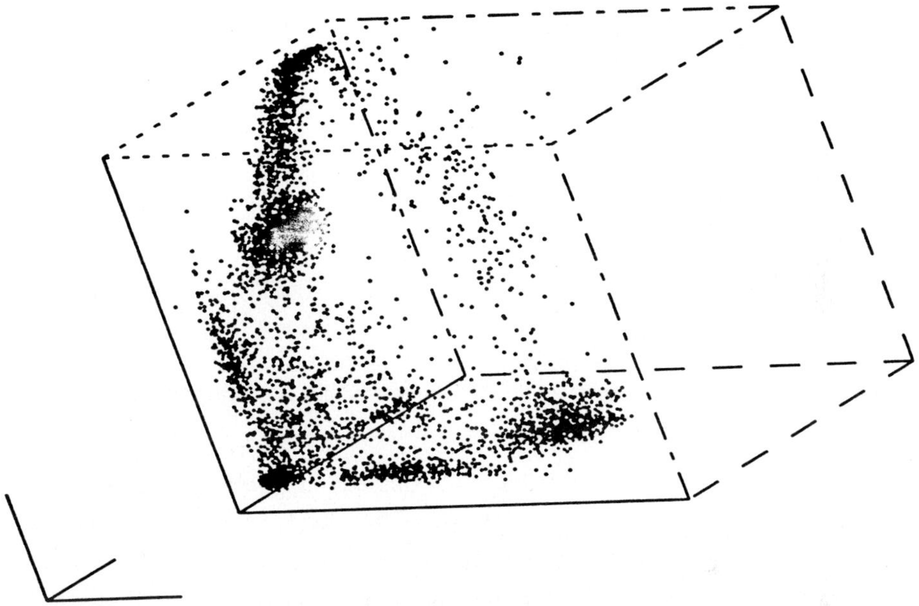

Figure 9: A trivariate dot plot of 3 parameters that demonstrates the difficulties in deciphering spatial relationships among the parameters.

The trivariate equivalent of contour plots is a cloud plot in which surfaces, as compared to contour lines, become the equal count surfaces. As is the case for trivariate dot plots, motion of the figure can bring out details that can be hidden from view in a single presentation. Cloud plots have been used by Kachel and Schneider (1986) to present trivariate FCM data.

When the data contains 4 or more parameters, for flow cytometric data, it is not possible to display the data set in its entirety in a single figure. In general, it becomes necessary to break the data set down into a number of bivariate or trivariate plots. However, there are some techniques, that are equivalent to dot plots, that show complete sets of measurements for a limited number of data points. Such plots are called ribbon plots, profile plots or N-plots. (Salzman 1975, Murphy

1985, Chambers 1983) Figure 10 displays such a plot of 4 parameter data of 150 cells. Each line corresponds to the set of measurements made on one cell. As can be seen, when too many lines are plotted, the figure saturates in a manner similar to the saturation phenomenon that limits bivariate dot plots. When single lines can be followed across the figure, the correlations between the the parameters for a single cell can be determined. When a group of lines clusters together, the signature of a subpopulation of cells can be apparent.

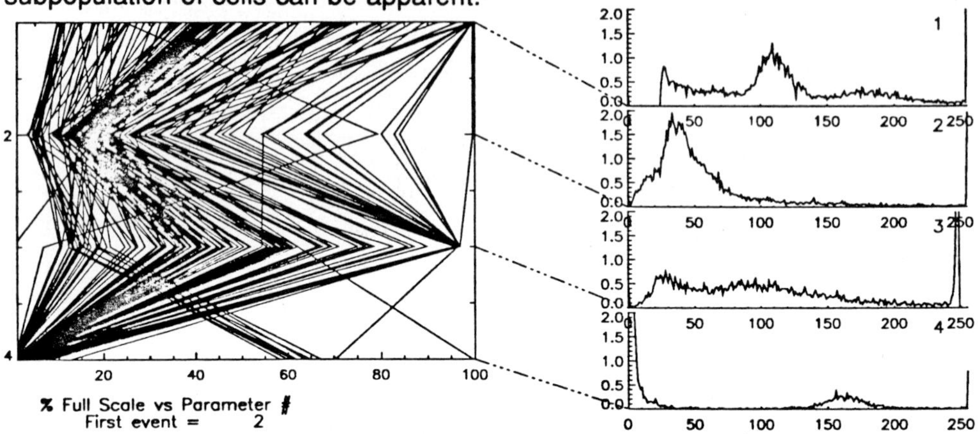

Figure 10. On the left side are profile plots for 4 parameters from an 8 parameter data set. Each line in the plot represents the measurements one one cell. The univariate histograms on the right of the figure show the distribution of each parameter for the complete data set.

Other plotting techniques for multivariate data of 4 or more parameters that are used in flow cytometry leave out some aspect of the data. While it is possible to generate all possible combinations of bivariate plots, that number increases as N(N-1)/2 and is equal to 28 for 8 parameter data. Even if bivariate plots of all possible combinations are generated, the correlations between parameters is hard to realize. However, it is possible to gain insight into the correlations by partitioning the data set. (Chambers 1983) This is accomplished by plotting a series of bivariates and defining a subset of the population in one bivariate space and replotting the data with that subpopulation plotted with a different symbol or color. Figure 4 demonstrates this technique. A region of the bivariate space of parameters 6 and 2 was outlined and all events falling inside the region denoted by a new plotting symbol on the other dot plots.

Box plots (Chambers 1983) can be used to display the individual parameters in a manner that gives one a feeling for the distribution of the parameters. Robinson, et al. (199`) have made extensive use of box plots to summarize immunophenotyping data for a single individual that was collected from multiple samples labeled with a panel of antibodies.

How is multivariate FCM data analyzed ?

Sometimes, the first step in the analysis of a multiparameter set of data involves the arithmetic manipulation of the data. This can be performed on either histograms or on the measurements of individual cells. Bivariate or trivariate histogram manipulation has exact parallels in the univariate world. Histograms can be added, subtracted, normalized, averaged or locally smoothed. An example of the use of histogram addition would be to reduce the statistical uncertainty in each bin by adding together or averaging a number of data sets from the same population of cells. Histograms can be normalized to display them as if they had been derived from some given number of events in the entire bivariate space or in some region of the bivariate space. Transforms developed for image analysis can be applied to sharpen edges and make the data more distinct for display purposes. Channel-by-channel comparisons can be made to determine regions in which two histograms differ significantly. Refer to Bagwell, et al. (1979) for a discussion of the univariate analysis.

The measurements on individual cells can be used to compute new parameters for the cells that are not directly measurable. For example, measurements of the vertically and horizontally polarized components of fluorescence emission are used to compute emission anisotropy. The density of cell surface antigens can not be directly measured. However, with a few assumptions it can be calculated from measurements of cell volume and the log of surface fluorescence (Jett 1980). For each cell, the ratio of cell surface fluorescence to cell surface area computed by raising the cell volume measurement to the two thirds power was computed on a cell-by-cell basis and added to the list mode file as a new parameter. The computed cell surface antigen density was then presented as a histogram or as a box plot. Comparison of surface antigen density distributions was made by comparing the box plots on logarithmic and linear scales.

One technique for the determination of the fractions of cells in S-phase of the cell cycle is by differential staining of cells that have incorporated bromodeoxyuridine (Crissman 1987). In this technique, the difference between the mithramycin and Hoechst 33342 staining determined electronically on a cell-by-cell basis identifies the cells in S-phase. Figure 11 demonstrated that the equivalent information can be obtained by calculation using the primary list mode data. Through digital determination of the difference, the need for a complicated analog difference circuit that is not readily available is eliminated.

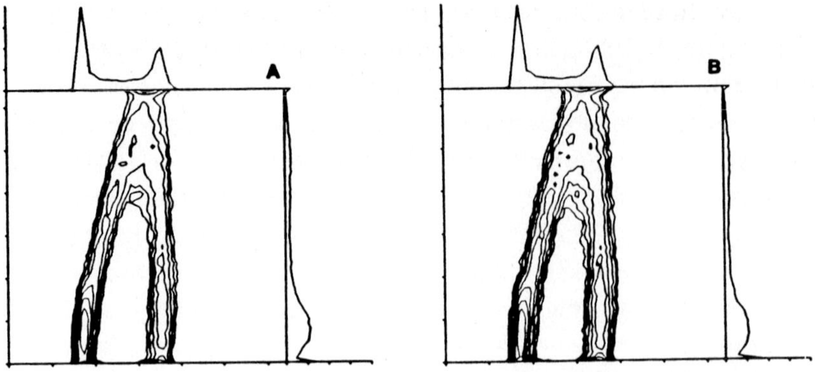

Figure 11. Contour plot A shows the analog difference recorded at the time of data acquisition between the Hoechst and mithramycin measurements versus the mithramycin values. Panel B shown the same parameters using the difference calculated from the Hoechst and mithramycin measurements for each cell. The distributions are visually very similar and yield the same values for the fraction of cells in S-phase. Univariate histograms are shown on the sides and the top of the contour plots.

Often a FCM data set represents measurements on a heterogeneous cell population. By displaying the multivariate data using techniques such as the one described above, it is possible to get a feeling for the number and characteristics of subpopulations. Salzman, et al. (1978) have described techniques for manually partitioning a large data set into smaller data sets that represent single populations. This type of processing or partitioning of a list mode file is actually accomplished by computer. The manual part of the procedure is the setting of limits that define subpopulations. Often the approach is iterative in that the data set is partitioned into 2 subsets based on a single parameter such as one that discriminates live cells from dead cells. The live cell data are then further processed to delineate the identifiable subpopulations.

The application of clustering algorithms is another approach to partitioning of a data set. In this volume Salzman discusses the state of current research into the applications of clustering to FCM data.

Finally, a few words need to be said about the use of computational statistics to analyze FCM data. Many traditional approaches in multivariate data analysis are not applicable to flow cytometric data. This is due to several factors. The distribution of individual parameters is not normal nor are the multivariate distributions normal. If one uses procedures that are designed for hypothesis testing of multivariate normal distributions, the tests will almost always fail. This is due to the fact that the distributions are not normal along with the large number of events in a data set. These combined factors mean that very small deviations from normality become significant.

In general, nonparametric approaches to FCM data analysis have to be used. Texts such as Conover (1980) and Hollander and Wolfe (1973) describe nonparametric approaches to statistical analysis. There are several statistical packages that run on a variety of computing platforms. Among them are SAS, BMDP, and many new programs that run on personal computers. Many of these packages use parametric approaches that assume normal distributions.

Acknowledgements: The helpful discussions with Gary Salzman and Richard Beckman are gratefully acknowledged. This work was supported by the National Flow Cytometry Resource (Grant RR 01315) and the Department of Energy.

References:

Ault K (1990) Applications in Immunology and Lymphocyte Analysis. In: Flow Cytometry and Sorting, Melamed MR, Lindmo T and Mendelsohn ML Editors, Wiley -Liss, New York, NY.

Bagwell CB, Hudson JL and Irvin GL II (1979) Nonparametric flow cytometry analysis. J Histochem. Cytochem. **27**:293-296.

Chambers JM, Cleveland WS, Kleiner B and Tukey PA (1983) Graphical Methods for Data Analysis. Wadsworth &Brooks/Cole Pub.

Conover WJ (1980) Practical Nonparametric Statistics, Second Edition. John Wiley and Sons, New York, NY.

Cleveland WS and McGill ME (1988) Dynamic graphics for statistics. Wadsworth & Brooks/Cole Advanced Books & Software, Belmont, California.

Crissman HA and Steinkamp JA (1987) A new method for rapid and sensitive detection of bromodeoxyuridine in DNA-replicating cells. Exp. Cell Res. **173**:256-261.

Crissman HA, Darzynkiewicz Z, Tobey RA and Steinkamp JA (1985) Normal and perturbed Chinese hamster ovary cells: correlation of DNA, RNA and protein content by flow cytometry. J. Cell Biol. **101**:141-147.

Dean PN, Bagwell CB, Lindmo T, Murphy RF and Salzman GC (1990) Introduction to flow cytometry data file standard. Cytometry **11**:321-322.

Donoho AW, Donoho DL and Gasko M (1988) In: Dynamic graphics for statistics. Wadsworth & Brooks/Cole Advanced Books & Software, Belmont, California.

Hollander M and Wolfe DA (1973) Nonparametric Statistical Methods. John Wiley and Sons, New York, NY.

Jett JH, Stevenson AP, Warner NL and Leary JF (1980) Quantitation of cell surface antigen density by flow cytometry. In: FlowCytometry IV, Laerum OD, LindmoT and Thorud E, Editors; Universitetsforlaget, Bergen, Norway

Johnson R, Bartholdi M, Hiebert R, Parson J and Cram LS (1985) A slit-scan flow cytometer for recording simultaneous waveforms. Rev. Sci. Ins. **56**:691-695.

Kachel V and Schneider H (1986) On-line three-parameter data uptake, analysis and display device for flow cytometry and other applications. Cytometry **7**:25-40.Lucas JN, Mullikin JC and Gray JW (1991) Dicentric chromosome frequency analysis using slit-scan flow cytometry. Cytometry **12**:316-322.

Mann RC (1985) On multiparameter data analysis in flow cytometry. Cytometry **8**:184-189.

Moore WA and Kautz RA (1986) Data analysis in flow cytometry. In: Handbook of Experimental Immunology, Volume 1. D. M. Weir, Editor, Blackwell Scientific Publications Oxford, England.

Murphy RF (1985) Automated identification of subpopulations in flow cytometric list mode data using cluster analysis. Cytometry **6**:302-309.

Robinson RD, Wheeless DM, Hespelt SG and Wheeless LL (1990) System for acquisition and real-time processing of multidimensional slit-scan flow cytometric data. Cytometry **11**:379-385.

Robinson JP, Durack G and Kelley S (1991) An innovation in flow cytometry data collection and analysis producing a correlated multiple sample analysis in a single file. Cytometry **12**:82-90.

Salzman GC, Crowell JM, Goad CA, Hansen KH, Heibert RD, LaBauve PM,

Martin JC and Mullaney PF (1975) A flow-system multiangle light-scattering instrument for cell characterization. Clin. Chem. **21**:1297-1304.

Salzman GC, Hiebert RD and Crowell JM (1978) Data acquisition and display for a high-speed cell sorter. Comput. Biomed. Res. **11**:77.

Salzman G (1992) Expert Systems in flow cytometry.(To be included in this volume.)

Steinkamp JA, Stewart CC, and Crissman HA (1982) Three color fluorescence measurements of single cells excited at three laser wavelengths. Cytometry **2**:226-231.

Watson JV, Hornsnell TS, Smith PJ (1988) Data compression: 8-dimensional flow cytometric data processing with 28K addressable computer memory. J Immunol Methods **113**:205-214.

EXPERT ASSISTANCE FOR FLOW IMMUNOPHENOTYPING

Gary C. Salzman, Richard J. Beckman, Carleton C. Stewart[1], and Sigrid J. Stewart[1]
Los Alamos National Laboratory, Los Alamos, New Mexico 87545, U.S.A.

INTRODUCTION

There is a serious data analysis bottleneck for flow immunophenotyping in the clinical laboratory when many parameters are required. Multiple tubes of cells are measured on the flow cytometer for each patient peripheral blood or bone marrow specimen. Each tube requires only a few minutes to measure five variates on each of 7000 or more cells. The data for each patient, however, may require 20 to 40 minutes to analyze with the manual methods now in common use. This paper describes the use of cluster analysis to replace the manual gating methods used to find the leukemic or other abnormal cell populations. It then discusses the use of an expert system to assist in describing the immunophenotype of the specimen. Since the literature on expert systems in cytometry has been reviewed recently (Salzman and Bartels, 1992), this article does not include such a review.

Cluster analysis is a mathematical technique used here for assigning multivariate flow cytometry events to a group in an n-dimensional space, in which n is the number of variates or parameters measured with the flow cytometer. The data are presented as a series of bivariate dot plots. A number is assigned to each group and used to associate a specific color with the dots in the group. The various populations of cells in the bivariate dot plots can then be visualized easily. A population of cells will have the same color in each of the dot plots. Clustering methodology is described in more detail below.

----

[1]Laboratory of Flow Cytometry, Roswell Park Cancer Institute
666 Elm Street, Buffalo, New York 14263, U.S.A.

NATO ASI Series, Vol. H 67
Flow Cytometry
Edited by A. Jacquemin-Sablon

MANUAL DATA ANALYSIS METHOD

In the Laboratory of Flow Cytometry at Roswell Park Cancer Institute, blood specimens are labeled with cocktails containing multiple antibodies. Ten panels, each containing three fluorescently-tagged monoclonal antibodies for a total of 30 antibodies, are used for the labeling. Panels for measuring autofluorescence and cell viability are also included. When the twelve tubes of labeled cells have been run on the flow cytometer, the analyst creates bivariate dot plots of the list mode data from each tube. He or she then searches through the bivariate fluorescence dot plots looking for the panel in which a possible leukemic cell population is best resolved from the other cell populations. A polygon region is then drawn around this population and the list mode data reanalyzed ("backgated") to produce a dot plot to see where the dots from these cells are located in the bivariate display of forward scatter (FSC) versus side scatter (SSC). Another polygon region called the scatter region is then drawn around the leukemic cell population in this bivariate display and the first polygon region is removed. The scatter region polygon is used to filter ("forward gate") the list mode data for all the other panels. The immunophenotype for leukemic cells can then be determined by inspecting the fluorescence dot plots for each panel.

CLUSTER ANALYSIS

The goal of cluster analysis is to partition a list mode data set into groups or clusters of points such that points in a cluster are closer to each other than to members of other clusters. The K-means algorithm used here partitions a dataset into disjoint subsets (Spath, 1985; Everitt, 1986). The user chooses K, the number of cluster means. Analytic methods for choosing K are discussed below. Cluster number labels are assigned at random to each event. An event is the vector of list mode data for one cell. The within cluster sum of squares (WSS) is then computed for each cluster. WSS is the sum of the squared distances from a cluster mean to each of the members of the cluster. The objective function is defined as the sum of all the WSS. A large value of WSS for a cluster indicates that the cluster is diffuse and may contain members that should be in other clusters. The algorithm then iterates over the dataset, changing the cluster labels of events and recomputing the WSS for each cluster and the objective function for the dataset. The algorithm terminates when only a few events are being successfully relabeled during each iteration.

The correct number of clusters for a dataset can be determined using one of the many cluster termination algorithms (Davies and Boulton, 1979; Milligan and Cooper, 1985). Many of these algorithms search for a minimum in the ratio of the objective function defined above to the between cluster sum of squares (BSS), which is the sum of the squared distances between the grand mean of the cluster centroids and each cluster centroid. The Davies-Boulton cluster separation measure is particularly useful for flow cytometry data.

K-means clustering often uses the Euclidean metric as a measure of the distance between events in the multivariate space. With this distance measure, K-means clustering tends to create "spherical" clusters even if the shape of a cluster is not spherical. The use of the Euclidean metric on two ellipsoidal shaped clusters lying close to and parallel to one another will result in both clusters being cut in half by the clustering routine. This problem can be dealt with by using the Mahalanobis distance (Everitt, 1986), which takes into account correlations among the variates. Since flow cytometry datasets often contain correlated events, this distance measure is frequently used.

CHOOSING THE DENSEST CLUSTER

In the manual data analysis method described above, the criterion for identifying the best-resolved cluster can be translated into a search for the cluster with the lowest coefficient of variation. This criterion often finds the leukemic cell population because it tends to be more homogeneous than the other cell populations. Figure 1 shows all the bivariate dot plot displays for one panel of a bone marrow specimen from a patient with acute myelogenous leukemia (AML). FSC stands for forward scatter and SSC stands for side scatter. FL1 is fluoresceinated CD16, which labels granulocytes or lymphocytes, FL2 is phycoerythrin-labeled CD64, which marks monocytes, and FL3 is Tandem-Conjugate labeled CD32, which marks granulocytes or monocytes. Cluster analysis using three clusters and the Mahalanobis distance produces a group of color bivariate dot plot displays. Figure 2 shows a group of bivariate dot plot displays for the densest cluster, which is that for the leukemic cell population. This technique can be applied to each of the 10 panels to identify the leukemic cell population. Then expert systems methods can assist in interpreting the data.

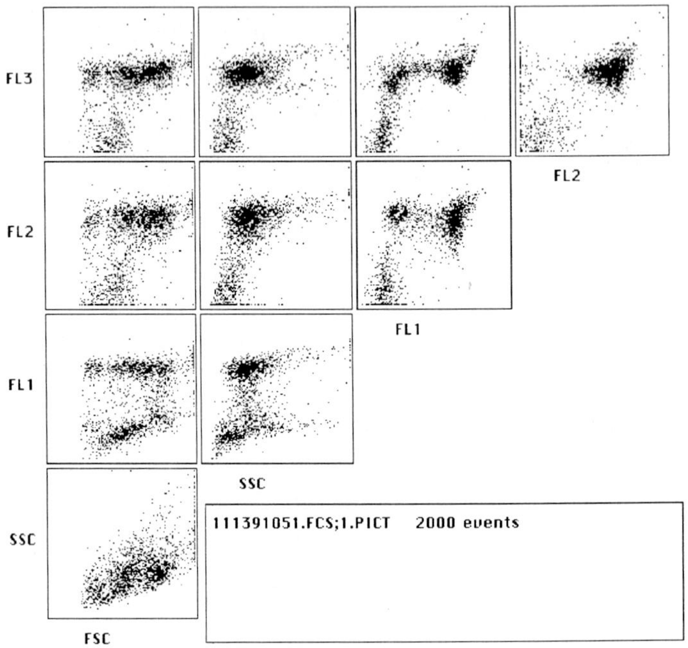

Fig. 1. All the bivariate dot plot displays for one panel of a bone marrow specimen from a patient with acute myelogenous leukemia (AML). The axes labels are described in the text.

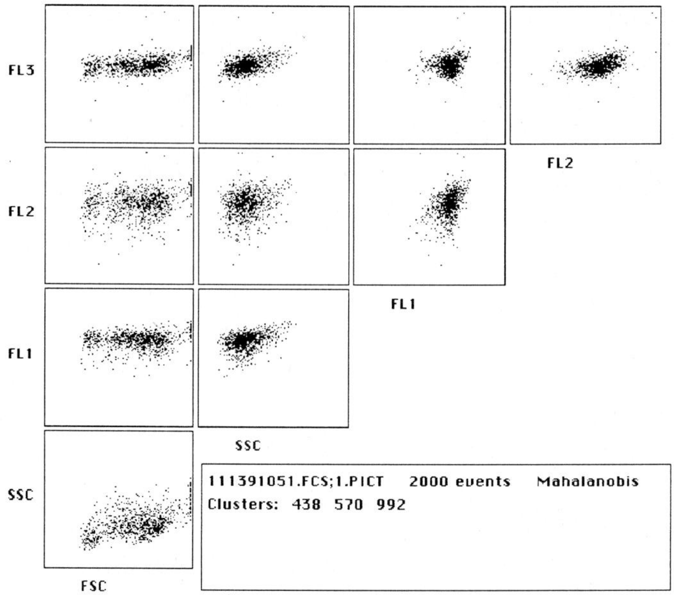

Fig. 2. All the bivariate dot plot displays for the densest cluster for the data shown in figure 1.

EXPERT SYSTEM

Expert systems are computer programs that attempt to emulate the problem-solving activities of human experts working in narrowly focused domains. These programs contain specific knowledge about the narrow domain in which they exhibit expertise. Expert systems can combine the problem-solving ability of a number of experts and make that capability broadly available at low cost. Expert systems should not be used for problems that can be solved algorithmically or for those that require creative invention or broad common sense knowledge. Expert systems can be expected to provide robust performance, to have the ability to do symbolic reasoning, to handle difficult problems with many complex rules, and to have self-knowledge so that they can explain their reasoning to a user.

In the medical area, there are four major types of expert systems. A diagnostic expert system provides a classification based on observed clues. A prognostic expert system predicts outcomes based on observed facts and data from previously diagnosed cases. An interpretative expert system provides evaluations based on input data. A design expert system works with constraints to calculate parameters for some action, such as radiation treatment planning.

An expert system performs symbolic reasoning by manipulating symbols that stand for concepts in the domain. The concept that a fluorescent dye emits strongly when excited by a laser is captured by the symbol "bright fluorescence" that can be manipulated by an expert system. The symbolic reasoning is facilitated by a variety of knowledge representation schemes, the most popular of which are rule-based systems. The three major pieces of an expert system are the knowledge base containing the rules, the working memory containing the facts, and the inference engine, which does pattern matching between the facts and the premises of the rules. When a fact in the working memory matches the premise of a rule, the rule "fires" and the conclusion of the rule is added to the working memory.

The C-Language Integrated Production System (CLIPS) expert system development tool is used for the work described here (Giarratano and Riley, 1989). It was developed by the U.S. National Aeronautics and Space Administration (NASA) and is continually being extended. Facts in CLIPS consist of a pattern of one or more fields enclosed in matching parentheses. A field is a character string with no spaces. For example, three facts extracted from list mode data containing a granulocyte population might be

```
(panel 11 CD 16 bright)
(panel 11 CD 64 negative)
(panel 11 CD 32 bright)
```

A rule contains a premise and a conclusion. A granulocyte identification rule in CLIPS is

```
(defrule conclude-grans "granulocytes"
        (panel ? CD 16 bright)
        (panel ? CD 64 negative)
        (panel ? CD 32 bright)
=>
        (cell-ID granulocytes))
```

The premise of this rule is the AND conjunction of the three fact patterns before the implication (=>). The question mark (?) is a wild card allowing a field in the appropriate position to match it. For the three facts above the question mark would be matched by the number 11. Since the facts above match the premise of this rule, the rule "fires" and adds the fact pattern of its conclusion to the list of facts in the working memory. The fact list then becomes

```
(panel 11 CD 16 bright)
(panel 11 CD 64 negative)
(panel 11 CD 32 bright)
 (cell-ID granulocytes)
```

For the diagnostic expert system that is being developed, cluster analysis is performed on the list mode data from each of the panels using an appropriate number of clusters. This number is determined based on the experience of the clinical laboratory experts. Rules for each panel identify the leukemic cell population. Other rules determine the immunophenotype.

ACKNOWLEDGMENTS

This work was performed under the auspices of the U.S. Department of Energy and was supported in part by U.S. National Cancer Institute Grant Number CA54518 (GCS), in part by the National Flow Cytometry and Sorting Research Resource (Grant Number P41-RR01315 (GCS)) and in part by U.S. National Institute of Allergy and Infectious Diseases Grant Number AI19490 (CCS).

REFERENCES

Davies DL, Boulton DW (1979) A cluster separation measure. IEEE Transactions on Pattern Analysis and machine Intelligence PAMI-1:224-227.

Everitt B (1986) Cluster analysis. 2nd edn. John Wiley & Sons, New York.

Giarratano J, Riley G (1989) Expert Systems. Principles and Programming. PWS-Kent Pub. Co. Boston. CLIPS is available from the NASA software distribution center, COSMIC, 382 E. Broad Street, Athens, GA 30602 USA. (404)542-3265.

Milligan GW, Cooper MC (1985) An examination of procedures for determining the number of clusters in a data set. Psychometrika 50:159-179.

Salzman GC, Bartels PH (1992) Expert systems for cytometry data analysis. In: Flow cytometry: principles and clinical applications. KE Bauer, RE Duque, TV Shankey, eds. Williams & Wilkins (in press).

Spath H (1985) Cluster dissection and analysis. Theory, FORTRAN programs and examples. Halsted Press. John Wiley & Sons, New York.

DNA CONTENT AND CELL CYCLE ANALYSIS BY FLOW CYTOMETRY IN CLINICAL SAMPLES : APPLICATIONS IN CANCER

F.Spyratos
Laboratoire de Biologie Tissulaire
Centre René Huguenin
35 rue Dailly
92211 St-Cloud
France

Flow cytometric DNA analysis has been established as a useful research tool for a number of years. The methods have been used for monitoring the stability of cell lines, for sensitivity testing, and for studying the action of anticancer drugs *in vitro.* Likewise the methods have been used to monitor the cell cycle perturbations produced by radiation, chemotherapy, and hormonal substances in murine and human tumors in experimental animals, and in malignant tumors in patients. Numerous papers have been published using DNA FCM measurements to characterize a variety of solid tumors. DNA analysis is currently assessed as a possible prognostic parameter in neoplastic disease. However, standard quality control parameters, are not universally in place in the clinical and research laboratories at the present time. Before standardization can occur, definitions for ploidy status, criteria for evaluability, and modeling parameters for S-phase need to be understood and well described.

Samples

Samples for FCM DNA analysis can be classified into two groups :
1- fresh tissue specimens (or frozen specimens) obtained by a surgical or biopsy procedure or by cytopuncture or fine-needle aspirate;
2- paraffin-embedded materials make it possible to do retrospective studies to investigate the relationship between DNA measurements and prognosis of patients with malignant tumors.
A perequisite for FCM measurement is to obtain single cell or nucleus in suspension, and the preparation of dispersed cells is a crucial procedure for FCM. Disagregated cells must be representative of the tissue in situ without preferential selection of particular cell sub-populations. Cells (or nuclei) in suspension should retain the original characteristics to be measured. In addition, it is desirable that the degree of cell recovery per unit volume of tissue be high. The ability to resolve subpopulations of cells with minor alterations in DNA content is dependent on technical factors, and is usually assessed through the

NATO ASI Series, Vol. H 67
Flow Cytometry
Edited by A. Jacquemin-Sablon

coefficient of variation (CV) of the $G_{0/1}$ peaks; the lower the CV, the better the resolution.

- Fresh tissue specimens

Several kinds of sample preparation methods, which permit suspended cells or nuclei to be obtained, are available for FCM measurements of the nuclear DNA content. The techniques widely employed for dispersing solid tumor cells are divided into two categories : first, enzymatic procedure which yields single-cell suspensions and makes it possible to collect information from cytoplasm : second, detergent-based method with which it is possible to obtain nuclear suspensions. In this case, DNA-unspecific fluorescence from the cytoplasm is avoided, but the possibility of simultaneous analysis of cytoplasmic or membrane-associated structures is thus sacrificed. The enzymatic dispersion methods are cumbersome and time consuming, compared to those using nonionic detergents, such as Triton X-100 and Nonidet P-40. The procedures developed by Vindelov, Krishan and Taylor are widely distributed throught the world because of simplicity in methodology. The analysis is performed on unfixed material. This is essential to avoid a potentially selective cell loss caused by centrifugation steps, and keeps the requirement of cells at a minimum. Clumping and staining artifacts caused by a fixative are avoided.
Samples can be long-term stored by freezing in a citrate buffer with dimethyl-sulfoxide (DMSO) (Vindelov1983).

- Paraffin-embedded tissus specimens

Retrospective studies including patients with long-term follow-up data have been of critical importance in investigating the relationship between DNA ploidy and prognosis of patients with malignant tumors. A sample preparation method for FCM DNA analysis using paraffin-embedded tissue specimens was introduced by Hedley et al. DNA histograms from paraffin-embedded specimens usually show good agreement with those from fresh specimens regarding DNA ploidy (Kallionemi 1988) but there are also reports that DNA aneuploid sub-populations detected in fresh tissue specimens cannot be found in paraffin-embedded specimens .
The common problems are excessive cell debris and broad peaks, and often these coexist. Pathological materials are usually fixed with formalin which provides nuclei with disadvantages for FCM DNA analysis. Neutral buffered formalin of formalin-acetone-acetic acid generally gives satisfactory results, while the use of more specialized fixatives such as Bouin's fluid or the mercury-based fixatives often results in uninterpretable DNA histograms. With increasing fixation time in formalin the coefficients of variation (CVs) detoriate and G_0/G_1 peaks shift to the left. Under such a condition, aneuploid peaks close to a diploid peak may result in a single peak with a large CV. Accordingly, DNA aneuploid sub-populations with minimal aberration of DNA content is potentially not detected in poor quality histograms

and misinterpreted as diploid. When paraffin-embedded specimens are used, data are unavailable in at least 10-15% of cases, independent of tumor type, because of poor resolution. A problem in the analysis of formalin-fixed material is the lack of a suitable reference for estimates of the ploidy level. Normal cells inherently includes in the tumor samples can be used as an internal reference, but reliable internal standards that can be externally added to the suspension samples are required when we are obliged to determine the diploid peak in histograms with multiple peaks. The position of the diploid sub-population in nuclear suspensions from paraffin-embedded specimens is frequently different even between samples processed in the same manner.

With respect to the resolution of DNA analysis, fresh specimens are superior to paraffin-embedded ones. Sometimes the validity of the data using paraffin-embedded materials may be questionable .

More improvements in methodology and data analysis, standardization of preparatory techniques, criteria for interpretation of histograms and quality control are required before clinical application of flow cytometric analysis to paraffin-embedded specimens.

Nomenclature

In order to distinguish results obtained by DNA flow cytometry from data obtained by cytogenetic techniques, it was recommended by the Convention on Nomenclature in 1984, that terms "normal" and "abnormal stemline" be used instead of diploidy and aneuploidy, respectively (Hiddemann et al., 1984). The term "DNA aneuploid" was agreed on as being a synonym for "abnormal stemline".

DNA ploidy status of tumors is defined based on the amount of DNA relative to normal. The DNA index (DI) is a value given to express the amount of DNA content relative to normal and is calculated by the following equation.

$$DI = \frac{\text{mean or modal channel no. of DNA aneuploid Go/G1 peak}}{\text{mean or modal channel no. of DNA diploid Go/G1 peak}}$$

DNA staining

Usually DNA is stained with propidium iodide routinely after addition of RNase for the possible elimination of nonspecific staining of double stranded RNA. This technique is applicable to various kinds of tumor specimens including biopsy samples as small as fine needle aspirates or cytopunctures (Briffod 1989, Remvikos 1989, Spyratos 1987, 1992).

A wide variety of fluorescence probes which bind stoichiometrically to DNA have been developed for measurement of nuclear DNA content. Propidium iodide which finds to double stranded polynucleotides (DNA

and RNA) is mostly popular for FCM measurement of nuclear DNA. Prior to DNA staining, RNA should be removed from nuclei. This dye can be excited at a wavelengh of 488 nm emitted from the argon ion laser with which the conventional type of flow cytometer is equipped. Propidium iodide is dropped into nuclear suspensions at a final concentration of 50 microliters/ml. However, the probes is too highly charged to cross the membrane of living cells. For practical use, Hoechst 33342 is the only available fluorescence dye employed in cell cycle analysis for living cells.
It should be emphasized that only a fraction of nuclear DNA is accessible to dyes and that the extent of DNA accessibility varies depending on the dye used, chromatin structure, and staining procedure.

Standards

A DNA reference standard is essential to the determination of relative DNA content. Both internal and external standards have been used as normal DNA content reference standards from a variety of sources: chicken red blood cells (CRBC), trout red blood cells (TRBC). This is achieved by adding a mixture of chicken (CRBC) and trout erythrocytes (TRBC) to the sample before staining (Vindelov et al., 1983). This peaks produced by the standards provide two points of reference at DI value of # 0.30 (CRBC) and 0.80 (TRBC), and allow DI determination independent of zero-point shift. However, occasionally, the dual standard of trout and chicken erythrocytes may lead to systematic inaccuracy in the DNA content measurement because of their low DNA content (Hoehn 1977). Other standards include commercially available fluorescent reference beads or fixed CRBCs to aid in instrument alignment and setting up gates or thresholds on the flow cytometer.
However, the further addition of internal standard cells to the sample is unecessary for most cases; indeed, human tumors are inherently contaminated with a large amount of non-tumor cells such as leukocytes, fibroblasts and endothelial cells.

Controls in DNA flow cytometry

Controls as well as defined criteria to interpret DNA histograms are essential components of DNA flow cytometry methodology.
- Cytologic examination of specimen is necessary to evaluate the quality of the specimen and to assess the percentage of tumor versus nontumor cells present.
- Microscopic review of the fluorescently stained cells of nuclei. provides information regarding specific uptake of dye into the nuclei , and the degree of clumping. Microscopic screening of histologic and fluorescent samples thus allows one to evaluate the quality of the specimen and may also be helpful in the interpretation of difficult DNA histograms.

Criteria for evaluation and interpretation of DNA histograms

The interpretation of DNA histograms is very important in flow cytometric analysis since the investigation of the relationship between DNA ploidy and the disease course is largely affected by DNA histogram analysis. Although guidelines for DNA ploidy analysis were proposed by the Society for Analytical Cytology, classification of histograms is made subjectively by each investigator, and many interlaboratory differences concerning the interpretation of histograms have also been reported (Joensuu 1989).
The use of certain criteria help the investigator to evaluate and interpret a DNA histogram. These criteria may not be applicable to all tumor types, and individual sets of criteria may be necessary for disease groups and specific studies, especially if the specimen has been obtained from a patient undergoing treatment.

A defined set of evaluation criteria are required that include
- limits and definitions of the coefficient of variation (CV) of the mean DNA content of the G0G1 cell subpopulation. Its value corresponds to the ability to resolve two G1 peaks in mixed-cell populations and helps to assess quality of sample preparation and instrument alignment. The larger the CV, the wider the peak and the greater the chance of missing and abnormal population. In addition, the wider the CV, the less accuracy will be achieved in trying to estimate the S-phase region of the histogram. Therefore criteria for CV are often set to allow for distinction between which histograms are evaluable and which are not. It is important to define in all published studies. Histograms with CV value of <5% are generally acceptable to most investigators .
- the DNA index (DI),
- specific definitions for DNA content and the definition of a DNA aneuploid peak (normal ie DNA diploid versus abnormal (DNA aneuploid) and subgroups of abnormal (hypodiploid, hyperdiploid, near diploid, near tetraploid, tetraploid, hypertetraploid, and multiploid). Investigators who have reported definitions for DNA aneuploid generally only consider an abnormal stemline present when two distinct G0/G1 populations are observed. In our laboratory, a peak is defined as being DNA aneuploid if at least 10% of the total events collected in the histogram are found in the G0/G1 peak and a corresponding G2M can be identified (Spyratos, 1987).
- criteria for evaluability of cell cycle analysis should also be defined.

In some cases, histograms may be doubtful (inability to resolve a shoulder, wide CV, discrete peak falling short of 10% of the total collected events) or uninterpretable (all debris, unsufficient number of cells for analysis, or poor resolution). In these cases, the assay should be repeated if there is sufficient material available.

In particular, such problems may be exacerbated in cases of analysis using formalin-fixed paraffin-embedded material in which not only large CVs are produced but also an extrinsic internal standard is unavailable. Therefore, the nuclear DNA analysis for paraffin-embedded material is not recommended with respect to the resolution of DNA analysis. Differences in processing of sample preparation do not largely affect the results if fresh specimens are available. DNA ploidy is judged based on the presence of absence of abnormal peaks with a normal diploid peak. In most tumors with DNA aneuploidy, abnormal and normal peaks are clearly separated from each other and it is easy to determine their DNA ploidies. The degree of DNA content aberration is expressed as the DNA index (DI). If CVs 0.01 are achieved and yet the difference in DIs between two peaks is greater than 0.02, their peaks are separated. It is usually difficult to discern a difference in DIs less than 0.02 between two different cell populations. Taking into account the fact that human cell nuclei contain approximately 7×10^9 base pairs DNA per diploid, a difference in DI of 0.02 would be equivalent to a difference in DNA content of 1.4×10^8 base pairs, which almost corresponds to the average DNA content of one chromosome. DNA aneuploidy detected by FCM indicates considerable aberration of nuclear DNA content. When there are peaks in different samples taken from the same tumor, and the differences in DNA indices between different histograms are more than 10% their DNA ploidies are considered different.
In addition to the determination of DNA ploidy, the estimation of S-phase fraction (SPF) is the major purpose of the FCM DNA measurement of tumor cells. The SPF is easily estimated by computer programs. Occasionally, however, these programs may cause confusion.

Now is the time to standardize sample preparation methods including nuclear staining, interpretational criteria of histograms and mathematical analysis.

Heterogeneity in DNA ploidy

Flow cytometric DNA analysis has revealed DNA aneuploid peak(s) in a variety of human malignant tumors, especially in solid tumors obtained from patients with clinically poor prognosis. Usually nuclear DNA ploidy has been determined in a single specimen obtained from tumors. There was a general assumption that most human tumors consist of a single DNA stemline, as the incidence of multiclonality was less than 10%. However, recent FCM studies have disclosed regional differences in DNA ploidy in the same tumor. The incidence of DNA ploidy heterogeneity was thought to be relatively low (5-20%).

Proliferative fraction

By FCM DNA measurement we can automatically obtain at least two parameters; DNA ploidy and three cell cycle phases. In general, the proportion of S and G2M phase is higher in actively growing tumors than in slowly growing tumors. It is well-known that the thymidine labelling index (TLI) reflects the prognosis of cancer patients; a high TLI indicates poor survival. Although it is doubtful as to whether the TLI is equal to the S-phase fraction (SPF) estimated by FCM, the SPF is used as an indicator for the aggressiveness (growth activity) of a tumor which influences the survival of patients. Tumors with DNA aneuploidy and high SPF are usually more aggressive than diploid tumors with low SPF. Considerable heterogeneity in SPF was found in multiple samples from the same tumor. There are many ways of analyzing DNA histograms and several computer programs for an automatic analysis of the cell cycle are now available from commercial or private sources. A tumor consists of two distinct populations; normal and tumor cells. The difficulty arises for cell cycle analysis in many cases, especially in tumors without DNA aneuploidy. In the cases, especially in tumors without DNA aneuploidy. In the cases of diploid tumor, a tumor cell population cannot be distinguished from normal cells on the basis of DNA content. However, if there is a considerable difference in DNA content between diploid and aneuploid populations, cell cycle analysis for each population is possible using the broadned polynomial method. Software packages are now also commercially available.

Clinical implications

Although clinical implications of nuclear DNA measurement remain controversial, many papers report that flow cytometric measurements of nuclear DNA content is useful to predict prognosis in several types of human cancer. In general, DNA aneuploidy determined by FCM has been considered to be associated with the low survival rate in cancer patients. Now clinical adoption of FCM nuclear DNA analysis is being performed for a variety of human tumors, with the aim of objective selection of treatment for patients with malignant tumor and prediction of their prognosis.

Tumor diagnosis
It is well establishen that numerical and/or structural abnormalities of chromosomes are found in almost all cases of malignant tumors. Cytogenetic abnormalities can be estimated by measuring nuclear DNA content with a flow cytometric technique. The presence of an abnormal DNA stemline, a subpopulation with DNA aneuploidy, is a reliable tumor marker available to flow cytometry, under the assumption that nonmalignant lesions have normal DNA ploidy. This rule can be applied to a variaty of human solid tumors. Although DNA aneuploidy was found in only 50-70% of malignant tumors as indicated by the earlier

studies, the frequency of DNA aneuploidy in malignant tumors may be much higher than these values. It must be remembered, however, that the absence of DNA aneuploidy does not preclude a diagnosis of malignant tumor. Nuclear DNA content can be easily measured on both biopsy and aspirated specimens with high reproducibility. DNA ploidy determined for a single sample has been considered to be representative of the entire tumor and DNA aneuploid clones are usually demonstrated in all parts of the tumor. However, there are cases in which subpopulations with DNA aneuploidy can be found only in the limited parts of the same tumor. Accordingly, multiple sampling from a tumor is recommended for cancer diagnosis using FCM measurement of nuclear DNA content.
The values of the DNA index in primary lesions are usually the same as those in metastatic sites. Accordingly, the measurement of DNA content is useful for establishin the relationship between primary and metastatic sites in cases with multiple tumors.

Prognosis
There are many studies focusing on the relationship of DNA ploidy to prognosis of patients with malignant tumors. Although earlier workd suggested that DNA ploidy was a major independent prognostic factor in a variety of tumors and that it was useful for predicting a clinical course for cancer patients, the relationship between DNA ploidy and patient outcomeis stil controversial. A technique for formalin-fixed paraffin-embedded specimens by Hedley and his colleagues facilitates large retrospective studies for archival material. It is easy to arrange homogeneous subjects with respect to tumor type, stage and treatment, when paraffin-embedded materials can be used. This is extremely important for examining the clinical relevance of DNA ploidy, because cancer patient outcome depends on a variety of factors such as stage of disease, histological types and therapy. However, we must always keep in mind the limitation in the analysis of formalin-fixed, paraffin-embedded samples, because resolution for them is inferior to that using fresh tissue specimens.
In markedly advanced tumors, such as Stage IV cases, DNA ploidy is not useful for predicting prognoses. On the other hand, in early stage cancer, DNA aneuploidy has great impact on prognosis ; DNA aneuploidy is associated with a poorer prognosis in a variety of malignant tumors.
It has been thought that the high percentage of S phase (SPF) indicates poor prognosis as well as DNA aneuploidy. The SPF data may be more useful than DNA ploidy as independant prognostic predictors in cancer patients. There is a tendency for SPF to be higher in aneuploid tumors than in diploid tumors, and the prognosis of cancer patients should be estimated on the basis of multiple parameters.

DNA content and S-phase fraction measured by flow-cytometry may also be useful as indicators of chemosensitivity to chemotherapy (Briffod 1989, Spyratos 1992).

DNA analyses and S-phase fraction are currently assessed as possible prognostic parameters in neoplastic diseases (Merkel 1987,1990). Final evaluation awaits the results of large multivariate prospective studies of uniformly staged and treated patients, examined with high resolution DNA analyses and quality control programs.

Briffod M, Spyratos F, Tubiana-Hulin M, Pallud C, Mayras C, Filleul A, Rouessé J (1989) Sequential cytopunctures during preoperative chemotherapy for primary breast carcinoma. Cytomorphology changes, initial tumor ploidy, and tumor regression. Cancer , 63, 631-637.

Clark GM, Dressler LG, Owens MA, Pounds G, Oldaker T and Mc Guire WL (1989) Prediction of relapse or survival in patients with node-negative breast cancer by DNA flow cytometry. New Engl J Med 320: 627-633

Dressler LG (1990) Controls, standards, and histogram interpretation in DNA flow cytometry, in : Darzynkiewicz Z, Crissman HA (Eds) Methods Cell Biol, Academic Press, San Diego, California, 33: 157-171

Gnant MFX, Blijham G, Reiner A, Reiner G, Reynders M, Schutte B, van Asche C, Steger G and Jakesz R (1992) DNA ploidy and other results of DNA flow cytometry as pronostic factors in operable breast cancer: 10 year results of a randomised study. Eur J Cancer 28: 711-716

Haag D, Feichter G, Goerttler K and Kaufmann M (1987) Influence of systematic errors on the evaluation of the S phase portions from DNA distributions of solid tumors as shown for 328 breast carcinomas. Cytometry 8: 377-385

Hedley DW (1990) DNA analysis from paraffin-embedded blocks. Methods cell biol.33: 139-147

Hiddemann W, Schumann J, Andreef M (1984) Convention on Nomenclature for DNA cytometry. Cytometry, 4 , 445-446

Hoehn H,Johnston P, Callis J. (1977) Flow cytogenetics : sources of DNA contentvariation among euploid individuals. Cytogenet Cell Genet 19, 94-107

Joensuu H, Kallionemi OP (1989) Different opinions on classification of DNA histograms produced from paraffin embedded tissue cytometry 10,711-717

Kallionemi OP. Comparison of fresh and paraffin-embedded tissue as strating material for DNA flow cytometry and evaluation of intra-tumor heterogeneity (1988) Cytometry 9, 164-169.

Kallioniemi OP, Blanco G, Alavaikko M, Hietanen T, Mattila J, Lauslahti K, Lehtinen M and Koivula T (1988) Improving the pronostic value of DNA flow cytometry in breast cancer by combining DNA index and S-phase fraction. Cancer 62: 2183-2190

Mc Guire WL, Tandon AK, Allred DC, Chamness GC, Clark GM (1990) How to use prognostic factors in axillary node-negative breast cancer patients. J. Natl. Cancer Inst. 82: 1006-1015

Merkel DE, Dressler LG and Mc guire WL (1987) Flow cytometry, cellular DNA content, and prognosis in human malignancy. J Clin Oncol 5: 1690-1703

Merkel DE and Mc guire WL (1990) Ploidy, proliferative activity and prognosis. Cancer 65: 1194-1205

Meyer JS and Coplin MD (1988) Thymidine labeling index, flow cytometrics S-phase measurement, and DNA index in human tumors. AJCP 89: 586-595

Remvikos Y, Magdelenat H and Zajdela A (1988) DNA flow cytometry applied to fine needle sampling of human breast cancer. Cancer 61: 1629-1634

Sahin AA, Ro JY, El-Naggar AK, Wilson PL, Teague K, Blick M and Ayala AG (1991) Tumor proliferative fraction in solid malignant neoplams. Anat Pathol 96: 512-519

Scott N, Cross D, Plumb MI, Dixon MF and Quirke P (1992) An investigation of different methods of cell cycle analysis by flow cytometry in rectal cancer. Br J Cancer 65: 8-10

Sigurdsson H, Baldetorp B, Borg A, Dalberg M, Fernö M, Killander D Olsson H and Ranstam J (1990) Flow cytometry in primary breast cancer: improving the prognostic value of the fraction of cells in the S-phase by optimal categorisation of cut-off levels. Br J Cancer 62: 786-790

Silvestrini R, Daidone MG, Valentinis B, Ferraris C, Di Re E, Raspagliesi F, Landoni F, Scarfone G and Bolis G (1992) Potentials of cell kinetics in the management of patients with ovarian cancers. Eur J Cancer 28: 386-390

Spyratos F, Briffod M, Gentile A, Brunet M, Brault C, Desplaces A (1987) Flow cytometric study of DNA distribution in cytopunctures of benign and malignant breast lesions. Anal. Quant. Cytol. Histol. , 9, 485-494.

Spyratos F, Briffod M, Tubiana-Hulin M, Andrieu C, Mayras C, Pallud C, Lasry S, Rouessé J (1992) Sequential cytopunctures during preoperative chemotherapy for primary breast carcinoma. II DNA flow cytometry changes during chemotherapy, tumor regression, and short-term follow-up. Cancer , 69, 470-475.

Toikkanen S, Joensuu H and Klemi P (1989) The prognostic significance of nuclear DNA content in invasive breast cancer-a study with long-term follow-up. Br J Cancer 60: 693-700

Tsugita M, Tsuru S, Takasaki T, Shinomiya N, Kobayashi S, Hanyu F (1991) Flowcytometric measurement of the cell cycle of experimental tumors: Some devices for accurate measurement of proliferative activity. Oncology 48: 343-346

Vindelov LL, Christensen IJ, Nissen NI (1983) A detergent trypsin method for flow cytometric DNA analysis. Cytometry 4, 317-322

IMMUNOMAGNETIC PARTICLES FOR CELL ISOLATION

C. Andreoni
Immunotech S.A.
Luminy Case 915
13288 Marseille Cedex 9
France

In vitro studies of cell-cell interactions and regulation of cellular functions require cell subsets of high purity. With conventional techniques for immunological cell separation, such as rosetting (Gold, 67), complement lysis (Loski, 83) or panning (Maage, 77), the amount of reagent and the time required for the separation is considerable and out of proportion to the quality of the separation. Fluorescence-activated cell sorting (FACS) (Parks, 86) is a powerful method for the isolation of subpopulations, but application is often limited by the small separation capacity of approximately 10^7 cells per hour. A recent, alternative cell sorting method involves immunomagnetic particles (Rembaum, 88). Why immunomagnetic particles ? Because the coated surface conserves the advantage of monoclonal selectivity. Because the mobile particle allows rapid reaction with its antigen. For example, the time needed for an antibody against placental alkaline phosphatase to bind the enzyme was only 5min, when the antibody was bound to particles, but it took 24hr to bind the same amount of enzyme when the same amount of antibody was bound to the wall of a microtiter well (Millan, 85). Because magnetic particles can be concentrated easily by a simple external permanent magnet. This property allows easy handling of large number of cells and introduces no problems regarding sterility. Since the first immunomagnetic cell separation (Molday, 77) different immunomagnetic devices have been developed (Lievestad, 86; Hebell, 89; Abts, 89; Autran, 91; Liberti, 91) and allows the optimization of the cell sorting efficiency. Besides giving details of our own experience in immunomagnetic sorting, essential properties of particles and caracteristics of particle-cell interactions are presented.

NATO ASI Series, Vol. H 67
Flow Cytometry
Edited by A. Jacquemin-Sablon

MAGNETIC PARTICLES

The particles used for immunomagnetic cell sorting are paramagnetic (Whitesides, 83). That means that they are not magnetic in the absence of an external magnetic field and can be easily dispersed as slowly settling suspensions. But, on application of an external magnetic field, particles are attracted to the region of highest magnetic intensity (typically close to the pole of the magnet). The force exerted on the particule is proportional both to the strenght of the magnetic field and to the volume of the particle (assumed to be spherical) :

$$Fx = V \chi_V H \frac{\delta H}{\delta x}$$

Fx=force exerted on the particle in direction x
V=volume of the particle, χ_V=magnetic susceptibility per unit volume
H=strenght of the magnetic field, $(\delta H/\delta x)$=magnetic field gradient

Large magnetic particles (diameter >0.5μm) are attracted by a simple permanent magnet, but smaller particles migrate with the same velocity only on application of a higher magnetic field gradient (figure 1). Extremely high gradients are easily obtained by merely placing stainless steel filaments between the two poles of a magnet. In this device, the magnetic field near the steel is the sum of the external magnetic field due to the magnet and the magnetic field induced by the the steel filaments. Particles are then retained by the filament and this is the principal disavantage of such a magnetic device because in cellular immunoisolation the cells are in contact with exogeneous material, which must be sterile.

IMMUNOADSORBENT PARTICLES

Monoclonal antibodies may either be physically adsorbed or chemically coupled to the surface of magnetic particles.

Physical adsorption. The hydrophobic surface properties of beads may suffice to bind monoclonal antibody firmly through hydrogen bonding, but the amount of antibody adsorbed and its conformation vary with the pH, the ionic strength, the temperature of the medium and the native characteristics of the immunoglobulin. It was found that some monoclonal IgM antibodies have better antigen-binding capacity, when only adsorbed to particles, as opposed to covalent binding (Funderud, 90).

Covalent coupling. Increasingly, the tendency is to use covalent coupling of the antibody to the particle in oder to obtain greater stability and greater sensitivity of the immunoadsorbent. The binding technique is adapted to the reactive chemical groups on the

FIGURE 1 : PRINCIPLE OF MAGNETIC PARTICLES SEPARATION

$$F_x = V \chi_v H \frac{\delta H}{\delta x}$$

Fx=force of the particle in direction x
V=volume of the particule, χ_v=magnetic susceptibility per unit volume
H=strenght of the magnetic field, ($\delta H/\delta x$)=magnetic field gradient

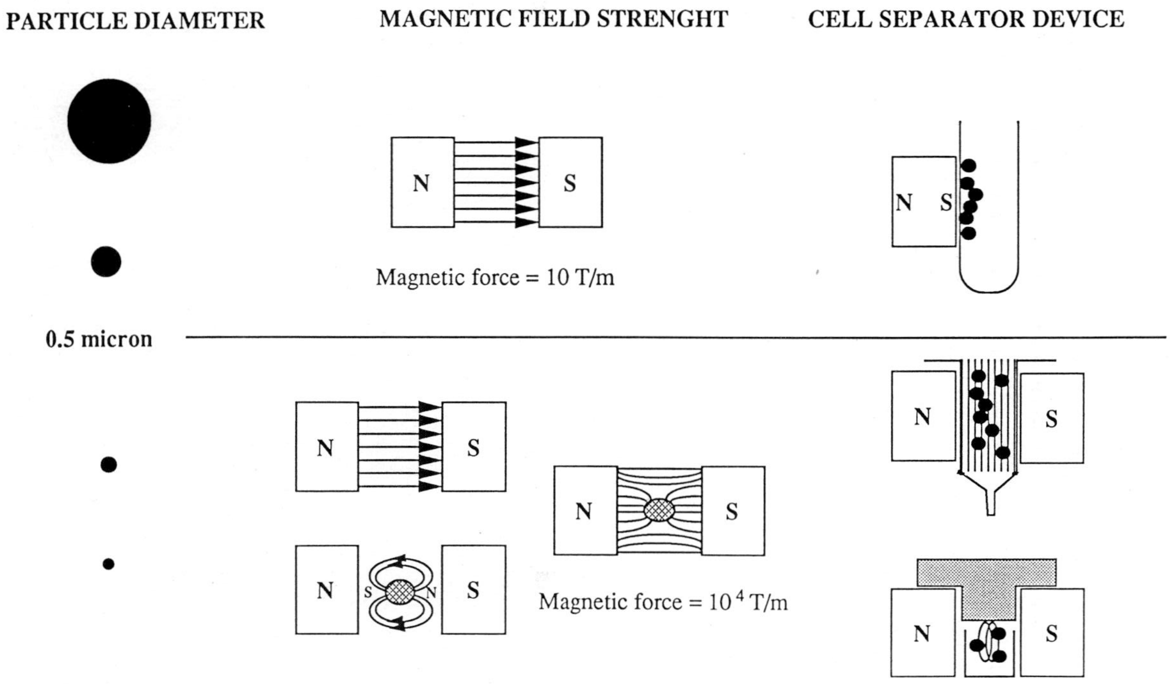

particle surface. Covalent coupling of a linker molecule, placed between the solid support and the specific antibody, enhances the flexibility of the specific antibody. Thus, a particularly attractive coating system consists of monoclonal antibodies coupled to the solid support via avidin-biotin complex (figure 2). The N-terminus amino group of avidin is coupled to carboxyl-terminated particles, employing carbodiimide under particularly mild conditions (Srere, 76), stabilizing the avidin layer. The avidin strongly reacts ($Kd=10^{-15}M$) (Green, 75) with any biotinylated monoclonal antibody providing a large sphere of immunomagnetic application. Finally, the orientation and flexibility of the monoclonal antibody is optimized by linking NHS-biotin via ε-aminocaproic acid and limiting the number of biotin molecule bound per molecule of antibody.

OPTIMAL CONDITIONS FOR PARTICLE-BOUND ANTIBODY AND CELLULAR ANTIGEN REACTIVITY

The antigen/antibody reactions with immobilize antibody exhibit specific features which are related to two principal phenomena : the antibody presentation and the antigen accessibility.

Antibody presentation. The reactive particles are covered with at least 800-1000 molecules/μm^2 (Gruenberg, 85) of specific monoclonal antibody. But more than the antibody density, the antibody orientation is important. Antibodies are properly oriented when the two paratopes are available for their target epitopes increasing the affinity of the cell-particle bond (Delaage, 90). To meet this requirement two factors are optimised : firstly, the number of binding sites must be as small as possible to inhibit "flattening" of the specific antibody against the solid support (the amount of linker required for maximum isolation is tested for each linker-specific antibody couple) ; secondly, the binding site of the specific antibody must be as far as possible from the active site. For particle-bound biotinylated antibodies, the use of the extended spacer arm (22.4Ä) analogue of NHS-biotin (13.5Ä) increases the capacity of coated beads to bind target cells. The more flexible system is obtained by interposing biotinylated anti-mouse IgG between the solid support and the specific antibody. The length of the linker (native IgG), the specificity of the linker (anti-Fc) and the ratio linker/specific antibody (10:1) lead to the most effective immunomagnetic beads because they catch antigens which are not accessible to specific biotinylated antibody-coated beads.

Antigen accessibility. The critical antigen density on the cell membrane is around 50 molecules/μm^2 (Gruenberg, 85) or about $5x10^3$-$1x10^4$ sites/cell (Gee, 91). As in the case

FIGURE 2 : COVALENT COUPLING OF MONOCLONAL ANTIBODIES TO MAGNETIC PARTICLES VIA THE AVIDIN-BIOTIN COMPLEX

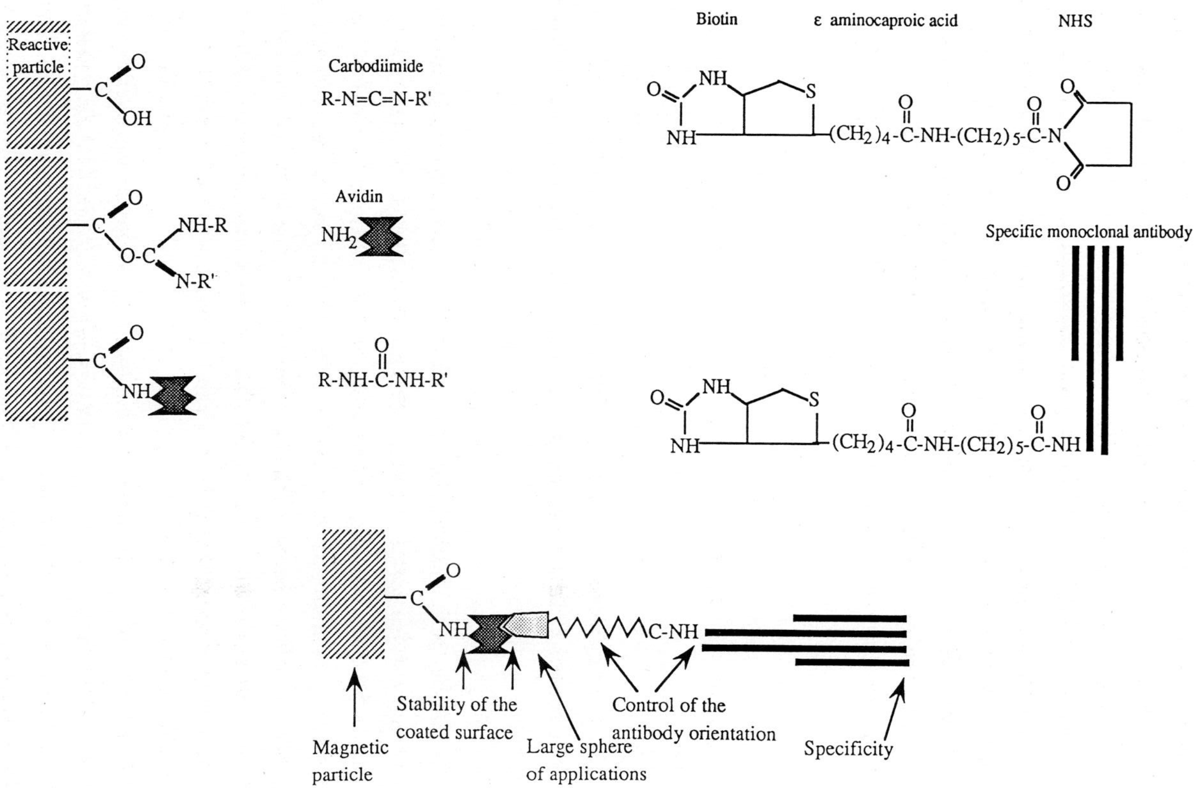

of the immunoadsorbent particle, the antigen expression level is less important than its cellular configuration. Soluble antibody can reveal membrane antigens (CD3+lymphocytes, CD4+monocytes) which remained inaccessibles for the particle-bound antibody. A particular conformation of the molecule such as the globular complexe TCR/CD3 (Kurrle, 89) or the mainly cytoplasmic CD4 monocyte antigen (Filion, 90) might prevent the accessibility of the coated beads. In the same way, previous labeling of cells with the specific monoclonal antibody reduces the reactivity of the beads (Gee, 91). The configuration of the antibody complexed with the cellular antigen is poorly accessible to the beads or steric hindrance interferes with the specificity of the beads.

INTERACTIONS BETWEEN IMMUNOMAGNETIC PARTICLES AND TARGET CELLS.

The ability of target cells to bind specifically immunomagnetic particles is directly related to the probability that they will collide with each other. This probability is determined by factors such as concentration, size and density of both the cells and the particles. Numerous, small beads with a density which is approximatly that of the cells'own density increase the opportunity for collisions and consequently accelerate the kinetics of particle/cell interaction. That is important because slow kinetics of interaction results in non specific binding and phagocytosis (Pilling, 89). As the particles and cells come into contact, their surface caracteristics are determinant. Electrostatically charged beads promote undesirable binding of cells having opposite net charges (Andreoni, 92). This can be suppressed by adding serum or NaCl to the incubation medium or by neutralizing the surface charge of the particle. The affinity of the particle-bound monoclonal antibody for the antigen becomes then determinant. High affinity monoclonal antibodies ($Kd=10^{-8}$ to $10^{-10}M$) are the most efficient for magnetic sorting.

The binding of the particle to the target cell is rapidly followed by a definite shallow depression in the cellular surface (figure 3). Very short incubation times (5min) at 4°C do not inhibit the membrane process. It is probably due to multiple antigen/antibody interactions between the cell membrane and the immunomagnetic bead by a zipper-like mechanism earlier described for macrophages (Silverstein, 77). This represents an atypical type of phagocytosis induced by antigen-antibody interaction whereas the "professional phagocytes", the monocytes, are not recruited by the particles under the same experimental conditions. The cell-particle interaction is very strong and cannot be disrupted by mechanical manipulation.

FIGURE 3 : THE ZIPPER-LIKE BINDING OF IMMUNOMAGNETIC PARTICLES TO THE MEMBRANE OF TARGET CELLS

Schematic diagram

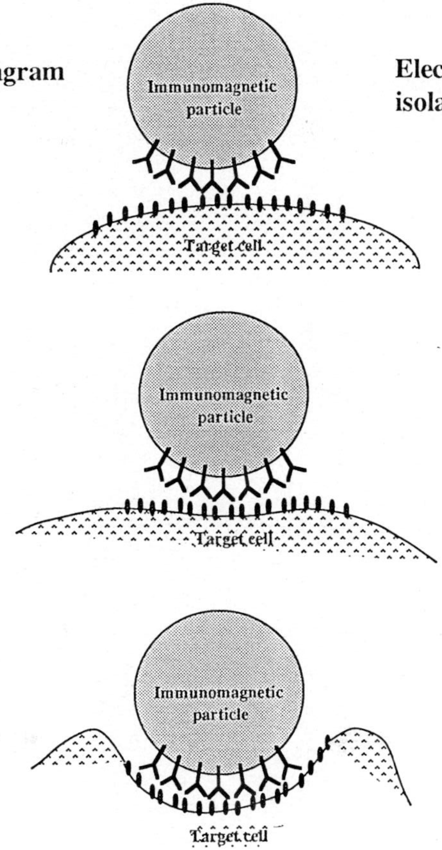

Electronic microscopy of T lymphocytes isolated with CD8 IObeads (x 19 600)

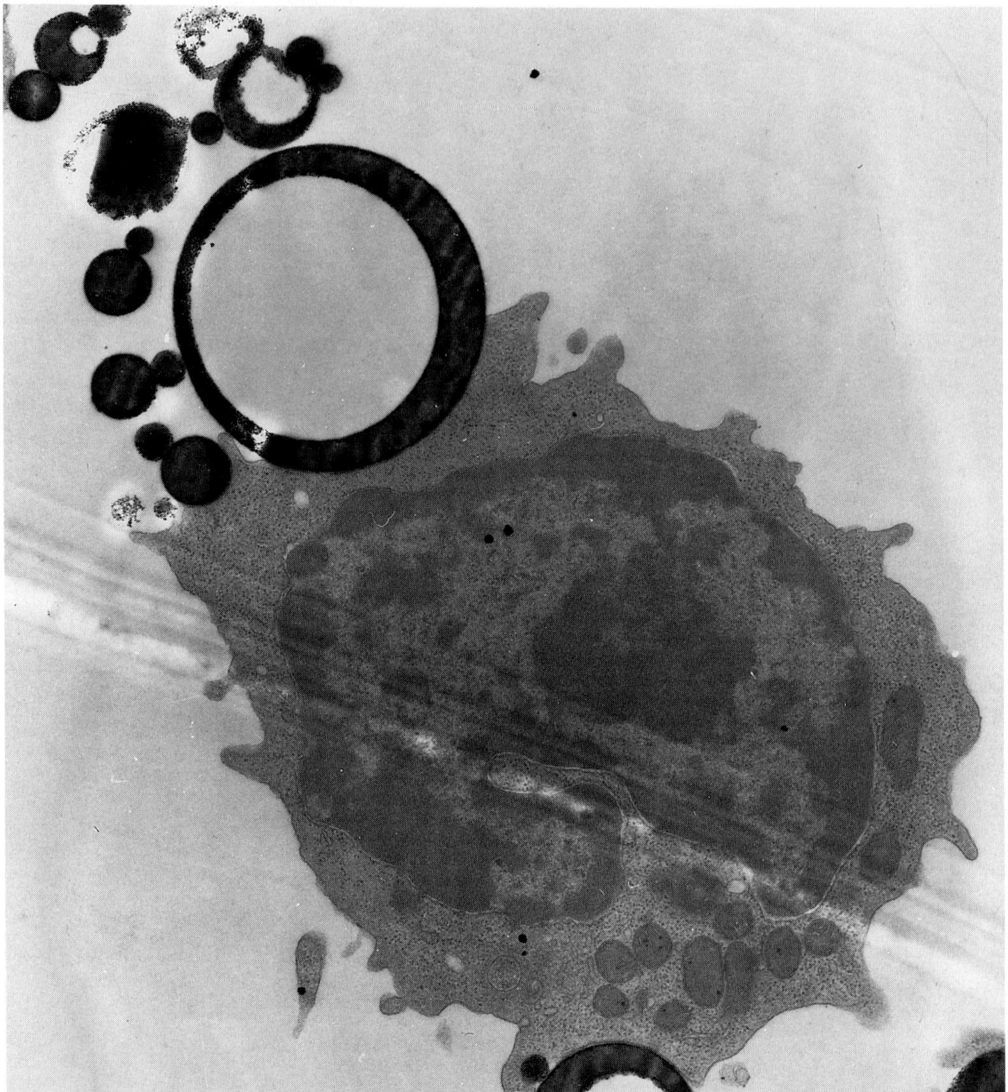

DETACHMENT OF THE IMMUNOMAGNETIC PARTICLES FROM THE POSITIVELY SELECTED CELLS

Several solutions are supplied for the detachment of the immunomagnetic particles from the positively selected cells : physiological release, antigenic competition, enzymatic or chemical cleavage.

Physiological release. Some antigens, such as CD19 (Lea, 86) and CD34 (Bagnara, 91), are versatile and contact with the ligand promotes the internalization of the molecule. The immunomagnetic beads are spontaneously released into the medium after an overnight incubation at 37°C. Unfortunatelly internalization of the marker expression after particle binding is unusual and an active detachment protocol must be initiated.

Antigenic competition. Incubation of the particle-cell complex with goat anti-mouse-Fab antiserum at room temperature for one hour interferes with and blocks the interaction between the mouse antibody coated onto the particle and the antigenic determinant on isolated cells (Rasmussen, 92). This detachment depends on the affinity of the particle-bound antibody and appears to be restricted to IgM antibody which have often low affinity, but enough avidity to be efficient when coupled to the magnetic particle. Sorting efficiency is at the expense of detachment efficiency which is obviously not ideal for the isolation of rare cells (Egeland, 91).

Enzymatic or chemical cleavage. As an alternative, the particle-cell complex may be treated with proteolytic enzymes. Antigen resistance to proteolysis is variable and limited both by antigen structure and the kind of protease used. Chymopapain partially damages the CD34 antigen, only the My10 epitope seems to be affected (Civin, 90). Conversely, the ph glycoprotease destroys 5 out of 7 CD34 epitopes (Marsh, 91). The application of such a system is restricted to certain antigens and the effects on other membrane proteins are not controlled. A promising option is to adapt the immunomagnetic particles with the aim of cleaving the particle-antibody bond. NHS-SS-biotin is coupled to the monoclonal antibody and coated onto avidin particles. The cell-sorting efficiency remains unchanged and the disulfide bonds is reduced with dithiothreitol for 30min at room temperature as was described previously for a different cell-sorting device (Thomas, 89).

HOW TO CHOOSE AN IMMUNOMAGNETIC CELL SEPARATION DEVICE ?

The recent development of numerous immunomagnetic cell devices allows the utilizer to choose (figure 4) : -the easiest manipulation (no maintenance of the cell separator device, no utilization of Rock-N-Roller, no incubation in ice) ; -the fastest protocol with highest sorting efficiency (no previous isolation of mononucleated cells, no previous elimination of monocytes nor removal of dead cells, short-time incubation of particles with cells, rapid attraction to the magnet) ; -the most conveniently selected cell suspension for a given application (designed for flow cytometric analysis, adapted for cell culture, approppriate for detachment) ; -the cost of the cell separation device (mono-size particles and high gradient magnetic cell separators are expensive).

PERFORMANCE OF IMMUNOMAGNETIC CELL SEPARATION

Peripheral blood was incubated with IObeads, for 5min, at room temperature, without rotative agitation. The negatively and positively selected cells were separated after 10min attraction on an external permanent magnet. The performance of the IObeads cell sorting is monitored by microscopic observation and by flow cytometric analysis of the positive and negative suspensions after direct immunofluorescent labeling.

Negatively isolated cells. The labeling of negatively isolated cells is performed under usual conditions. Cells are analysed on a fluorescence-activated cell sorter (FACScan). The standard instrument settings are adjusted for analysis of FITC- and PE- stained peripheral blood leucocytes. The cytofluorometric estimation of the residual target cell population in the negative suspension defines the sorting efficiency (figure 5). It is proportional to the quantity of IObeads used for the separation. Under optimal conditions (1mg of particles per ml of whole blood) the sorting efficiency reaches 95-100%(figure 5).

Positively selected cells. Cellular viability of the positively selected cells is monitored by exclusion dyes (ethidium bromide and acridine orange) and by microscopic observation. It is usually superior to 95%. The cell recovery is proportional to the quantity of IObeads used for the separation. A maximum of target cells is isolated with 0.5mg to 1mg of IObeads per ml of whole blood (figure 6), the amount of free particles in the suspension is minimal with 0.5mg particles. The positively selected cells do not lend themselves to indirect labeling because fluorescent conjugates specifically bind to particle-bound antibodies giving the cells, coated by immunomagnetic particles, a bright fluorescent signal. The particle-bound cells are detetected in flow cytometry because they increase the

FIGURE 4 : THE CHOICE OF AN IMMUNOMAGNETIC CELL SEPARATOR DEVICE

IMMUNOMAGNETIC CELL SORTING	PARTICLE DIAMETER (μm)				RESTRICTING UTILIZATION
	5	1	0.1	0.03	
Magnetic force on the particle	high	high	low	low	Cost and maintenance of the particle separator
Particle density	high	low	null	null	Homogeneisation using a Rock-N-Roller
Chemical stabilisation of coated proteins	yes	yes	yes	yes	
Kinetics of specific cell/particle interaction	slow	fast	fast	fast	Incubation 4°C to inhibit phagocytosis
Unspecific trapping	weak	null	important	important	Previous elimination of monocytes and dead cells
Alteration of cellular physical caracteristic	important	weak	null	null	Flow cytometric analysis of particle/cell impossible
Biodegrability	no	no	yes	yes	Presence of exogeneous material in the culture
Particle detachment	restrict	universal	no	no	Choice for the target antigen limited

FIGURE 5 : DEPLETION OF B LYMPHOCYTES FROM PERIPHERAL BLOOD WITH CD19 IObeads

Immunofluorescent labeling (CD19-FITC) and flow cytometric estimation of the target cell population before and after magnetic cell sorting.

control sample

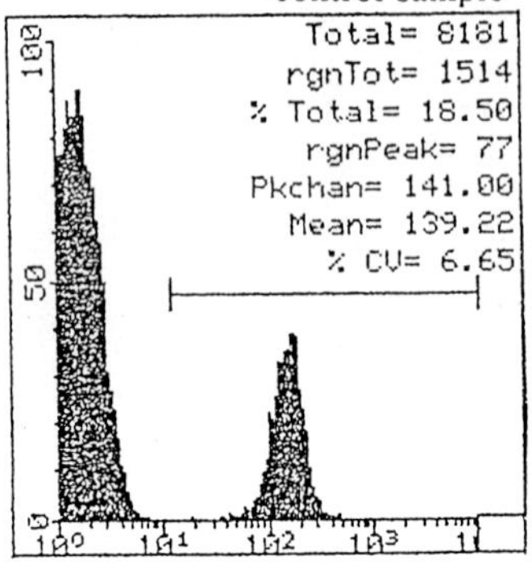

negative fraction

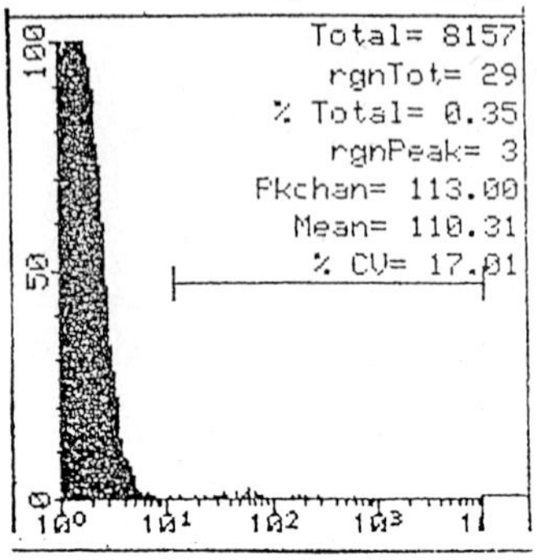

$$\text{Depletion rate} = 1 - \frac{\%\text{positive cells}/\%\text{negative cells in the control sample}}{\%\text{positive cells}/\%\text{negative cells in the negative fraction}} \times 100$$

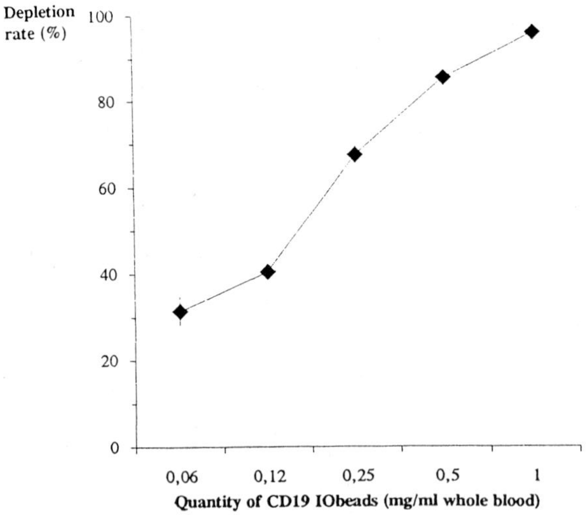

90° side scatter signal. The magnitude of amplification is satisfactory for quantitative discrimination between positive and negative cells (figure 7). The autofluorescence of positive cells is higher than that of the negative cells but specific staining is easily distinguishable and the level of antigen expression is not affected. This demonstrates that immunomagnetic cell isolation combined with ordinary two-color staining may be employed for the simultaneous detection of cells expressing three different surface markers (figure 6 and 7). However, it is important to note that additive fluorescent labeling of the particle-bound antigen is unsuccessful even for the cognate epitope. It seems like all the target antigens are recruited in the interaction with the immunoparticle and are inaccessible because of the zipper-like binding. The purity of the positively selected cells is determined by prior fluorescent labeling of the target cells followed by immunomagnetic sorting or by employing a marker corresponding to approximately the same cellular distribution. Sorting purity is estimated to 100%.

In conclusion, with the IObeads, cells can be isolated, in a single step directly from blood or other biological fluid. The sorting efficiency and purity obtained approximate 100% and makes possible studies of homogeneous populations.

APPLICATIONS OF IMMUNOMAGNETIC CELL SEPARATION

High purity cell subsets are advantageous in the descriptive studies related to cellular phenotype as well as in functional studies related to cell communication and stimulation.

The immunophenotyping of rare cell populations represent a challenge due to their inherent low frequency. The immunomagnetic isolation of rare cells permit the identification of subsets corresponding to differentiation state and functional particularity (Egeland, 91; Gore, 91).

Concerning cellular functions, it is generally accepted that interaction with soluble factors (33) as well as cellular secretion (Autran, 91; Jones, 90) are not influenced by the bound particles. Opinion is divided, however, about the recovery of functions requiring cell/cell adhesion or cellular motility. There are several reported cases of particle/cell complexes that are active in culture (Lievestad, 86; Abts, 89; Morgan, 90) but usually workers prefer to eliminate undesirable cells by magnetic cell sorting and to culture the enriched remaining population (Verfaillie, 90) suggesting that the particule-binding cells may not be altogether normal.

FIGURE 6 : ENRICHMENT OF B LYMPHOCYTES FROM PERIPHERAL BLOOD WITH CD19 IObeads

Microscopic numeration of cells coated with beads

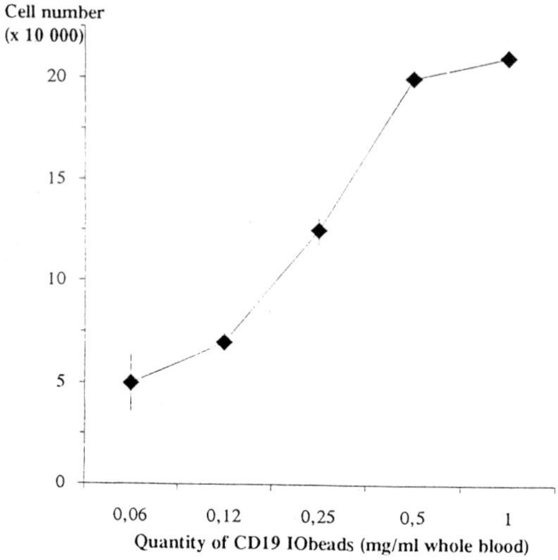

Flow cytometric analysis of the positive fraction

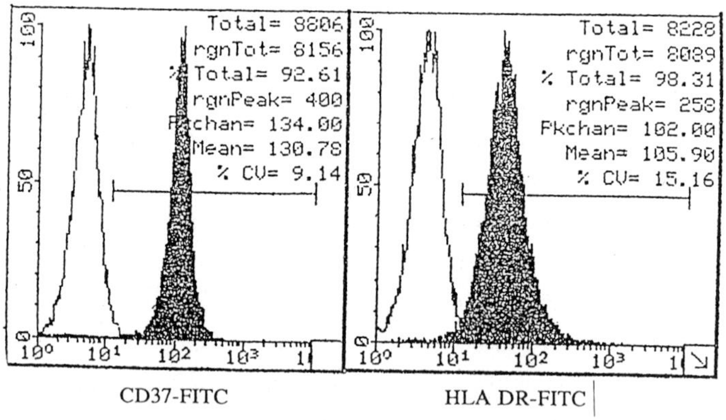

FIGURE 7 : FlOW CYTOMETRIC ANALYSIS OF T LYMPHOCYTES SEPARATED FROM PERIPHERAL BLOOD WITH CD2 IObeads

Control cell suspension

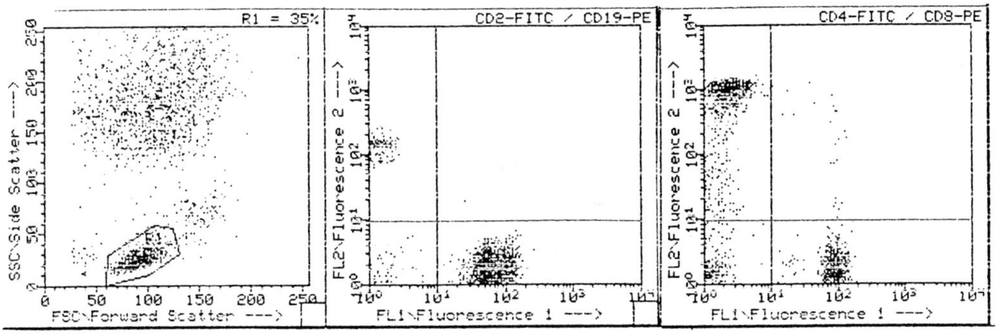

0.5mg CD2 IObeads per ml of whole blood, incubation for 5min at room temperature

Negatively selected cells

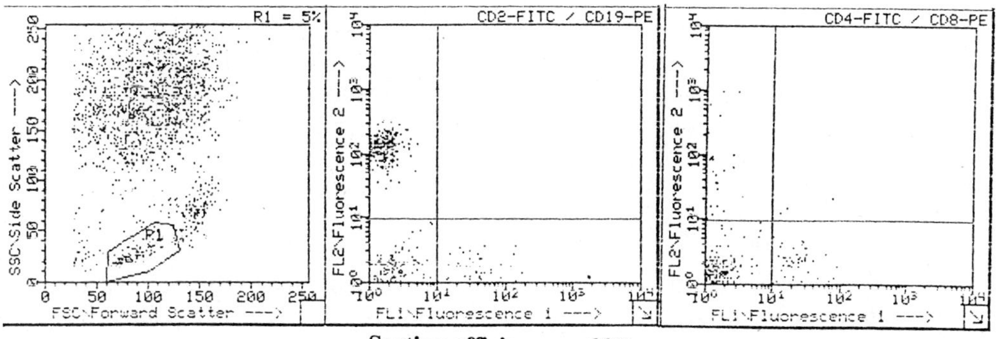

Sorting efficiency = 92%

Positively selected cells

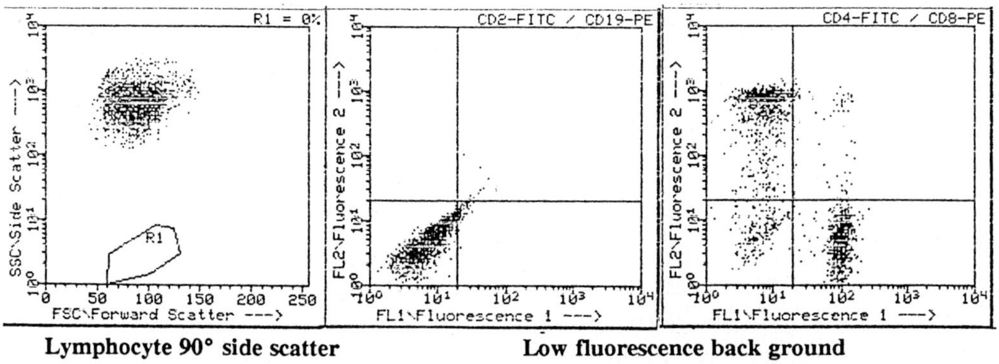

Lymphocyte 90° side scatter signal highly increased

Low fluorescence back ground
Target antigens saturated
Level of antigenic expression respected

REFERENCES

ABTS H, EMMERICH M, MILTENYI S, RADBRUCH A, TESCH H (1989) CD20 positive human B lymphocytes separated with the magnetic cell sorter (MACS) can be induced to proliferation an antibody secretion in vitro. J Immunol Methods 125:19-28

ANDREONI C, DELAAGE M (1992) Avidin-binding to peripheral blood B lymphocyte and monocyte results from cellular charge display. J. Immunol. Methods (in press)

AUTRAN B, LEBLOND V, SADAT-SOWTI B, LEFRANC E, GOT P, SUTTON L, BINET JL, DEBRE P (1991) A soluble factor released by CD8+CD57+ lymphocytes from bone marrow transplanted patients inhibits cell-mediated cytolysis. Blood 77:2237-2241

BAGNARA GP, ZAULI G, VITALE L, ROSITO P, VECCHI V, PAOLUCCI G, AVANZI GC, RAMENGHI U, TIMEUS F, GABUTTI V (1991) In vitro growth and regulation of bone marrow enriched CD34+ hematopoietic progenitors in Diamond-Blackfan anemia. Blood 78:2203-2210

CIVIN C, STRAUSS LC, FACKLER MJ, TRISCHMANN TM, WILEY JM, LOKEN MR (1990) Positive stem cell selection - Basic science. In : Bone Marrow Purging and Processing. LISS A.R., Inc., 387-402

DELAAGE M, PRINCE P, BRAILLY H, LE DOUSSAL JM, MOREL A, BARBET J (1990) Antigen-antibody reactions on surfaces. Orientation and environmental effects. Immunoanal Biol Spec 24:15-20

EGELAND T, STEEN R, QUARSTEN H, GAUDERNACK G, YANG YC, THORSBY E (1991) Myeloid differentiation of purified CD34+ cells after stimulation with recombinant human granulocyte-monocyte colony-stimulating factor (CSF), granulocyte-CSF, monocyte-CSF, and interleukin-3. Blood 78:3192-3199

FILION LG, IZAGUIRRE CA, GARBER GE, HUEBSH L, AYE MT (1990) Detection of surface and cytoplasmic CD4 on blood monocytes from normal and HIV-1 infected individuals. J Immunol Methods 135:59-69

FUNDERUD S, NUSTAD K, LEA T, VARTAL F, GAUDERNACK G, STENSTAD P, UGELSTAD J (1990) Fractionation of lymphocytes by immunomagnetic beads. In : Lymphocytes : a pratical approach. KLAUS GGB (ed), Oxford University Press, New York, 55-61

GEE AP, MANSOUR VH, WEILER MB (1991) Effects of target antigen density on the efficacy of immunomagnetic cell separation. J Immunol Methods 142:127-136

GOLD ER, FUNDENBERG HH (1967) Chromium chloride as a coupling reagent for passive hemagglutination reactions. J Immunol 99:859-863

GORE SD, KASTAN MB, CIVIN CI (1991) Normal human bone marrow precursors that express terminal deoxynucleotidyl transferase include T-cell precursors and possible lymphoid stem cells. Blood 77:1681-1690

GREEN NM (1975) Adv. Protein Chem. 29:85-133

GRUENBERG J, HOWELL KE (1985) Immuno-isolation of vesicles using antigenic sites either located on the cytoplasmic or exoplasmic domain of an implanted viral protein. A quantitative analysis. Eur J Cell Biol 38:312-321

HEBELL T, GÖTZE O (1989) The isolation of B lymphocytes from human peripheral blood using antibodies coupled to paramagnetic particles and rosetting techniques. J Immunol Methods 123:283-291

JONES BM (1990) Evaluation of CD5 and other differenciation antigens on human immunoglobulin-secreting cells using a combination of immunobead rosetting and reverse haemolytic plaque formation. J Immunol Methods 132:119-126

KANDZIA J, ANDERSON MJD, MULLER-RUCHHOLTZ W (1981) Cell separation by antibody-coupled magnetic microspheres and their application in conjunction with monoclonal HLA antibodies. J Cancer Res Clin Oncol 101:165-170

KURRLE R (1989) Cluster report CD3 In : Leucocyte typing IV, White cell differentiation antigens. KNAPP W, DÖRKEN B, GILKS WR, RIEBER EP, SCHMIDT RE, STEIN H, VON DEM BORNE AEGKr (eds) Oxford University Press, New York, 290-293

LEA T, SMELAND E, FUNDERUD S, VARTDAL F, DAVIES C, BEISKE K, UGELSTAD J (1986) Characterisation of human mononuclear cells after positive selection with immunomagnetic particles. Scand J Immunol 23:509-512

LIBERTI PA, FEELEY BP (1991) Analytical and process scale cell separation with bioreceptor ferrofluids and high gradient magnetic separation. In : Cell separation science and technology. KOMPALA D, TODD P (eds) American chemical society, Washington D.C., chap. 17

LIEVESTAD T, GAUDERNACK G, QVISTADT E, UGELSTAD J, THORSBY E (1986) Isolation of functionally active T-cell subsets by monoclonal antibodies conjugated to monodisperse magnetic microspheres (M-450). Transplantation Proceedings 18:731-732

LOSKI CL, RAMM LE, HAMMER CH, MAYER MM, SHIN ML (1983) Cytolysis of nucleated cells by complement cell death displays multihit characteristics. Proc Natl Acad Sci USA 80:3816-3821

MAAGE MG, McHUGH LL, ROTHSTEIN TL (1977) Mouse lymphocytes with and without surface immunoglobulin : preparative scale separation in polystyrene tissue culture dishes coated with specifically purified anti-immunoglobulin. J Immunol Methods 15:47-56

MARSH JCW, SUTHERLAND DR, DAVIDSON J, MELLORS A, KEATING A (1991) Retention of progenitor cell function in CD34+ cells purified using a novel o-sialoglycoprotease. Blood 78 suppl. 1, 300a.

MILLAN JL, NUSTAD K, NORGAARD-PEDERSEN B (1985) Highly sensitive solid phase immunoenzymometric assay for placental and placental-like alkaline phosphatases with a monoclonal antibody and monodisperse plolymer particles. Clin Chem 31:54-58

MOLDAY RS, YEN SPS, REMBAUM A (1977) Application of magnetic microspheres in labeling and separation of cells. Nature 268:437-438

MORGAN A, BURKE LM (1990) Comparison of mitogen responses of lymphocyte subpopulations separated using coated metal beads and by traditional methods. J Immunol Methods 126:1-6

PARKS DR, LANIER LL, HERZENBERG LA (1986) Flow cytometry and fluorescence activated cell sorting (FACS). In : Handbook of experimental immunology, 4th edition, WEIR DM (ed). Blackwell Scientific Publications, Oxford, chap 110

PILLING D, KITAS GD, SALMON M, BACON PA (1989) The kinetics of interaction between lymphocytes and magnetic polymer particles. J Immunol Methods 122:235-241

RASMUSSEN AM, SMELAND EB, ERIKSTEIN BK, CAIGNAULT L, FUNDERUD S (1992) A new method for detachment of Dynabeads from positively selected B lymphocytes. J Immunol Methods 146:195-202

REMBAUM A, TÖKES ZA (Eds) (1988) Microspheres : medical and biological applications. CRC Press, Inc. Boca Raton, Florida. 234p

SILVERSTEIN SC, STEINMAN RM, COHN ZA (1977) Endocytosis. Ann Rev Biochem 46:669-672

SRERE PA, UYEDA K (1976) Functional groups on enzymes suitable for binding to matrices. In : Methods in enzymology, vol XLIV, Immobilized enzymes. MOSBACH K (ed) 11-19

THOMAS TE, SUTHERLAND HJ, LANSDORP PM (1989) Specific binding and release of cells from beads using cleavable tetrameric antibody complexes. J Immunol Methods 120:221-231

VERFAILLIE C, BLAKOLMER K, McGLAVE P (1990) Purified primitive human hematopoietic progenitor cells with long-term in vivo repopulating capacity adhere selectively to irradiated bone marrow stroma. J Exp Med 172:509-520

WHITESIDES GM, KAZLAUSKAS RJ, JOSEPHSON L (1983) Magnetic separations in biotechnology. Trends in Biotechnology 1:144-148

Optical Trapping: Instrumentation and Biological Applications

T. N. Buican
Cell Robotics, Inc.
2309 Renard Pl. SE
Albuquerque, NM 87106
USA

Optical trapping, a new technique for the manipulation of microscopic particles, was invented in the late '60s by Arthur Ashkin of AT&T Bell Labs. This technique, examples of which were first reported by Ashkin (1970), relies on the pressure created by one or more laser beams that are scattered by a microscopic object in order to trap, levitate, and move that object. As opposed to other trapping techniques, optical traps are intrinsically stable and very localized in their effects. As such, they can be incorporated into relatively simple devices that allow single cells, chromosomes, and other cell organelles to be accurately positioned and transported. Furthermore, optical trapping only requires low-intensity laser beams and can be operated at wavelengths at which absorption by the trapped particle is minimized. As reported in Ashkin and Dziedzic (1987) and Ashkin *et al.* (1987), the trapping laser beams seem to have negligible biological effects on most cells that have been optically trapped.

Cell separation based on optical trapping belongs to the class of active, single-cell sorting techniques, together with flow sorting and mechanical micromanipulation. As such, automated optical trapping instruments analyze single cells and, after converting the measurement results into a sorting decision, physically remove selected single cells from the original sample. Although passive cell separation based on optical trapping has also been proposed by Ashkin (1973), the complex dependence of separation properties on the physical properties of the cell and the existence of beam-mediated interactions between cells reported by Buican *et al.* (1987) make this technique unattractive. By contrast, active separation based on optical measurements of single-cell properties and automated beam control can be very precise both in terms of its analysis of single cells and in the physical separation of these cells. Furthermore, automation applies in a natural way to optical trapping as the trapping optics can easily be used for light collection and imaging, and the direction and intensity of the laser beams used for trapping can easily be brought under automated control.

The range of cell properties on which automated cell separation can be based in an optical trapping system includes labelling fluorescence intensity, light scattering intensity distribution, and cell morphology. As the speed at which cells travel through the analysis volume can be accurately controlled and is much lower than the typical speeds encountered in flow sorting, high-resolution measurements of cell properties can be made. Furthermore, since cell position can be controlled with very high accuracy, other, nonoptical measurements can be performed.

The purely optical nature of cell manipulation by optical trapping means that cell analysis and manipulation can be performed inside *completely enclosed* sample systems. Thus, separation based on optical trapping has a clear advantage over mechanical micromanipulation,

NATO ASI Series, Vol. H 67
Flow Cytometry
Edited by A. Jacquemin-Sablon

which can only be performed in open containers. The small diameter of the trapping beams and the highly localized character of the optical trap allow manipulation to be performed inside commensurately small compartments. Consequently, a single optical manipulation chamber can contain a large number of compartments and interconnecting channels. The former can be used not only for collecting cells that have been separated, but also for performing further analysis and processing of those cells. Thus, complex experimental protocols can be carried out at the level of the single cell inside an optical manipulator, without the need for recovering the cells after each step and transporting them between separate instruments.

We begin this article with a brief presentation of the principles of optical trapping. Following this, we discuss the applications of two-dimensional (2-D) and three-dimensional (3-D) optical traps in cell separation. This discussion follows that presented in Buican (1991). For each type of trap we discuss instrument optics, manipulation chambers, control electronics, and the procedures for automated control. We conclude with a brief discussion of future developments in the field of cell separation and processing by optical trapping.

For the interested student, we recommend the reviews of biological optical trapping by Block (1990) and Wright *et al* (1990).

A Qualitative Discussion of Optical Trapping

Whenever light is scattered by an object, the change in the momentum of the scattered photons leads to a momentum of equal magnitude but opposite direction being transferred to the scattering object. If one considers a system consisting of a scattering object and the incident and scattered beams (Figure 1), and if P_0 and P_1 are, respectively, the momentum fluxes (momentum transported by the beams per unit time) of the incident and scattered beams, then, because of conservation of total momentum, a force, $F = P_1 - P_0$, is produced that acts on the scattering object. This is the *radiation pressure force*, and we will refer in what follows to its components along the direction of the incident beam (axial force) and perpendicular to that direction (radial force).

The momentum carried by a light beam in a given direction can be computed from the far-field angular distribution of the beam intensity, ρ. As shown in Bohren and Huffman (1983), the total beam momentum is given by

$$P = \frac{W}{c} \int_{4\pi} \rho \cos\alpha \, d\Omega$$

where W is the total beam power, c is the speed of light, α is the angle relative to the given direction, and $d\Omega$ is the solid angle element. For an axially symmetric beam, the factor, $\cos\alpha$, shows that, at constant total beam power, beam momentum decreases as the divergence of the

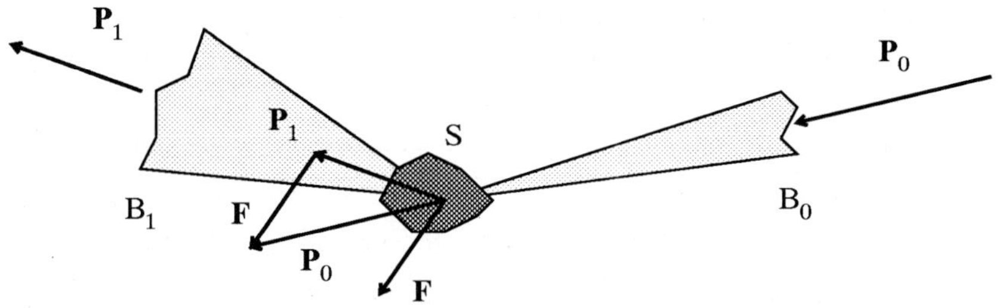

Figure 1. Conservation of momentum in a light scattering system: B_0, B_1—incident and scattered beams; S—scattering object; $\mathbf{P}_0$, $\mathbf{P}_1$—momentum fluxes of the two beams; **F**—light pressure force due to scattering of the incident beam by S. (Reproduced with permission from Buican (1991), Copyright 1991 Am. Chem. Soc.)

beam increases. It follows immediately that beam momentum is decreased by (i) absorption (which reduces beam power); and (ii) diffuse scattering (which spreads out the intensity distribution). Absorption produces a purely axial force in the direction of the incident beam. Diffuse scattering, if symmetric relative to the direction of the incident beam, also contributes a force in the direction of the incident beam.

It can be shown that the momentum flux of a beam with axial symmetry is parallel to the beam axis and has a magnitude given, for small θ, by $P = W(1 - \theta^2)/c$, where θ is an angular parameter describing beam divergence. This relationship shows that the beam has maximum momentum when it is maximally collimated ($\theta = 0$), and its momentum decreases as its divergence increases. The magnitude of the force acting on the scattering particle is thus given by $F = W\,\Delta(\theta^2)/c$. If the scattering object behaves like a lens or mirror, the beam angle, θ, can either increase or decrease, depending on the geometry of the object and incident beam. Consequently, we may expect both the magnitude and direction of the light pressure force acting on an optical element to depend on the optical properties of the object, as well as its position relative to the incident beam.

The Lens Model. It was shown by Salzman (1982) that, under most experimental conditions, live cells have a refractive index close to that of the suspension medium and, therefore, at least in a first approximation, partial reflection does not play an important role in biological optical trapping. An adequate qualitative description of the optical trapping of biological particles can thus be obtained by treating the particles as lenses that refract the incident beam.

Radial Trapping. It is easy to see that, whenever an object that behaves like a convergent lens (internal index of refraction higher than that of the surrounding medium) is moved away from the axis of a light beam incident on it, the light beam is deflected away from the initial direction (Figure 2a). Therefore, the radial component of the light pressure force acting on the object points in a direction opposite to particle displacement (toward the beam axis), and a stable, 2-D trapping effect is created. One can similarly show that the radial force acting on a divergent lens (object with a lower refractive index than that of the medium) points in the

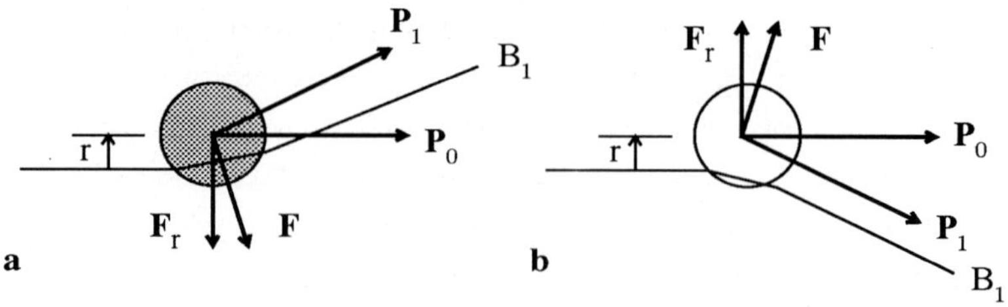

Figure 2. The orientation of the radial force: (a) particle with higher refractive index than that of the surrounding medium; and (b) particle with lower refractive index. B_0, B_1—incident and scattered beams; **F**—light pressure force; $\mathbf{F}_r$—radial component of **F**; $\mathbf{P}_0$, $\mathbf{P}_1$—momentum fluxes for the incident and scattered beams; r—radial particle displacement. Only the axes of the beams are shown. (Reproduced with permission from Buican (1991), Copyright 1991 Am. Chem. Soc.)

same direction as the displacement (Figure 2b) and thus there is no radial trapping effect. Radial trapping applies to both spherical and nonspherical particles. In the case of the latter, torques may develop which give the particles preferential orientations relative to the incident beam.

Axial Trapping. In the lens model of optical trapping, the particle in the laser beam is regarded as a lens of focal length, f (Figure 3). In a purely geometrical optics approximation, one can easily see that there are always two points along the beam axis for which the angle of the output ("scattered") beam equals that of the incident beam. Thus, the momentum fluxes of

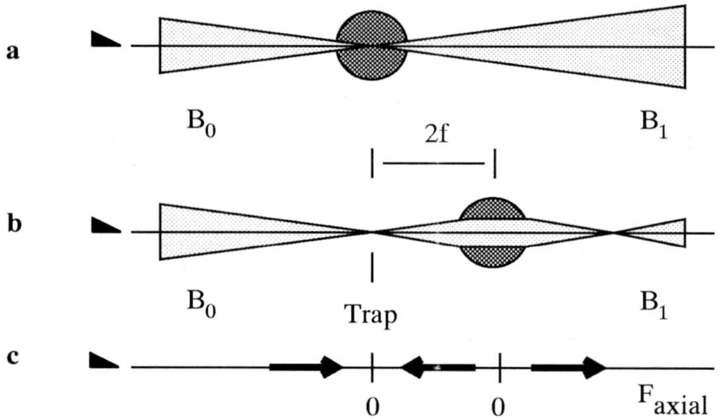

Figure 3. Lens model of optical trapping. Two particle positions for which the axial force vanishes: (a) particle at beam focus; (b) particle at a distance 2f from the beam focus; (c) diagram showing the direction of the axial force (F_{axial}) for various positions of the particle. The position of the stable axial trap relative to the trapping beam is indicated. The focal length of the particle is denoted by f. The incident and scattered beams, B_0 and B_1, travel from left to right. (Reproduced with permission from Buican (1991), Copyright 1991 Am. Chem. Soc.)

the two beams have the same magnitude and, for distances of 0 and 2f between the lens and the beam focus, the axial force vanishes (Figure 3c). It is also easy to see that, for any distance between 0 and 2f, the beam angle decreases after passing through the object and, therefore, the beam momentum increases and the light pressure force acting on the lens is directed opposite to the beam. For all points outside this interval, the beam momentum decreases and the lens is pushed in the direction of the beam. As a consequence of this dependence of the axial force on the distance between lens (object) and beam focus, a stable axial trap always exists in a purely geometrical optics model and is situated at the beam focus. However, if one considers a Gaussian beam and the corresponding imaging relationships described by Kogelnik and Li (1966), one can show that an axial trap exists only if $\theta^2 > \lambda/\pi f$, where λ is the wavelength and θ is the asymptotic angle of the Gaussian beam. Thus, for weakly convergent beams (small θ), the inequality above is not satisfied. Consequently, the axial force always points in the direction of the beam and there is no axial trapping. One can thus define two trapping regimes, depending on the degree of convergence of the beam: (i) 2-D trapping, which applies to weakly convergent beams and in which there is a trapping effect only in a plane perpendicular to the axis of the trapping beam; and (ii) 3-D trapping, in which a strongly convergent laser beam creates a full, 3-D optical trap.

Beyond the qualitative description we have presented, the interested student will find more elaborate treatments of optical trapping in Roosen (1979), Bakker Schut *et al.* (1991), Visscher and Brakenhoff (1991b), and Wright *et al* (1989).

2-D Optical Trapping and Cell Separation

We showed in the previous section that weakly collimated laser beams produce 2-D optical traps and that trapped particles are propelled along the beam axis. Such traps may be used to transport microscopic particles over macroscopic distances and can thus be the basis for both remote delivery and sorting systems.

As the particles in a 2-D trap move along the beam axis, hydrodynamic forces which depend on particle shape and size will affect both the velocity at which the particles travel and the stability of the trapping effect. Thus, during the 2-D trapping of blood cells, we observed that lymphocytes, which are roughly spherical, are stably trapped and can be transported over relatively large distances (Buican *et al*, 1987). By contrast, erythrocytes are only briefly held in the beam and are readily pushed out by orientation-dependent hydrodynamic forces. This phenomenon could be used for the passive separation of spherical from nonspherical cells.

The velocity at which stably trapped cells travel along weakly focused beams depends on both the optical and the hydrodynamic properties of the cells. The use of differences in travel velocity for passive cell separation has been proposed by Ashkin (1973). However, the travel velocity of a given particle is further affected by the scattering of the beam by closely following, trapped particles. We have reported (Buican *et al*, 1987) that the interaction between simultaneously trapped particles may lead to the formation either of particle clumps or of stable systems of particles which move at the same velocity while maintaining a constant separation. In either case, passive separation is no longer possible, as travel velocity now depends on the properties of more than one cell. The simultaneous presence of several particles in the trapping beam can be avoided by lowering the rate of particle injection and thus limiting the throughput of any passive sorting system of this kind.

The use of weakly focused laser beams to transport, rather than sort, microscopic particles has the important advantage that proper alignment of the transportation beam with the collection volume is sufficient to ensure precise delivery of the trapped particles. Thus, a transportation system based on a weakly collimated beam requires minimal adjustment. Furthermore, such a system can accurately deliver microscopic particles through apertures with diameters comparable with the diameter of the trapping beams, and can thus move selected particles between separate compartments in a complex, integrated instrument.

Laser Sorting. The macroscopic transportation capabilities of weakly focused laser beams can be used for the automated sorting of microscopic particles. A particle travelling along a weakly focused beam can be transferred to another, intersecting beam, provided the intensity of the latter is above a threshold value that depends on the intensity of the first beam and the

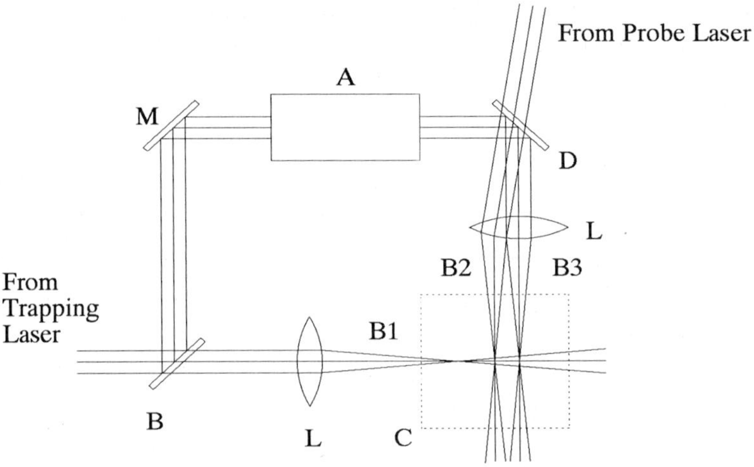

Figure 4. Simplified diagram of the laser sorter optics: A—acoustooptic modulator; B—beam splitter; B1—propulsion beam; B2—probe beam; B3—deflection beam; C—manipulation chamber; D—dichroic mirror; L—focusing lenses; M—mirror. The light collection optics are not shown. (Reproduced with permission from Buican (1991), Copyright 1991 Am. Chem. Soc.)

relative positions of the beam waists. Intersecting beams can thus be used to direct trapped particles to one of several collection volumes. An automated sorting system based on light scattering measurements was described by Buican *et al* (1987) and a diagram of the instrument optics is shown in Figure 4.Naturally, other optical, as well as nonoptical, parameters can be used to control sorting. A manipulation chamber for laser sorting is shown in Figure 5. Channels machined into the top of the chamber are enclosed by a window that allows the particles in the chamber to be visualized. Light scatter measurements are also made through this window. The laser beams enter the channels through side windows, and the sample is injected into the propulsion beam through a sample port. There are two ports just ahead of each side window that are used to create localized fluid flows. Individual cells are trapped in the propulsion beam and start travelling along its axis. As each cell intersects the probe beam (of wavelength different from that of the trapping beams and of lower intensity), the scattered light intensity is measured and a sorting decision is made. When a cell reaches the deflection beam (typically about 100 μm downstream from the probe beam), the intensity of this beam is modified by an acoustooptic modulator so as to direct the particle to the appropriate elution volume. Localized fluid flows just ahead of the beam output windows are used to elute the cells.

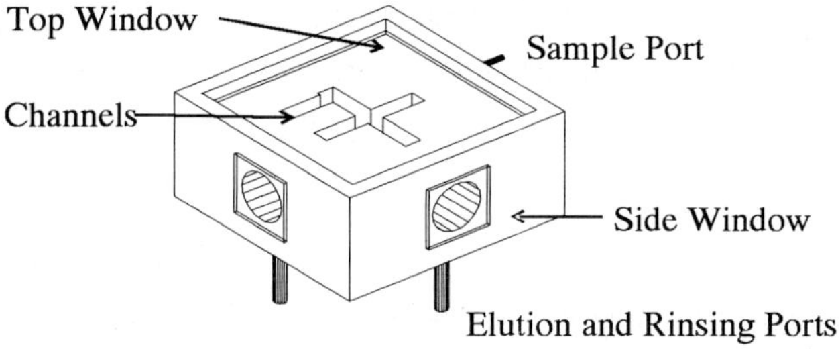

Figure 5. View of the laser sorter manipulation chamber (see Figure 4 for the geometry of the beams inside the chamber). The side windows allow the three laser beams to enter and exit the chamber. The elution and rinsing ports are used to create local fluid flows near the side windows. (Reproduced with permission from Buican (1991), Copyright 1991 Am. Chem. Soc.)

The beam focusing optics on the laser sorter described by Buican *et al.(1987)* had a focal length of 50 mm. At a power of approximately 300 mW per beam, typical particle velocities reached several hundred μm/s. For a separation between probe and deflection beams of 100 μm, this resulted in a maximum sorting frequency of a few cells per second. Although this rate is too low to make the instrument a practical sorter, its simplicity and the relatively large distances travelled by the cells (over 6 mm) may allow a similar system to be used as a microscopic gating device that prevents unwanted cells from drifting through narrow apertures connecting the compartments of an integrated analysis and preparation instrument, while letting through selected cells.

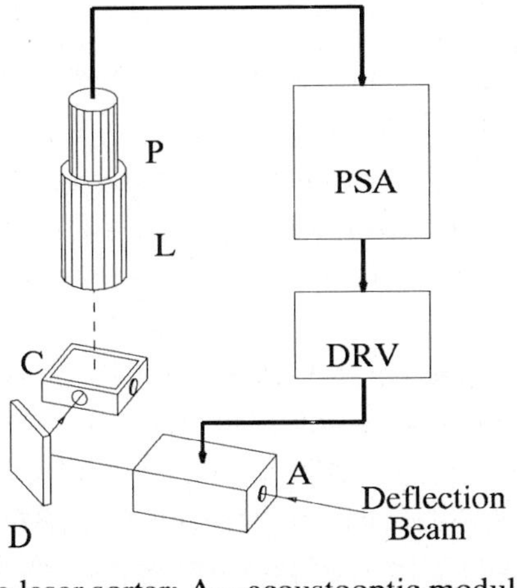

Figure 6. Automated control of the laser sorter: A—acoustooptic modulator; C—manipulation chamber; D—dichroic mirror; DRV—modulator driver; PSA—pulse shape analyzer; L—lens; P—photomultiplier. The dashed line represents the light collection path, while the wide, solid line represents the signal and control path. The deflection beam focusing lens is not shown. (Reproduced with permission from Buican (1991), Copyright 1991 Am. Chem. Soc.)

The elements of the sorting control system are shown in Figure 6. The light scattered by each cell as it passes through the probe beam is collected by a lens and falls on a photomultiplier. The pulse produced by the detector is analyzed by a pulse shape analyzer and a sorting decision is made on the basis of the properties of this pulse.

The parameter of choice for controlling the laser sorter is pulse amplitude. Pulse area, which is a commonly used sort control parameter in flow sorters, is not suitable for controlling laser sorting because pulse widths are not constant. This is due to the fact that particle velocity in the propulsion beam depends on many particle properties and is also affected by the presence of other particles in the beam .

Secondary parameters, such as pulse width and time interval between pulses, are also derived by the pulse shape analyzer. These parameters, the values of which can indicate a separation between successive particles too small to allow individual analysis and/or separation, are used to disable deflection and thus avoid compromising sort purity.

Microrobotic Manipulation

A different approach to cell separation is made possible by the use of a 3-D optical trap. While the laser sorter is operationally equivalent to a flow sorter (in the sense that it separates cells into a small number of collection volumes and that measurements are performed on particles moving along a well-defined axis), the 3-D optical trap can be used for microscopic robotic manipulation of single cells and cell organelles. This technique is well suited for the automated separation of rare cells and can also be easily integrated into complex instruments.

Optical Manipulator Designs. Two basic optical manipulator designs, both originating in Ashkin's work (Ashkin, 1970, Ashkin and Dziedzic, 1974), have been proposed and implemented. The two designs differ in the number of laser beams used to produce a 3-D optical trap (Figure 7).The single-beam optical manipulator (Figure 7a) relies upon the use of a high numerical aperture focusing lens and has been implemented on commercial microscopes (Ashkin and Dziedzic, 1974). The double-beam system (Figure 7b) uses lower numerical aperture lenses, and a stable 3-D trap is produced by two coaxial, counterpropagating beams (Buican *et al.*, 1989). In both cases, the lenses used for focusing the trapping beam(s) are also used for imaging, and dichroic mirrors are used to separate the trapping and imaging wavelengths.

The single beam optical trap illustrated in Figure 7a can be implemented on any microscope with only minor modifications to the microscope optics. The illumination source is the microscope condenser, while imaging is done through the existing microscope optics. The role of the additional divergent lens is to expand the trapping laser beam in order to fill the input aperture of the microscope objective and thus maximize the convergence of the trapping beam. The position of the trap within the field of view of the microscope can be modified by moving the divergent lens (Ashkin *et al.*, 1987). Longer-range motion of the trap can be achieved by moving the manipulation chamber relative to the trapping beam. A variant of

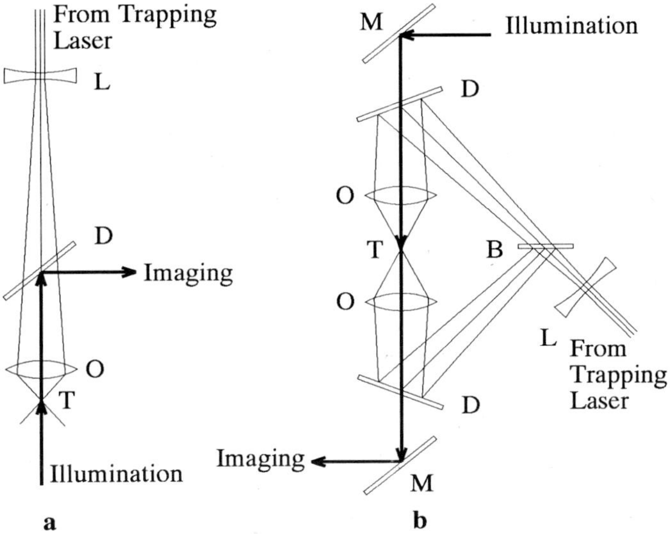

Figure 7. Simplified diagram of single-beam (a) and double-beam (b) optical trapping and imaging systems. B—beam splitter; D—dichroic mirror; L—divergent lens; M—mirror; O—microscope objective; T—optical trap. The arrows represent the imaging optical paths. (Reproduced with permission from Buican (1991), Copyright 1991 Am. Chem. Soc.)

Ashkin's single-beam design was proposed in (Visscher and Brakenhoff, 1990), where an optical trap was added to a confocal microscope by means of a second microscope objective. The second objective, which is used exclusively for trapping, is mounted on an xyz stage and can thus control the position of the optical trap.

The trapping beams in the double-beam system are derived from a single laser beam by means of a beam splitter. The use of a divergent lens for beam expansion is optional, as a high degree of convergence of the trapping beams is no longer critical in this system. The optics for this type of manipulator are more complex than those of the Ashkin design and attention must be paid to the proper alignment of the trapping beams. However, beam alignment can be monitored by a video camera that images the beams after a complete round trip through the system and can thus be automated (Buican *et al.*, 1989).

Because of their different optical characteristics, the two optical manipulator designs have different applications. The single-beam manipulator, which uses a high numerical aperture lens, is capable of high-resolution imaging and can produce small-diameter, high-intensity traps. This manipulator is therefore most suitable for the analysis and manipulation of small particles such as cell organelles. At the same time, the use of a high numerical aperture lens leads to a small field of view and a very small working distance. The small field of view restricts the use of such a manipulator in an automated separation system, where the performance of automated navigation algorithms depends to a large extent on the size of the field of view. Furthermore, the very small working distance makes difficult the use of complex manipulation chambers.

By contrast, the dual beam system provides lower-resolution imaging but has a wider field of view and larger working distance. This system is eminently suited for automated cell and chromosome separation. In applications such as chromosome separation, the lower resolution of the images provided by the double beam manipulator can be compensated for through the use of specific fluorescent probes. The large working distance which, for modern ultralong working distance objectives, can be around 1 cm, allows the optical manipulator to be easily integrated into complex systems

In certain applications, it is important to create and use several, independently controlled, optical traps. Such applications generally involve the manipulation of long and thin microscopic objects, such as certain bacteria, chromosomes, and DNA. While multiple laser beams can be used to create multiple traps, an interesting system that uses a single, scanned and modulated laser beam, was developed by Visscher and Brakenhoff (1991a). In this system, the trapping beam focus is moved in three dimensions along a polygonal path that has the desired trapping points as its vertices. The trapping beam is turned on only as it reaches a vertex. If the scanning is periodic and each vertex is revisited before a particle could drift out of the trapping region, then the net effect is to produce multiple optical traps at the vertices of the scanning path. As the scanning system is computer-controlled, the positions of the simultaneous traps can be independently controlled by the user through an appropriate interface.

Visscher and Brakenhoff (1991b) have also demonstrated a combined optical trapping instrument and confocal microscope. Confocal imaging is particularly attractive on optical trapping instruments, as it makes possible three-dimensional imaging and navigation. Of course, both the scanning optics and the image processing electronics must be quite fast.

Another interesting optical manipulator system that uses a scanned laser beam was described by Misawa *et al.* (1991) and Sasaki *et al.* (1991a, 1991b). In this system, a constant-intensity trapping beam is scanned in such a way that the beam focus moves along a curve at a speed too high for particles to become trapped and to follow the trap. However, it can be shown that the perturbation created by the trapping beam moving across a particle results in the particle undergoing a small displacement in the direction of the scan (Fig. 8). If the scanning is periodic, than a "flow" pattern results, with particles moving at various speeds along the scanning path. Stable flow and positional patterns of considerable complexity have been created by this method.

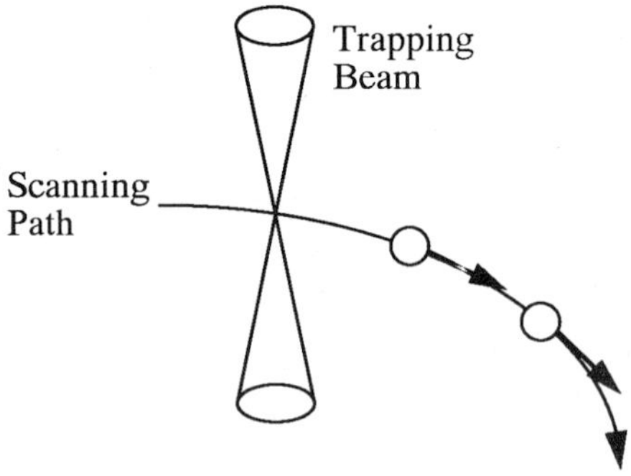

Figure 8. Fast scanning of a trapping beam. Two particles are shown as they are propelled along the scanning path.

Optical Manipulation Chamber Designs. Optical manipulation chambers may incorporate features that facilitate cell separation. For instance, multiple compartments inside the chamber allow selected cells to be physically separated from the original cell sample. It is also desirable for the manipulation chamber to have ports through which the sample can be introduced and the separated cells can be eluted. If the chamber has multiple compartments, then it is preferable for each compartment to be connected to separate external ports. Finally, if the cell sample is highly concentrated (as is the case, for example, with blood samples), it is desirable for the manipulation chamber to be shallow. Low chamber depth ensures that, after sedimentation of the cells, the surface concentration is sufficiently low to permit the trapping and removal of single cells.Manual cell separation through optical manipulation was demonstrated by Ashkin (*3*) in a rather crude chamber that contained a hollow, transparent fiber. We described a multicompartment chamber design (*11*) that is well suited for cell separation, while also containing features that make it attractive for complex work with live cells. This chamber (Figure 9) consists of three layers that are cemented together. The outer layers are

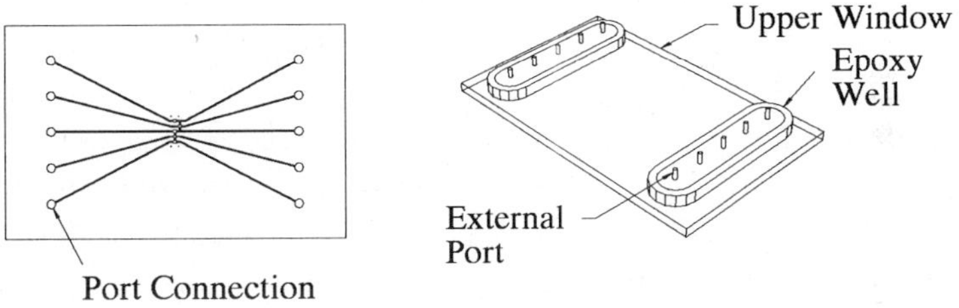

a **b**

Figure 9. Multicompartment manipulation chamber: (a) Top view of the central layer, showing compartments (central region) and interconnecting channels; and (b) perspective view of the upper window, showing the external ports. (Reproduced with permission from Buican (1991), Copyright 1991 Am. Chem. Soc.)

good quality glass of 170-μm and 1-mm thickness respectively. The central layer, which contains the pattern of compartments and channels, can be fabricated either from stainless steel shim stock or from a photosensitive ceramic (Fotoceram, Corning). Small compartments and channels can be accurately cut through the central layer by laser machining (stainless steel) or photoetching (Fotoceram). A view of the central section is shown in Figure 9a, where one can see five compartments (300 μm by 30 0 μm) and channels connecting the compartments with each other and with the external ports. The external ports are connected to internal channels through holes in the 1-mm-thick glass window, to which they are cemented (Figure 9b). The small thickness and large surface area of this manipulation chamber lead to efficient heat exchange and temperature control. This characteristic is particularly important when separating and processing live cells, as well as for performing biochemical assays inside the chamber. Furthermore, the channels connecting each compartment to external ports allow reagents and culture media to be circulated through the compartments, and complex, long-term protocols can thus be carried out inside the chamber.

Automated Control Based on Machine Vision. The ability of optical manipulators to image and manipulate single cells makes this type of instrument amenable to machine-vision-based automation. The control problem in such an instrument consists of the following: (i) finding the cells of interest; and (ii) separating those cells without collecting unwanted particles. In order to simplify image acquisition, we assume that the particles injected into the chamber have been allowed to settle onto the bottom window. Obviously, a shallow chamber has the advantage that particle settling occurs quickly and the delay between sample injection and the beginning of the search procedure is short. The search procedure consists of scanning through the sample and analyzing each particle in the instrument's field of view. Once a cell of interest has been identified, it is trapped and raised to a certain height above the bottom window, fol-

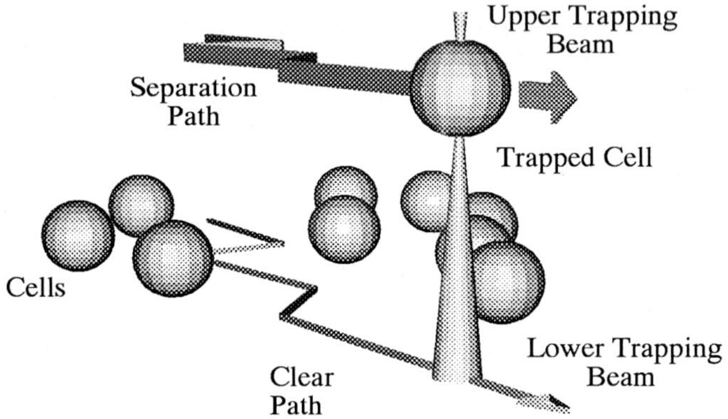

Figure 10. Geometry of cell separation by optical manipulation. The trapped cell is carried along a separation path that avoids other cells on the bottom of the manipulation chamber. The cells in the chamber are imaged along the common axis of the trapping beams. The clear path, which is the projection of the separation path onto the bottom of the chamber, is constructed so as to avoid the trapping of unwanted cells. (Reproduced with permission from Buican (1991), Copyright 1991 Am. Chem. Soc.)

lowing which it is transported to an area where it is either recovered or further analyzed and processed.

Given the initial position of the trapped cell and the coordinates of the target position, separation must proceed along a path that does not collide either with the chamber walls or with unwanted particles. For a simple chamber consisting of a single, convex compartment, the first requirement is easily satisfied. The simplest path satisfying the second requirement depends on the depth of the chamber and the degree of convergence of the trapping beam(s). As optical traps are generally better localized in a plane perpendicular to the trapping beams than along the beam axis, a shallow chamber used with a double beam trap may require the trap to be moved around unwanted particles in order to avoid trapping them. This is illustrated in Figure 10, where we distinguish between the *separation path* followed by the trap, and the *clear path*, which is the projection of the former onto the bottom of the chamber. Finding a clear path from a knowledge of the positions and sizes of the settled cells is the classical collision avoidance problem of robotics. In the case of an optical manipulator, a clear path must be computed gradually, on the basis of a succession of overlapping video frames. Automated navigation is not required for chambers that are deeper than the axial dimension of the trap, and the separation path is, consequently, a straight line at a height above the bottom window that is sufficient to prevent the trapping of unwanted cells. This advantage is generally enjoyed by single-beam traps.

Automated Control System. The main elements of the control system for microrobotic cell separation are shown in Figure 11. The imaging system, which is no more than a video microscope, provides a video signal which is digitized by a frame grabber and made available to a processor. The processor performs a series of operations on the digitized video frame which lead to the computation of a local clear path. Once the clear path for the current frame has been computed, the processor downloads the coordinates of the path vertices to a servo controller, and the servo actuators on an xyz stage move the manipulation chamber accordingly.

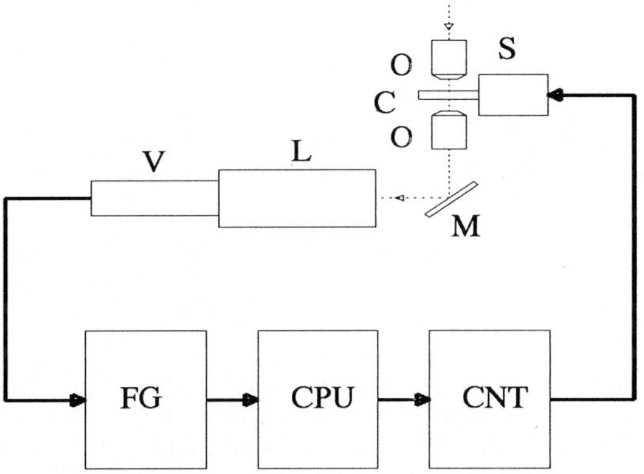

Figure 11. Automated control of the optical manipulator: C—manipulation chamber; CNT—servo controller; CPU—central processing unit; FG—frame grabber; L—imaging lens; M—mirror; O—microscope objective; V—video camera; S—xyz stage with servo actuators. The dotted line represents the imaging optical path, while the solid line represents the data and control path. (Reproduced with permission from Buican (1991), Copyright 1991 Am. Chem. Soc.)

Given the computation-intensive nature of the image processing and machine vision algorithms, it is obvious that the speed at which cells can be separated depends, to a large extent, on the computing power available for controlling the optical manipulator. As shown in Figure 11, the robotic manipulator described by Buican *et al.* (1989) uses a single processor both for running the image analysis and machine vision algorithms, and for controlling the relative position of the optical trap. Furthermore, the same processor operates the rest of the hardware, as well as a high-resolution, interactive user interface. Consequently, separation of trapped cells does not proceed at the maximum possible speed, which is only limited by viscous drag and beam intensity. The data processing bottleneck can be overcome by using real-time, multiprocessor systems, which are now available at reasonable cost.

Commercial Instrumentation

The first commercial optical manipulator, the LazerTweezers™ 1000, has recently been introduced by Cell Robotics, Inc. This instrument (Figure 12) is of the single-beam type, with a moving stage that positions the manipulation chamber relative to the trap. The instrument consists of a set of modules that can be assembled around an inverted microscope. In its basic configuration, the instrument consists of the following: (i) a trapping laser module that mounts on the microscope's side camera port; (ii) a self-contained, servo-controlled, xyz stage that replaces the microscope's original stage; (iii) a 3-D trackball used to control the instrument; and (iv) a system controller. In addition to these modules, a video camera and col-

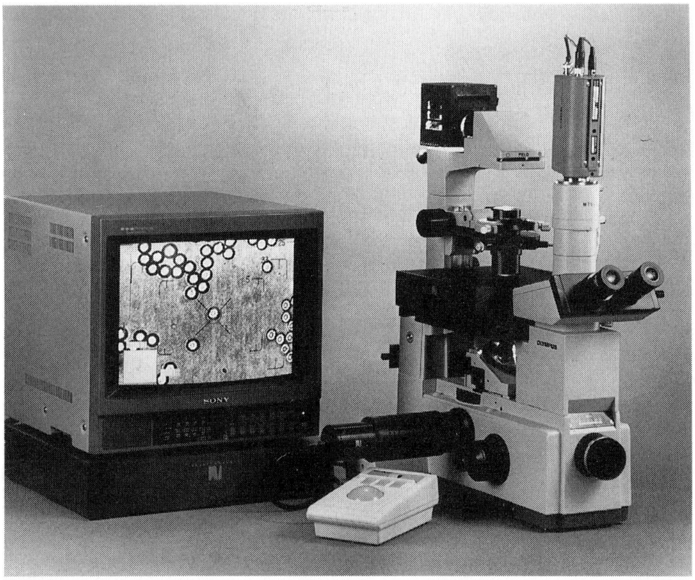

Figure 12. The LaserTweezers™ 1000 optical trapping system from Cell Robotics, Inc. The system consists of an inverted microscope, a trapping laser module (mounted on the side camera port of the microscope); a servo-controlled xyz stage (it replaces the original microscope stage); a trackball for user control of the instrument; and a system controller (seen under the video monitor. A video camera and color monitor are also included.

or monitor are also required. A variety of trapping laser modules, all containing diode lasers emitting in the 820 nm - 850 nm range, will be available. These lasers will range in power from 40 mW to 250 mW. Higher-power lasers will be available on a custom basis, as well as lasers emitting at other trapping wavelengths.

The xyz stage carries the manipulation chamber and controls its position relative to the trap. The stage has a positional accuracy of t he order of 1μm in all three axes. The same stage provides temperature regulation for the chamber, which is an essential element in many biological applications. Manipulation chambers can be easily moved in and out of the stage. As described elsewhere, the chambers are completely enclosed. They are wafer-shaped, with microscopic compartments and channels. Particle suspensions, culture media, and reagents can be introduced into and circulated through the compartments. An automated fluidic system, which will automatically introduce samples, media, and reagents, will be available in the near future.

The system controller controls the xyz stage, trapping laser, and will also control the fluidic system. The controller provides the operator with information about trap position, laser power, and status of the fluidic system by means of a set of a graphic overlay superimposed over the live video image. A set of menus can be opened when special functions are required. Thus, the live video image of the particles inside the manipulation chamber is kept uncluttered most of the time. Special control functions include storage of certain important positions inside the chamber and the ability of the instrument to move the trap automatically to any of these positions.

The basic system will be supplemented in the near future by a 486 PC that will use machine-vision software to automate the cell manipulation. This addition will be particularly important in applications where rare cells are to be identified and separated, or where relatively large numbers of cells must be isolated and processed.

Biological Applications

Optical trapping instrumentation will give the biologist a spectrum of capabilities that cannot be matched by any single technique that is already in use. As a manipulation tool, optical trapping can be used both to separate biological particles and to bring them together. In either case, sophisticated computer control has been demonstrated by several teams, including Buican *et al.* (1987, 1989), Visscher and Brakenhoff (1990, 1991a, 1991b), and Sasaki *et al.* (1991a, 1991b). Buican *et al.* (1987) have also demonstrated computer control based on machine vision, thus taking advantage of the very detailed morphological information that can be easily obtained in most optical trapping systems. With or without sophisticated computer control, optical trapping has already been used as a very selective sorter to separate natural killer cells (Seeger *et al.*, 1991), spermatozoa (Tadir *et al.*, 1989, Wright *et al.*, 1990), chromosomes (Buican *et al.*, 1989), and chromosome fragments (Seeger *et al.*, 1991). It should be emphasized at this point that optical trapping can be used to achieve absolute purity in sorting. Indeed, as both the operator and a machine vision control computer can monitor the contents of the trap during separation, it is possible to ensure that all desired particles are collected and that all other particles are rejected. This makes optical trapping the method of choice for the separation of rare cells, such as stem cells and fetal cells in peripheral blood.

The use of an optical trap as a tool for bringing together selected particles in order to perform microscopic cell biology experiments has been demonstrated in several labs. For example, Wiegand Steubing *et al.* (1991) have used a trap in order to bring and hold together cells in cell fusion experiments. The possibility of performing carefully controlled, cell-cell interaction experiments was demonstrated by Seeger *et al.* (1991), who brought together NK and tumor cells with an optical trap, and were subsequently able to monitor the interaction between these cells. These experiments define two potential ranges of applications for optical trapping instruments, namely (i) rapid production of hybrid cell lines with carefully controlled properties, and (ii) automated assays of cell function at the single cell level.

The optical trap has also been used by several investigators as a probe of the mechanical properties of various intracellular mechanisms. For example, Berns *et al.* (1989) have initiated studies of the interaction between individual chromosomes and the mitotic spindle by trapping chromosomes and perturbing their position. Block *et al.* (1989, 1991) have measured the compliance of bacterial flagella by tethering the flagellum and rotating the bacterium with an optical trap. Block *et al.* (1990) have also used the trap in studies of single kinesin molecules.

Indirect manipulation of single DNA molecules has been demonstrated by Chu (1991), who has used coated microspheres attached to the opposite ends of DNA chains as "handles" that can easily be manipulated by optical traps. These experiments suggest that optical traps may play an important role in the various genome projects currently under way or under consideration.

Optical traps as biological tools exhibit the following important features: (i) they can be used in a wide variety of techniques and on scales ranging from whole cells to single DNA molecules; (ii) they can be easily integrated with many other preparative and analytical techniques; and (iii) the manipulation techniques made possible by optical trapping are well suited for large-scale automation and integration. Therefore optical trapping has the potential to provide the technological basis for complex, powerful, and efficient "microscopic laboratories" for cell biology, cytogenetics, and molecular biology. Such instruments would perform complex analytical and preparative procedures on single cells and chromosomes, while requiring a minimum of human intervention. This approach may open up new avenues in experimental biology and biotechnology.

Bibliography

Ashkin A (1970) Acceleration and trapping of particles by radiation pressure. Phys. Rev. Lett. 24:156-159

Ashkin A (1973) U. S. Patent No. 3,710,279

Ashkin A, Dziedzic JM (1974) Stability of optical levitation by radiation pressure. Appl. Phys. Lett. 24:586-588

Ashkin A, Dziedzic JM (1987) Optical trapping and manipulation of viruses and bacteria. Science 235:1517-1520

Ashkin A, Dziedzic JM, Yamane T (1987) Optical trapping and manipulation of single cells using infrared laser beams. Nature 330:769-771

Bakker Schut TC, Hesselink G, de Grooth BG, Greve J (1991) Experimental and theoretical investigations on the validity of the geometrical optics model for calculating the stability of optical traps. Cytometry 12:479-485

Berns MW, Wright WH, Tromberg BJ, Profeta GA, Andrews JJ, Walter RJ (1989) Use of a laser-induced optical force trap to study chromosome movement on the mitotic spindle. Proc. Natl. Acad. Sci. USA 86:4539-4543

Block SM, Blair DF, Berg HC (1989) Compliance of bacterial flagella measured with optical tweezers. Nature 338:514-518

Block SM (1990) Optical tweezers: A new tool for biophysics. In: Foskett JK, Grinstein S (eds) Noninvasive techniques in cell biology. Modern Cell Biology 9:375-402, Wiley/ Liss, New York
Block SM, Goldstein LSB, Schnapp BJ (1990) Bead movement by single kinesin molecules studied with optical tweezers. Nature 348:348
Block SM, Blair DF, Berg HC (1991) Compliance of bacterial polyhooks measured with optical tweezers. Cytometry 12:492-496
Bohren CF, Huffman DR (1983) Absorption and scattering of light by small particles. John Wiley & Sons: New York, NY, p. 72
Buican TN, Smyth MJ, Crissman HA, Salzman GC, Stewart CC, Martin JC (1987) Automated single-cell manipulation and sorting by light trapping. Appl. Opt.26:5311-5316
Buican TN, Neagley DL, Morrison WC, Upham BD (1989) Optical trapping, cell manipulation and robotics. In New Techniques in Cytometry, Salzman GC (ed) Proceedings of SPIE; SPIE: Bellingham, Washington, 1063:190-197
Buican TN (1991) Automated cell separation techniques based on optical trapping. In: Kompala DS and Todd P (eds) Cell separation science and technology. ACS Symposium Series, 464:59-72
Chu S (1991) Laser manipulation of atoms and particles. Science 253:861-866
Kogelnik H, Li T (1966) Laser beams and resonators. Appl. Opt. 5:1550-1567
Horn, BKP (1986) Robot vision. The MIT Electrical Engineering and Computer Science Series; MIT Press, Cambridge, MA
Misawa H, Koshioka M, Sasaki K, Kitamura N, Masuhara H (1991) Spatial pattern formation, size selection, and directional flow of polymer latex particles by laser trapping technique. Chem. Lett.469-472
Roosen G (1979) La levitation optique de spheres. Can. J. Phys. 57:1260-1279
Sasaki K, Koshioka M, Misawa H, Kitamura N, Masuhara H (1991a) Laser-scanning micromanipulation and spatial patterning of fine particles. Jpn. J. Appl. Phys. 30:L 907-L 909
Sasaki K, Koshioka M, Misawa H, Kitamura N, Masuhara H (1991b) Pattern formation and flow control of fine particles by laser-scanning micromanipulation. Opt. Lett. 16:1463-1465
Salzman GC (1982) In Catsimpoolas, N (ed) Cell Analysis, vol. 1. Plenum Publishing Co., New York, p 111
Seeger S, Monajembashi S, Hutter K-J, Futterman G, Wolfrum J, Greulich KO (1991) Application of laser optical tweezers in immunology and molecular genetics. Cytometry 12:497-504
Tadir Y, Wright WH, Vafa O, Ord T, Asch RH, Berns MW (1989) Micromanipulation of sperm by a laser generated optical trap. Fertility and Sterility 52:870-873
Visscher K, Brakenhoff, GJ (1990) Cytometry Supplement 4, p. 17
Visscher K, Brakenhoff GJ (1991a) 3 Dimensional multiple trap imaged by confocal microscopy. Abstract 10, Cytometry Suppl. 5:27
Visscher K, Brakenhoff GJ (1991b) Single beam optical trapping integrated in a confocal microscope for biological applications. Cytometry 12:486-491
Wiegand Steubing R, Cheng S, Wright WH, Numajiri Y, Berns MW (1991) Laser induced cell fusion in combination with optical tweezers: the laser cell fusion trap. Cytometry 12:505-510
Wright W, Sonek GJ, Berns MW (1989) Optical trapping of microspheres as a model for biological cells. IEEE Engineering in Medicine and Biology Society 11th Annual International Conference, 1203-1204
Wright W, Sonek GJ, Tadir Y, Berns MW (1990) Laser trapping in cell biology. IEEE J. Quant. Electron. 26:2148-2157

Subject Index

Printing: Druckhaus Beltz, Hemsbach
Binding: Buchbinderei Schäffer, Grünstadt

NATO ASI Series H

NATO ASI Series H

Vol. 22: **Neural Development and Regeneration. Cellular and Molecular Aspects.**
Edited by A. Gorio, J. R. Perez-Polo, J. de Vellis, and B. Haber. 711 pages. 1988.

Vol. 23: **The Semiotics of Cellular Communication in the Immune System.**
Edited by E.E. Sercarz, F. Celada, N.A. Mitchison, and T. Tada. 326 pages. 1988.

Vol. 24: **Bacteria, Complement and the Phagocytic Cell.**
Edited by F. C. Cabello und C. Pruzzo. 372 pages. 1988.

Vol. 25: **Nicotinic Acetylcholine Receptors in the Nervous System.**
Edited by F. Clementi, C. Gotti, and E. Sher. 424 pages. 1988.

Vol. 26: **Cell to Cell Signals in Mammalian Development.**
Edited by S.W de Laat, J.G. Bluemink, and C.L. Mummery. 322 pages. 1989.

Vol. 27: **Phytotoxins and Plant Pathogenesis.**
Edited by A. Graniti, R. D. Durbin, and A. Ballio. 508 pages. 1989.

Vol. 28: **Vascular Wilt Diseases of Plants. Basic Studies and Control.**
Edited by E. C. Tjamos and C. H. Beckman. 590 pages. 1989.

Vol. 29: **Receptors, Membrane Transport and Signal Transduction.**
Edited by A. E. Evangelopoulos, J. P. Changeux, L. Packer, T. G. Sotiroudis, and K.W.A. Wirtz. 387 pages. 1989.

Vol. 30: **Effects of Mineral Dusts on Cells.**
Edited by B.T. Mossman and R.O. Begin. 470 pages. 1989.

Vol. 31: **Neurobiology of the Inner Retina.**
Edited by R. Weiler and N.N. Osborne. 529 pages. 1989.

Vol. 32: **Molecular Biology of Neuroreceptors and Ion Channels.**
Edited by A. Maelicke. 675 pages. 1989.

Vol. 33: **Regulatory Mechanisms of Neuron to Vessel Communication in Brain.**
Edited by F. Battaini, S. Govoni, M.S. Magnoni, and M. Trabucchi. 416 pages. 1989.

Vol. 34: **VectorsasTools for the Study of Normal and Abnormal Growth and Differentiation.**
Edited by H. Lother, R. Dernick, and W. Ostertag. 477 pages. 1989.

Vol. 35: **Cell Separation in Plants: Physiology, Biochemistry and Molecular Biology.**
Edited by D. J. Osborne and M. B. Jackson. 449 pages. 1989.

Vol. 36: **Signal Molecules in Plants and Plant-Microbe Interactions.**
Edited by B.J.J. Lugtenberg. 425 pages. 1989.

Vol. 37: **Tin-Based Antitumour Drugs.** Edited by M. Gielen. 226 pages. 1990.

Vol. 38: **The Molecular Biology of Autoimmune Disease.**
Edited by A.G. Demaine, J-P. Banga, and A.M. McGregor. 404 pages. 1990.

Vol. 39: **Chemosensory Information Processing.**
Edited by D. Schild. 403 pages. 1990.

Vol. 40: **Dynamics and Biogenesis of Membranes.**
Edited by J. A. F. Op den Kamp. 367 pages. 1990.

Vol. 41: **Recognition and Response in Plant-Virus Interactions.**
Edited by R. S. S. Fraser. 467 pages. 1990.

NATO ASI Series H

Vol. 42: Biomechanics of Active Movement and Deformation of Cells.
Edited by N. Akkas. 524 pages. 1990.

Vol. 43: Cellularand MolecularBiologyof Myelination.
Edited by G. Jeserich, H. H. Althaus, and T. V. Waehneldt. 565 pages. 1990.

Vol. 44: Activation and Desensitization of Transducing Pathways.
Edited by T. M. Konijn, M. D. Houslay, and P. J. M. Van Haastert. 336 pages. 1990.

Vol. 45: Mechanism of Fertilization: Plants to Humans.
Edited by B. Dale. 710 pages. 1990.

Vol .46: Parallels in Cell to Cell Junctions in Plants and Animals.
Edited by A. W Robards, W. J . Lucas, J . D. Pitts, H . J . Jongsma, and D. C. Spray. 296 pages. 1990.

Vol. 47: Signal Perception andTransduction in Higher Plants.
Edited by R. Ranjeva and A. M. Boudet. 357 pages. 1990.

Vol. 48: Calcium Transport and Intracellular Calcium Homeostasis.
Edited by D. Pansu and F. Bronner. 456 pages. 1990.

Vol. 49: Post-Transcriptional Control of Gene Expression.
Edited by J. E. G. McCarthy and M. F. Tuite. 671 pages. 1990.

Vol. 50: Phytochrome Properties and Biological Action.
Edited by B. Thomas and C. B. Johnson. 337 pages. 1991.

Vol. 51: Cell to Cell Signals in Plants and Animals.
Edited by V. Neuhoff and J. Friend. 404 pages. 1991.

Vol. 52: Biological Signal Transduction.
Edited by E. M . Ross and K . W. A. Wirtz. 560 pages. 1991.

Vol. 53: Fungal Cell Wall and Immune Response.
Edited by J. P. Latge and D. Boucias. 472 pages. 1991.

Vol. 54: The Early Effects of Radiation on DNA.
Edited by E. M. Fielden and P. O'Neill. 448 pages. 1991.

Vol. 55: The Translational Apparatus of Photosynthetic Organelles.
Edited by R. Mache, E. Stutz, and A. R. Subramanian. 260 pages. 1991.

Vol. 56: Cellular Regulation by Prote in Phosphorylation.
Edited by L. M. G. Heilmeyer, Jr. 520 pages. 1991.

Vol. 57: MolecularTechniques in Taxonomy.
Edited by G . M . Hewitt, A. W. B. Johnston, and J. P. W. Young . 420 pages. 1991.

Vol. 58: Neurocytochemical Methods.
Edited by A. Calas and D. Eugene. 352 pages. 1991.

Vol. 59: Molecular Evolution of the Major Histocompatibility Complex.
Edited by J. Klein and D. Klein. 522 pages. 1991.

Vol. 60: Intracellular Regulation of Ion Channels.
Edited by M. Morad and Z. Agus. 261 pages. 1992.

Vol. 61: Prader-Willi Syndrome and Other Chromosome 15q Deletion Disorders.
Edited by S. B. Cassidy. 277 pages. 1992.

NATO ASI Series H